FLOW VISUALIZATION

Proceedings of the International Symposium
on Flow Visualization, October 12-14, 1977,
Tokyo, Japan

Edited by

Tsuyoshi Asanuma

Professor of Mechanical Engineering
Tokai University
Hiratsuka, Japan

⬤HEMISPHERE PUBLISHING CORPORATION

Washington   New York   London

IN COOPERATION WITH

McGRAW–HILL INTERNATIONAL BOOK COMPANY

New York   St. Louis   San Francisco   Auckland   Beirut   Bogotá
Düsseldorf   Johannesburg   Lisbon   London   Lucerne   Madrid
Mexico   Montreal   New Delhi   Panama   Paris   San Juan   São Paulo
Singapore   Sydney   Tokyo   Toronto

207408

FLOW VISUALIZATION

1 2 3 4 5 6 7 8 9 0 L I L I 7 8 3 2 1 0 9

Library of Congress Cataloging in Publication Data

International Symposium on Flow Visualization,
  Tokyo, 1977:
  Flow visualization.

  Bibliography: p.
  Includes index.
  1. Flow visualization—Congresses.   I. Asanuma,
Tsuyoshi.   II. Title.
TA357.I582   1977   620.1'064   79-12407
ISBN 0-89116-155-4

CONTENTS

TUFT AND WALL TRACING METHODS

CHEMICAL REACTION AND ELECTRICAL CONTROL METHODS

OPTICAL METHODS

CAVITATION

PREFACE

While most of the papers presented here deal with direct injection and optical methods, a number of other methods are also reported. Several papers on wall tracing and chemical reaction methods are introduced. Further, work employing shadowgraph and schlieren methods is reported, as is the electrical control method.

Considerable attention is given here to a newly developed tracer method, which employs such tracers of material as optically activated powder suspended in the working fluid, small ice particles in humid air, or cavitation bubbles produced in water. It should be noted that the development of interferometry and holography is yielding a remarkable trend in the field of flow visualization because of noncontacting nature and multidimensional information, while recent progress in the electrical control method shows the most promising quantitative measurement of a flow.

The International Symposium on Flow Visualization (held October 12-14, 1977, in Tokyo) was the forum for presentation of these papers. This international meeting grew out of a strong interest in the subject developed in the Japanese scientific and engineering communities through five years of domestic symposia on flow visualization, sponsored by the Institute of Space and Aeronautical Science of the University of Tokyo.

As evidenced by the large number of participants and contributed papers, the international response to this topic was beyond our expectations.

Contributors from outside Japan introduced papers especially on tracer methods of the direct injection and optical methods. All the work employing shadowgraph and schlieren methods was presented by non-Japanese participants. On the other hand, the bulk of the papers on the electrical control and tuft methods were reported by Japanese participants, who introduced as many papers on the direct injection methods as did the non-Japanese participants. We consider these facts to evidence the need for international meetings and publication of the proceedings to an international audience.

I would like to thank the members of the organizing committee, especially those from outside Japan. I am also grateful to the Japan Society for the Promotion of Science for financial support and to all the sponsors who contributed to the symposium's success.

T. Asanuma

ORGANIZING COMMITTEE

Chairman: Prof. T. Asanuma *(Tokai University, Japan)*
Secretary: Prof. Y. Tanida *(University of Tokyo, Japan)*
Members:

Mr. K. Akashi *(Mitsubishi Heavy Ind., Ltd., Japan)*
Prof. S. Asaka *(Ochanomizu University, Japan)*
Prof. R. S. Brodkey *(Ohio State University, United States)*
Prof. J. Hanawa *(Ibaraki University, Japan)*
Prof. T. Hayashi *(Chuo University, Japan)*
Prof. R. L. Hummel *(Toronto University, Canada)*
Prof. T. Ishihara *(University of Tokyo, Japan)*
Prof. T. Kobayashi *(University of Tokyo, Japan)*
Prof. W. Merzkirch *(Ruhr University, West Germany)*
Prof. H. Murai *(Tohoku University, Japan)*
Prof. H. Nakaguchi *(University of Tokyo, Japan)*
Prof. Y. Nakayama *(Tokai University, Japan)*
Prof. H. Ohashi *(University of Tokyo, Japan)*
Dr. Y. Okamoto *(Japan Atomic Energy Research Inst.)*
Prof. R. Řezniček *(Academy of Science, Czechoslovakia)*
Prof. T. Tagori *(University of Tokyo, Japan)*
Prof. S. Taneda *(Kyushu University, Japan)*
Dr. C. Véret *(ONERA, France)*
Prof. F. J. Weinberg *(Imperial College, United Kingdom)*
Prof. W.-J. Yang *(University of Michigan, United States)*

SUPPORTING SOCIETIES

The Japan Society for Aeronautical and Space Sciences
The Japan Society of Mechanical Engineers
The Japan Society of Civil Engineers
The Society of Naval Architects of Japan

FINANCIAL SPONSOR

The Japan Society for Promotion of Science

GENERAL LECTURES

FLOW VISUALIZATION TECHNIQUES IN JAPAN

TSUYOSHI ASANUMA*

INTRODUCTION

There are a great number of cases in history in which scientists happened to obtain an idea of elucidating a pysical phenomenon from visual observation. In case of a fluid moving process we often need to devise certain means to enable direct observation of the corresponding fluid motion. We call the above devices on the whole as the flow visualization techniques which have been long studied and exploited for various applications. One famous example is the experiment carried out by Osborne Reynolds (1879) in which the transition from a laminar flow to a turbulent flow has been shown to take place in a horizontal circular pipe by visualizing the behaviour of the dye injected into a water flow. A variety of techniques have been established by not only the specialists in fluid mechanics but also many investigators studying other fields in science and engineering who therefore need to improve or develop suitable visualization methods. The achievements in those techniques for the first time led to "the Symposium on Flow Visualization" held by ASME in 1960. The results of this meeting considerably stimulated many research people all over the world.

Flow visualization techniques upheld the primary intention of acquiring the following observation and measurement, (i) a streak line, a path line and/or a stream line which all show the flow direction, (ii) flow velocity, its gradient and/or acceleration and the corresponding distributions, (iii) generation and decaying of vortices, transition from a laminar flow to a turbulent one or vice versa, and separation and reattachment of a flow, (iv) density and temperature distributions of a fluid. Those remain to be true up to the present time.

Owing to the further development and improvement of the flow visualization techniques, however, the number of research fields as well as the objectives are growing to a large extent. In the next place I would like to present a summary of the trends for each of the flow visualization techniques which have been recently developed in Japan.

CURRENT TRENDS OF FLOW VISUALIZATION TECHNIQUES IN JAPAN

The use of flow visualization in the fields of sciences and engineering in Japan dates back to many years ago. In order to study the gas exchanging process of a   two-cycle engine (1927-1932) the tracer method using smoke or burning carbon powder for the tracer, as well as a technique which resulted in the colouring due to the interface reaction, were applied. The flow pattern around the intake valve of an engine was also visualized using the suspended tracer of aluminum powder (1929). With the help of the flow visualization techniques, other

---

* Professor of Mechanical Engineering, Tokai University, Hiratsuka, Japan

distinguished research work has been also conducted in a smoke tunnel in the aeronautical field or a water channel in the naval architecture field. After World War II, many efforts have been made to improve and develop the flow visualization techniques amongst the individual research worker of many fields. Unfortunately those were conducted rather independently and the communication between the research workers was quite poor. Since 1973, however, the symposium on flow visualization has been held every year by the Institute of Space and Aeronautical Science, the University of Tokyo. Each year the attendants to the meeing have been over seven hundreds, which is more than the expected initially, while the number of the papers presented exceeds thirty. This year we have the fifth domestic symposium as well as "the International Symposium on Flow Visualization" succeeding the former on this day.

Judging from the papers presented to the domestic symposia for the past five years, the subjects taken up are concerned with quite a wide range of specific fields, e.g. the aeronautical, the naval architectural, the mechanical, the chemical, the civil and the architectural fields, which together represent almost the whole of the engineering fields, and the natural science fields including physics, meteorology and oceanography, as well as the medical field which is showing much interest in the visualization topics recently. As shown in Table 1, the flow visualization techniques presented to the past domestic symposia have been classified into the following four categories according to the principle. The first is the wall tracing method which mainly investigates the flow near or on a solid wall surface. The second is the tuft method which indicates only the stream direction near the wall or in a flow. The third is the tracer method, the most representative of all the visualization methods, which uses some kinds of tracers in a flow. Finally the optical method depends upon the optical phenomena due to, for instance, the density deviation of a flow. The tracer method may be further divided into three types as follows: the direct injection type, the chemical reaction type and the electrical control type. In Table 2 there is shown the distribution of flow visualization techniques which are employed in the papers presented to the past five Symposia in our country. According to the figures in Table 2 almost half of the total number concerns the direct injection type of the tracer method, while one third is for the electrical control type and the optical method. The rest are the contributions from the wall tracing method, the chemical reaction type and the tuft method. Consequently it can be concluded that generally we are more or less interested in these six groups of the flow visualization techniques in Japan. It can be also seen from this table that over thirty papers were presented before three or four hundreds of participants at each meeting. Table 3 shows the distribution of flow visualization techniques in various application fields and almost half of the total number of the techniques are found to be concentrated to the basic scientific field.

In the following a brief review is made on the development of each visualization technique which has been and is now being undertaken in Japan. Since the results of investigations have been presented to the domestic Symposia and are described in detail in the "Handbook on Flow Visualization", which has been edited by all the Japanese members of Organizing Committee on International Symposium and was published last year. I would like to refer to these publications considerably, though purposely unspecifically, throughout much of my presentation. I would like to make a survey of the most pertinent of the current and newly developed techniques on flow visualization in our country. From the same reason, specific reference will be made neither to the investigators nor the laboratories which are responsible for the work. Such information can be readily obtained from the back numbers of the domestic Symposia or the references in the "Handbook on Flow Visualization".

4

1) WALL TRACING METHOD

This method visualizes the stream near a body surface placed in a flow, observing the flow pattern of some material on the surface due to either the wall shear stress or the difference of heat transfer depending upon the conditions of the surface boundary layer.

The oil film method belongs to the former type. It yields the flow condition, direction and speed of the stream near the boundary from the flow pattern being traced by the mixture of oil and pigment painted on a body surface. Examples of application are for the surface flow on a swept back blade or a cascaded blade of a turbomachine and the swirling air flow on a piston crown of an internal combustion engine. This method has been furthermore applied for a high turbulence flow in an actual turbomachine, the working fluid being both gas and water. There is a detailed examination on the composition of oil, pigment and an additive and also the optimum exposure time of the oil film in a flow.

The oil dot method employs colour oil poured from a particular oil hole or a small tube. The trace of the oil shows more easily the stream direction and the distribution of the shear stress. This method can be used for both gas and water, and can be applied to the flow pattern on an aircraft fuselage and also at the nose cone in a supersonic flow. Since both the oil film and oil dot methods rely upon some time mean values, the measurements are in general carried out only for a stationary flow.

Next the latter type of the wall tracing method, which depends upon the conditions of the boundary layer, includes the sublimation method, the soluble chemical film method and the thermosensible paint method, the last of which has been studied in detail and applied for an air flow to visualize the boundary layer transition on a delta wing, as shown in Fig.1. In addition the electric etching method has been recently developed in which the body surface with homogeneous solder coating is electrolysed in water, resulting in the surface erosion and the white colour electrolyte. This method requires a few of minutes to obtain a clear picture of the flow pattern so that it is appropriate only for a stationary water flow of considerably low speed, 0.01 - 0.10 m/s.

2) TUFT METHOD

A primary objective of this method is to visualize the flow direction indicated by tufts or streamers of fine silk or wool yarn for an air flow and those of nylon yarn or wooden stick for a water flow. There are three types of application.

The surface tuft method follows up the movements of the tuft fixed at one end on a body surface. This method has been used to investigate a surface flow of an aircraft or an automobile and the vortex formed at the stern of a ship. When the flow speed is low we can not neglect the tuft rigidity and the effect of the gravity force, so that extremely fine and light tuft material like a single silk fiber should be used. There have been examined in detail the characteristics and the material of a tuft in order to obtain faster response to a change in stream direction for visualizing an unsteady flow. Particularly to the surface tuft method, care must be taken not to disturb the boundary layer with the attached tufts.

The depth-tuft method is also used to detect the flow direction at some distance from a body surface which is placed in a flow. It has been applied to observe the flow pattern in the volute chamber of a centrifugal pump and around the bottom surface of a ship model from stem to stern.

As the last type, the tuft grid method employs a grid of tufts placed in a plane within a flow, usually perpendicularly to the stream direction, in order to visualize the wake behind a delta wing or a building, and the flow pattern in an impeller of a radial water pump.

Therefore the tuft method can be applied for a wide range of flow speed without special installation, so it can be employed frequently for a practical turbomachine and testing models.

3) TRACER METHOD I (DIRECT INJECTION TYPE)

The visualization technique which injects some foreign material into a flow as a tracer is the most popular one and has been long and widely used up to now for both air and water flows. This may be classified into five types according to the way of providing the tracer, e.g. streak line, path line, particle suspension, surface floating and time line. These types are in general applied for a low speed air flow (0 - 20 m/s) or a water flow (0 - 2 m/s) by using various sorts of material as the tracer, e.g. gases, fluids and solids.

Streak line method: As a tracer in an air flow, the mist consisting of kerosene or fluid paraffine is recently of general use and various kinds of mist generators are developed. The mist is used in so called a "smoke tunnel" in which the flow pattern around an airfoil with some high lift device is studied to propose an approximate method to obtain the lift force upon it. While mist can be also produced by sublimation of dry ice placed in warm water, as well as white smoke of titanium tetrachloride.

In case of a water flow a water colour paint, a cooking pigment and milk can also be used. The dye stream line method is employed for the visualization of a three dimensional behaviour of a swirling flow in a circular pipe or an impeller of the centrifugal compressor model. On the other hand, behaviour of the boundary layer and the wake after the parallel plate can be obtained by gradually dissolving into water the condensed milk which has been painted on the plate surface or a delta wing. Fig.2 shows the observation of an unsteady flow in an impeller, supplying dye tracers intermittently.

Path line method: In this method we get quantitative information of the local flow speed as well as the flow direction from the tracer particle displacement or the path line during a short interval. In general the tracer which does not much diffuse should be injected intermittently. For an air flow the soap bubble method has led to the improvement of producing equipments, and the detailed discussions are carried out on the producing conditions of the soap bubbles. There is also a quantitative study on the path of the scavenging flow in a two-cycle engine and the flow around a building model. For a water flow air bubbles are used widely to observe the turbulent flow in a turbomachine. Owing to the large difference in density between the tracer and the working fluid, however, one should choose the tracer of the least diameter. While the oil particles possess density pretty close to that of water, buy it is difficult to control the size and the number of particles produced in a unit time. The use of the luminous particles is restricted only to a large scale model despite of the merit which needs no special lighting from outside, since their size is about 10 mm.

Particle suspension method: In some cases air bubbles can be used in a water flow, but the buoyant speed is too large and it is too difficult to get all of the bubbles of a uniform size. It is preferable to use the tracer of a liquid type such as carbon tetrachloride, benzen, toluen, xylene, since their density can be adjusted to the same as that of water. Presently solid tracer like aluminum powder and polystyrene particles is most often used, for instance, to study the formation of vortex rings by a mechanical impulse upon a simulated arterial wall as illustrated in Fig.3. While, for an air flow, methaldehide (sublimation temperature 120°) and stealync and zinc (melting point 140 °C) are employed as the tracer, the path line of which is examined under the high intensity polarized strobo-light in order to find a flow pattern of ventilating air in a building or a velocity distribution around a heated plate.

Surface floating method: Since a free surface becomes necessary, this method applies only for a water flow. But it is so easily handled that it has been long used at almost every occasion, for instance, in the observations of the flow around a promotor of a turbulent flow, the flow in a divergent channel and the vertical vortex in a flooded river and so on. As the tracer we can choose aluminum, bakelite, sawdust, foamed styrene and skim milk.

Time line method: This is also mainly used for a water flow. A velocity distribution is conveniently obtained by means of the two successive time lines shown by the dye tracer.

Recently a new method has been proposed to obtain the shadow of an air flow on a fluorescent plate by using the absorption of ultraviolet rays in air containing a small amount of ozon as the tracer. Similarly there is another method in which luminescence particles are uniformly diffused in a fluid flow, for instance, water dissolving glycerine, and pulse light is radiated to provide the luminescence. A photograph is then taken and the flow speed can be calculated from it. Both techniques are newly proposed and in principle quite similar to the electrical control method which will be refered to later on. We have a great expectation of developing these techniques in future.

4) TRACER METHOD II (CHEMICAL REACTION TYPE)

According to Table 2 there are not many papers employing this type of the tracer method but they are the same in number as those of the wall tracing method for the past five years. The chemical reaction method belongs to the tracer method and employs colouration, colour change, colour vanishing or luminescence due to a chemical reaction. This may be classified into the interface reaction method, which applies for a chemical reaction due to the encounter of reactants and also electrolytic colouration technique, which uses electrolysis phenomenon. In the former case a colour comes out anywhere at the body surface, in the fluid or the tracer material. While the latter includes two types, one using the coloured small particles produced by electrolysis, which is called electrolysis precipitation, and the other in which the fluid around the electrode is coloured or becomes luminous. Generally speaking, most of those chemical reaction types show neither significant density change arising from luminosity nor the gravity force effect, because the coloured fine particles being produced are extremely small. Therefore it is best suited for the subjects in low speed flow, such as smoke diffusion from a chimney, exclusion of tracers from the front of a bluff body, generation of twin vortices near a circular cylinder and boundary layer around a cylinder starting oscillation, the last of which is shown in Fig.5.

5) TRACER METHOD III (ELECTRICAL CONTROL TYPE)

The topic here represents all of the tracer methods which can be controlled electrically. The most typical types are the hydrogen bubble method and the spark tracing method. Both need not a tracer foreign to the working fluid. The smoke wire method, on the other hand, employs oil painted on a metal wire which changes into white colour mist. We have included the smoke wire method in the present category, since the production rate of the tracer can be controlled electrically.

Hydrogen bubble method: In this technique the working fluid of water changes itself into the tracer due to electrolysis phenomenon. The electrode is a wire with the diameter 5 - 100 μm, which does not affect the flow. The initiation, the period, the quantity and the time interval of producing the bubbles can be controlled electrically. Streak lines, path lines and time lines can be obtained as desired, so that the method applies for some quantitative measurement of flow velocity. A joint use of the kinked electrode for D.C. and the straight one for pulse current allows the simultaneous measurement of the velocity components of a two dimensional flow in the direction along and normal to the stream. Further, the hydrogen bubbles, as the tracer, neither diffuse in water nor stain with an observation window. Therefore, the hydrogen bubble method can be used for visualization of not only a stationary flow but also an unsteady flow. The examples for analysis are a flow around a stationary circular cylinder, as well as a cylinder starting impulsively, a flow within element fluidics, the decay of a swirling flow inside a cylinder, velocity profile of a pulsating duct flow, the vortices behind an cylinder and a velocity distribution around an oscillating wing, the last of which is illustrated in Fig.6.

So many papers were presented already concerning the quantitative measurement of an unsteady flow by means of this type of visualization method, whence the following points are emphasized to be taken into consideration. First, the density ratio between the tracer of hydrogen bubbles and water is about $1 : 10^4$, so the buoyant speed of the bubbles with the optimum diameter of 60 - 200 μm for visualization amounts to 10 - 40 mm/s, which is not suited for a low speed flow. Second, a consideration for the virtual mass of a bubble in water concludes that the smaller the bubble size is, the better is the response to a periodically oscillating flow. In order to reduce the latter error within 2 % the bubble diameter should be less than 10 μm. Further there is a report stating that if one places a fine electrode of diameter d at a distance x away from a body (for instance, a cylinder) for observation within the range of x/d < 70, the bubbles can not move down to the body surface (the leading edge) against the stagnation pressure ahead of it. This is caused by the phenomenon similar to that of the electrolysis precipitation method.

Spark tracing method: In this technique the tracer is a spark line ionized by an electric pulse with high voltage and high frequency. The life time of the ion is about 0.1 - 1.0 msec, so a series of pulses with high frequency over 10 kHz at least have to be continuously produced. Since nothing but the electrodes is put into the air flow for measurement, there is almost none to disturb it. The method can be applied for a wide range of flow speed from several meters per second to supersonic speed. The generation of a spark can be electrically controlled as desired. Therefore, we can easily obtain a spark line, that is, a time line which enables to make visualization of an unsteady flow, as well as to yield quantitative flow velocity measurement. In addition the spark itself is bright enough to dispense with the illumination for photography, and it has also an advantage of not staining with the flow and the observation windows. This technique has been used since quite a long time before for a number of actual machines or experimental models, for instance, the flow through an intake valve and in the swirl chamber of an internal combustion engine, and the velocity profile of a flow around an stationary wing as well as a wing in pitching oscillation, which is shown in Fig.7.

In order to apply this type of visualization technique for the measurement of velocity quantitatively, a couple of points to prevent the errors should be taken into consideration. First the proper electrode geometry for the flow field needing visualization have to be chosen. Second we have to point out two errors particular to the spark tracing method. One is due to the diffusion effect of ionized particles forming a spark which enlarges the radius of curvature of the subsequent spark line in the boundary layer, hence the latter appears to be thickened rapidly. The other error arises from the following reason. It may be well imagined that the density of the ionized gas is fairly small compared with that of the air surrounding it, because the temperature of ionized gas is presumably over one thousand degrees centigrade due to rapid heating by an electric spark.

Further the spark tracing method needs an electric source accommodating electric pulses with high voltage and high frequency, and it can not detect the flow inside the boundary layer, because of the turbulence and the heating due to the discharge.

Nevertheless this technique has been widely used as a method to provide quantitative velocity measurement amongst many specific research fields. This tendency will further continue in future in a variety of fields.

Smoke wire method: A metal wire is painted with fluid paraffine or its mixture with a small quantity of machine oil or kerosene. The wire is placed perpendicularly to an air stream. A strong instantaneous electric current through it produces white colour mist which is suited for a photograph. The visualization method which employs this mist as the tracer is called "the smoke wire method". A series of photographs of the white mist movement illuminated by a strobo-light not only reveal the flow movement but also yield a velocity distribution. Compared with the other two types of the electrical control method described previously, the present one is quite simple and easy to handle in controlling the generation of mist. On the other hand it takes a couple of milliseconds after applying an electric current till the mist is produced, as well as the switching on the electric current is generally allowed after all only once at each operation. The mist is easily diffused into a flow, so that it should be used for observation of a stationary flow with relatively low speed (below 10 m/s). Some of the examples of use are found in the investigations of the formation of Kármán vortex behind a circular cylinder, the flow within the boundary layer around a spinning cylinder and the growth of a vortex ring out of a circular orifice, which is illustrated in a series of photographs of Fig.8.

6) OPTICAL METHOD

The flow visualization method which employs optical phenomena such as refraction, interference, birefringence, glow discharge, surface reflection, is called on the whole as the optical method. It has a great advantage of visualizing a flow without any disturbance on it. For instance a circulating flow of glycerine and shock waves are observed by the shadowgraph method which relies upon the difference of the fluid density gradient, while the transonic flow through turbine cascade blades and the flow near the trailing edge of an oscillating airfoil are investigated by the Schlieren method which depends upon the fluid density gradient. The Mach-Zehnder interferometry is used for taking a photograph of the interference fringes, showing the equi-density lines. The nozzle flow of a supersonic turbine, e.g. the Mach number distribution inside the turbine nozzle, is analysed as shown in Fig.9. Besides those, the Lazer holography technique is also being developed recently to reproduce the information of the three dimensional field, for instance, the natural convection flow from a heat spot and a supersonic jet flow. The Moiré method enables the visualization of the displacement changes of a fluid free surface, thereby the eccentric vortex in an impeller and the wave pattern due to a ship model are visualized successfully, the latter of which is illustrated in Fig.10. The deformation of melted polymer which is pushed out through a slit is observed by means of the birefringence method, but generally its application examples are quite few, because of the limitation of using a special liquid as the working fluid.

FINAL REMARKS

The major tendency of the recent development of the flow visualization techniques in our country has been summarized as follows:

1) The range of flow speed for visualization has been widening as we see from the examples of utilizing the tracer method of a chemical reaction type for an extremely low speed air flow due to the natural convection and also the optical method for a supersonic flow.

2) The objectives of the flow visualization techniques have been mainly to visualize qualitatively the flow pattern change and also the stream direction. Presently, however, the data recorded from the visualized experiments allow to provide some quantitative evaluation. For instance, we can easily reduce the velocity distribution of a flow field from either the tracer path lines or the time

lines.

3) Most of the flow visualization techniques up to now have handled a stationary flow. However, a remarkable improvement of the techniques of the illumination and the high speed photography which enable to provide some quantitative measurement is presently arousing considerable possibility to visualize a nonstationary flow.

Next, a brief summary for each of the flow visualization techniques is given as follows:

1) New tracer method of pulse luminescence type which employs the optically activated tracers will acquire considerable attentions in the near future.

2) It should be noted that the development of interferometry and holography of the optical method is yielding a remarkable trend of flow visualization, because of noncontacting nature and multi-dimensional information.

3) In order to visualize the extremely low speed flow, both the electrolysis etching and precipitation types of the electro-chemical method are newly developed. They will arise the interest of the concerning researchers.

4) The recent progress of the electrical control method, including the hydrogen bubble, spark tracing and smoke wire types, show the most promising possibility of the quantitative measurement of a flow.

Finally, I would like to classify the visualization techniques employed in both domestic and foreign papers which are presented to this international symposium. The results is shown in Table 4. The distribution of the techniques used in the domestic papers is similar to that of Table 2. The distribution for the foreign papers, however, shows a remarkable difference, that is , the methods of dominant use in foreign countries are the tracer method of the direct injection type and the optical method.

According to Table 4, the tracer method of direct injection type is found to be most popularly used in various research field, whether in the West or the East. While, the work, employing the tuft, wall tracing, chemical reaction and electrical control methods, is given by Japanese participants only, except for one from the U.S. On the contrary, all the papers of the optical method are contributed by foreign participants, while only three papers on holography and birefringence are presented from our country. Consequence is rather maldistribution in the numbers of paper and contributing countries over each technique which has been classified according to the principle. Here, the point is that the above fact should be the very reason for us to organize the present International Symposium.

I sincerely wish that the present International Symposium would provide an opportunity for exchanging the information as well as stimulating mutual interests and understandings upon the visualization techniques.

REFERENCES

1) N.Kawai and Y.Oguni, NAL TR-353 (1973), JAPAN.

2) Y.Seno and M.Yamaguchi, Turbomachinery, 3-3 (1975), 28.

3) T.Fukushima and T.Azuma, 3rd Sympo. on Flow Visualization, ISAS, Univ. of Tokyo, (1975), 89.

4) N.Nakatani, M.Matsumoto, Y.Ohmi and T.Yamada, 3rd Sympo. on Flow Visualization, ISAS, Univ. of Tokyo, (1975), 99.

5) S.Taneda, Recent Research on Unsteady Boundary Layer, ed.e. Eichelbrenner, 2 (1972), 1165.

6) Y.Tomonari, Private communication

7) T.Asanuma, Y.Tanida, T.Kurihara and T.Tanikatsu, Bulletin of Insti. of Space & Aeron. Scie., Univ. of Tokyo, 7-2 (1971), 491.

8) H.Hamada, 3rd Sympo. on Flow Visualization, ISAS, Univ. of Tokyo, (1975), 11.

9) T.Yano and N.Ukeguchi, Technical Review of Mitsubishi Shipbuilding & Engineering, 9-37 (1961), 2.

10) T.Tagori and K.Masunaga, Bulletin Socie. Naval Archi. Japan, 495(1970).

METHOD		TYPE	AIR	WATER
WALL TRACING		OIL FILM	0	0
		OIL DOT	0	0
		SUBLIMATION	0	
		THERMOSENSIBLE PAINT	0	
		SOLUBLE CHEMICAL FILM		0
		ELECTROLYTIC ETCHING		0
TUFT		SURFACE TUFT	0	0
		DEPTH TUFT	0	0
		TUFT GRID	0	0
TRACER	DIRECT INJECTION	STREAK LINE	0	0
		PATH LINE	0	0
		PARTICLE SUSPENSION	0	0
		SURFACE FLOATING		0
		TIME LINE	0	0
	CHEMICAL REACTION	INTERFACE REACTION	0	0
		SURFACE		
		INTER-FLUID		
		INTRA-TRACER		
		ELECTROLYTIC COLOURATION		0
		ELECTROLYSIS PRECIPITATION		
		COLOURATION		
	ELECTRICAL CONTROL	HYDROGEN BUBBLE		0
		SPARK TRACER	0	
		SMOKE WIRE	0	
OPTICAL		SHADOWGRAPH	0	0
		SCHLIEREN	0	0
		MACH-ZEHNDER INTERFEROGRAPH	0	0
		LASER HOLOGRAPHY	0	0
		MOIRÉ		0
		BIREFRINGENCE		0

Table 1. Flow visualization techniques

YEAR / TECHNIQUE			1973	1974	1975	1976	1977	TOTAL
WALL TRACING METHOD			6	4	3	3	1	17
TUFT METHOD			3	2	3	5	0	13
TRACER METHOD		DIRECT INJECTION	21	22	17	15	5	80
		CHEMICAL REACTION	6	1	1	4	2	14
		ELECTRICAL CONTROL	9	5	5	7	4	30
OPTICAL METHOD			4	3	7	6	2	22
TOTAL (NUMBER OF PAPERS)			49 (36)	37 (34)	36 (35)	40 (34)	14 (15)	176 (154)
TOTAL NUMBER OF PARTICIPANTS			697	731	720	747		

Table 2. Classification of flow visualization techniques presented to the symposia in Japan

12

	WALL TRACING	TUFT	TRACER DIRECT INJECTION	CHEMICAL REACTION	ELECTRICAL CONTROL	OPTICAL	TOTAL
BASIC SCIENCE	2 (3)	2 (2)	25 (13)	(10)	13 (9)	8 (4)	50 (41)
FLUID MECHANICS	(3)	(2)	1 (5)		(1)	3 (4)	4 (15)
HEAT & ENGINE	1	1	2 (9)		(2)		4 (11)
NAVAL ARCHITECTURE	1 (3)	(4)	(5)	(3)			1 (15)
AERONAUTICAL ENGINEERING	2 (1)		2 (4)	(1)	1	3	8 (6)
ARCHITECTURAL ENGINEERING		1	3 (4)	(1)			4 (5)
CIVIL ENGINEERING			1 (2)		1 (1)		2 (3)
MEDICAL ENGINEERING	1	1	(4)		(1)		2 (5)
TOTAL	7 (10)	5 (8)	34 (46)	0 (14)	15 (15)	14 (8)	75 (101)

NB In this table figures without brackets show the number of cases in which the working fluid is air, while those inside brackets are for the cases in which water is used as the working fluid.

Table 3. Distribution of flow visualization techniques versus relevant applied fields

TECHNIQUE		FOREIGN	DOMESTIC	TOTAL
WALL TRACING METHOD		1	1	2
TUFT METHOD		1	6	7
TRACER METHOD	DIRECT INJECTION	15	14	29
	CHEMICAL REACTION	0	1	1
	ELECTRICAL CONTROL	1	9	10
OPTICAL METHOD		16	7	23
TOTAL		34	38	72

Table 4. Comparison of flow visualization techniques between foreign and domestic presentation in the International Symposium, Tokyo, 1977

Fig.1. Boundary Layer Transition on a Delta Wing (Thermo-
        sensible Paint, Wall Tracing M.)
        A: M = 1.3, Re = 1.8 x 10^6.
        (N.KAWAI, Ref.1)

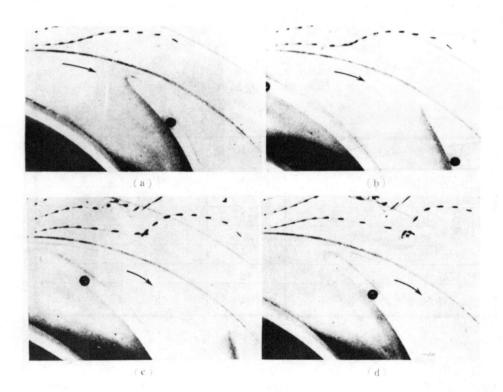

Fig.2. Unsteady Flow in an Impeller Visualized by Intermittent
        Supply of Dye Tracer (Dye, Stream Line M.)
        W: 0.04 m/s, Re = 560.
        (Y.SENO, Ref.2)

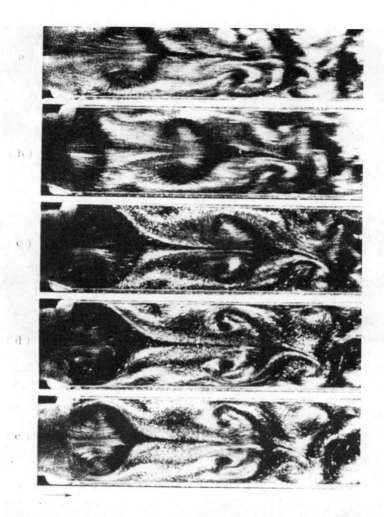

Fig.3. Formation of Vortex Rings by a Mechanical Impulse on
Simulated Artery Wall (AL.Dust, Particle Suspension M.)
W: 0.008 m/s, Re = 220.
(T.FUKUSHIMA, Ref.3)

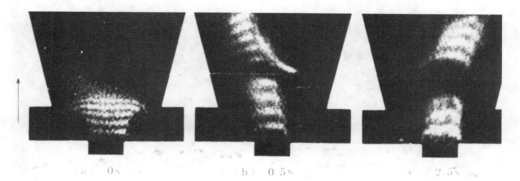

Fig.4. Flow Pattern in a Fluidic Element (Pulse Luminescence,
Time Line M.)
W: 0.34 m/s, Re = 3.4 x 10^3.
(N.NAKATANI, Ref.4)

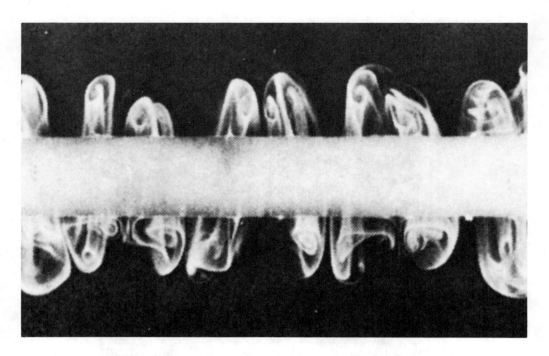

Fig.5. Boundary Layer on a Cylinder Starting Oscillation
      (Electrolysis Precipitation, Chemical Reaction M.)
      W: rest.
      (S.TANEDA, Ref.5)

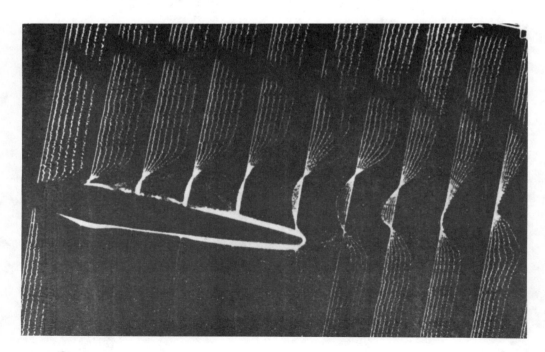

Fig.6. Velocity Profile of Flow around an Oscillating Aerofoil
      (Hydrogen Bubble, Electrical Control M.)
      W: 0.01 m/s,
      (Y.TOMONARI, Ref.6)

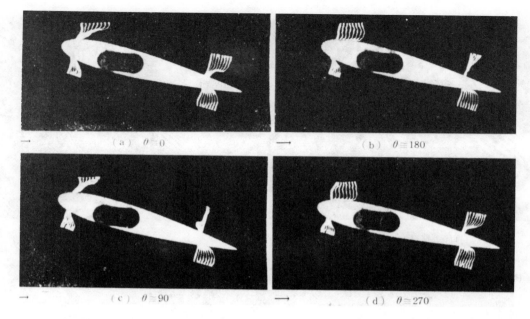

Fig.7. Velocity Profile around an Oscillating Aerofoil
(Spark, Electrical Control M.)
A: 20 m/s, Re = 2 x 10^5.
(T.KURIHARA, Ref.7)

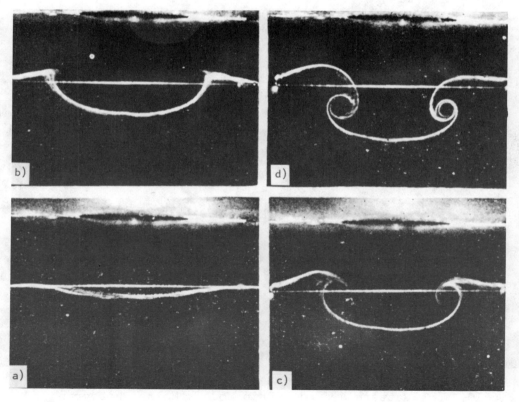

Fig.8. Formation of a Vortex Ring through a Circular Orifice
(Smoke Wire, Electrical Control M.)
A: 0.4 m/s, Re = 1066.
(H.YAMADA, Ref.8)

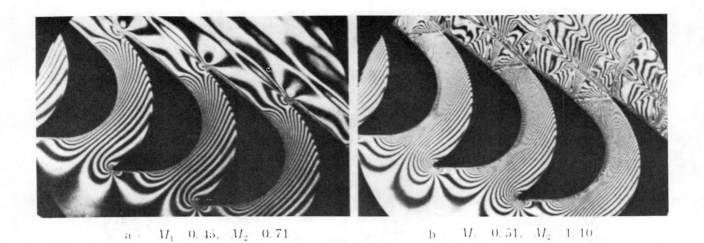

a M_1 0.45, M_2 0.71   b M_1 0.51, M_2 1.10

Fig.9. Transonic Flow through Cascaded Blades of a Turbine
        (Mach-Zehnder Interferometer, Optical M.)
        A: M_1 = 0.51, M_2 = 1.10, Re = 5 $\sim$ 8 x 10^5.
        (T.YANO, Ref.9)

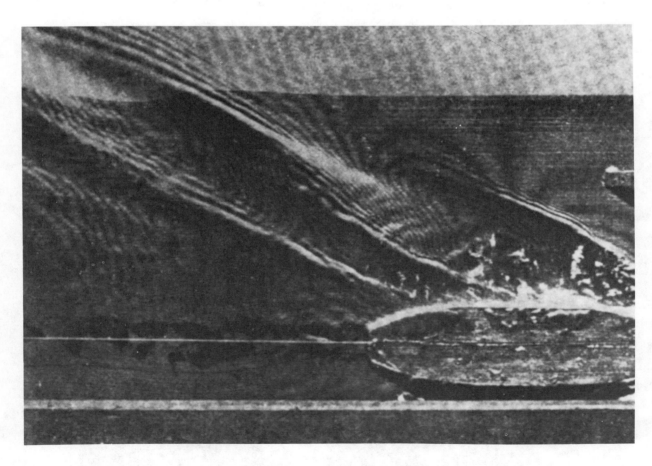

Fig.10. Wave Pattern around a Ship Model (Moiré Photography,
        Optical M.)
        W:
        (T.TAGORI, Ref.10)

FLOW VISUALIZATION IN CZECHOSLOVAKIA AND SOME COUNTRIES IN CENTRAL AND EASTERN EUROPE

R. REZNICEK*

Some memories concerning the application of flow visualization methods in Czechoslovakia. Remarks of general character on the importance and application of flow visualization methods. Survey of methods applied in Czechoslovakia, the Soviet Union, Poland, German Democratic Republic and Hungary. Proposal of the foundation of an international commission for the selection and distribution of educational films utilizing flow visualization.

I would like, first of all, to express my admiration and appreciation to the workers of the Institute of Space and Aeronautical Science of the University of Tokyo for their organization of this Symposium and, at the same time, even my gratitude for being able to take part in it. My participation here will, no doubt, become an unforgetable experience, both from the point of the expected high professional level and for the admirable enthusiasm, effort and devotedness of the initiators and organisers of the Symposium under the leadership of professor Asanuma and professor Tanida and, of course, also for the world-famous precision and hospitality of our Japanese hosts and, last but not least, for the beautiful atmosphere of the Symposium in the City of Tokyo which has always been a legend for me. All this is a good reason which the Institute of Space and Aeronautical Science can be proud of.

The problems of Flow Visualization in Czechoslovakia are being dealt with by quite a number of people, particularly from the point of application of several methods in the solution of technical development problems. The number of research workers working systematically in the field of research and development of visualization methods is, however, much smaller. I have the honour of representing these scientists at theis Symposium. It is beyond discussion that many of them, mostly my colleagues and friends, would love to be present here - unfortunately, Tokyo is much too far away from Prague.

The application of flow visualization in the field of hydrodynamics and aerodynamics has been put through in our country by professor Pešek, the Vice-President of the International Federation of Astronautics, chairman of the Section of Engineering Sciences of the International Academy of Astronautics. Professor Pešek, Head of Department of Aerodynamics, included in the specialised study of aerodynamics

---

* Professor, Head, Physics Dept. Faculty of Mechanization, VSZ, Prague, Czechoslovakia

and aeronautics at Prague Technical University even an exposition concerning several methods of flow visualization and he took part in putting them into practice. I was lucky enough to have been one of his students at that time and even later, during my PhD studies, it was professor Pešek again who encouraged me in 1960 to write a monography containing a survey and explanation of all, at least in my opinion, methods of flow visualization that were known at that time and I could examine a considerable part of them. The manuscript of this book was finished in 1966 and the Publishing House of the Czechoslovak Academy of Sciences published the book in 1972. It was also thanks to professor Pešek and the experimental and theoretical processes introduced by him, including flow visualization, that aero-dynamics and aircraft construction in our country reached a higher standard. Earlier, but already at the time of the beginning of a successful period in our aeronautics, the aircraft constructors based their designs more on their sense for a technical subject rather than on objectively obtained experimental facts. The late professor Hajn, head of the Institute of aircraft Construction at the above mentioned Prague Technical University, told me an interesting story in this connec-tion. The story was about the AVIA BH-10 aircraft of which professor Hajn was co-constructor in the year 1924 and which was an extremely progressive design at that time. It was an elegant single-engined low-wing airplane. Professor Hajn said that the chief constructor, dr. Beneš, pictured the basic proportions and shape of the aircraft and the airfoils of its wings completely off-hand, in the so called "live-line" like an artist and the aircraft built according to this basic design performed really well. Indeed, aircraft designing, much more than any other tech-nical fields, is subject to aesthetic feelings of the designer - and it is not without sense when we hear that a nice-looking aircraft also flies well. And we, the people dealing with flow visualization here, come across this factor of aesthetics in technology quite often. Just imagine-some of the pictures of flow visualization look very much like a piece of art. I can tell from my own teaching practice that attractive pictures of the flow not only give the students unforget-table technical knowledge but they also introduce interesting aesthetic experiences.

But our task is to deal with the problem of discovering new technical infor-mation in the field of fluid mechanics by using optical methods, that is, as professor Merzkirch writes in connection with Hubert Schardin - to make the invisible visible. Yes, indeed, the importance of the methods of flow visualization lies in the fact that they help to elucidate the physical principle of processes which take place in various instances of the flow and thus they facilitate the creation of a physical model approaching the reality of the process, which is the necessary basis of theoretical analysis. This makes evident even the importance of these methods in the field of engineering, agricultural engineering and medical technology; these fields abound with problems connected with fluid flows which cannot avoid the appli-cation of flow visualization. These are, for instance, the problems in the con-struction of new types of hydraulic, steam and gas turbines, combustion chambers, steam generators and engines, parts of aircrafts, ships and rockets, then comes pneumatic and hydraulic transportation, separation of particles of different density, capillary elevation of liquids in porous matter, heating, ventilation, cooling, making spray aerosols and the movement of their particles in the atmosphere, under-ground water flow, the function of wind-protecting belts, etc. Theoretical solution of problems in the above mentioned technical disciplines is very difficult in most cases, if not completely impossible, and that is why designers often rely on their long experience and technical commonsense. But the application of flow visualiza-tion makes it possible for them to understand the problems in greater detail, to obtain new data for the construction work and thus helps them to raise the technical standard of their work.

A number of different visualization methods are now applied in Czechoslovakia, the Soviet Union, German Democratic Republic, in Poland **and** Hungary. They can be

classified into 3 groups:
1) Methods based on introduction of particles into fluids, the particles having different properties than the fluid;
2) Methods which investigate the changes of special surface finish of the flown-around objects;
3) Methods exploiting natural changes of optical properties of a fluid which originate during the flow.

All these groups include methods which are applicable both for the study of a liquid flow as well as a gas flow. For instance in Group 1 - in the study of a liquid flow it is the matter of application of particles forming continuous filaments or larger continuous areas. That means the use of various dyes, the application of chemical reactions with colour effects, the electrolytical method, electrochemoluminiscence. In the application of separated particles which do not form continuous filaments or larger continuous areas the matter of application concerns sawdust, filedust, oil droplets and air-bubbles inside a liquid. A frequently used method is that of applying aluminium dust on the liquid surface, or the use of lycopodium powder. These methods are applicable in the investigation of a liquid flow. In gas-flow visualization, in the case of continuous filaments or larger continuous areas, the most frequently applied methods are those of various kind of smoke and vapor, thin flame filaments, methods of electrically heated wires, spark discharge, luminiscent methods, vapor-screen method for visualization of shock wave. Aluminium dust, fine balsawood sawdust, spark-discharge heating of air particles and tuft screen are used in the case of particles that do not form continuous filaments.

In fact, this first group also includes the investigation of cloud movement which is recorded by satellites, for example. And so, after all, flow visualization in this field enjoys the widest possible application.

In the second group, the methods can be divided according to the type of process into physical, chemical and mechanical. Special surface coatings are used for liquid flows, and chemical methods based on chemical reactions of gas and the surface of the flown around body are applied for the flow of gases. Among physical methods is the method of sublimation or evaporation with the application of kaolin coating; finally, among mechanical methods applied are such as tufted-yarn probes, or the methods based on tracer praticles which are blown away from the surface or deposit on the same, depending on each particular problem, then various oil and varnish coatings, fluorescence of an oil film in ultraviolet radiation.

I have no information of the surface-finishing method by means of liquid crystals to have been used in the countries I am referring to.

Methods of the third group employing natural changes of optical properties of fluids originating during the flow are represented in cases of liquid flows, by the method of flow birefringence. Recently it has been possible to include in this group even the processes connected with cavitation. The study of cavitation by optical methods leads to the formation of a separate group among the frequently used methods in the field of research of cavitation. These methods belong therefore, from the wider aspect of the sense of flow visualization and also from the aspect of the sort of method applied, into the field of flow visualization, although the study of cavitation has not appeared so far in various surveys and accounts of visualization methods. Of the methods of the third group which are applicable in gas-flow visualization and which are commonly used in our part of the world I should mention the shadow method, the Schlieren method also with coloured image, interferometric method including the application of holography. In the study of gas flows at very low pressures experiments have been made with the application of the principle of absorption of electromagnetic or corpuscular radiation.

All the metioned methods have been studied or applied at various depth and scope of application in Czechoslovakia, the Soviet Union, German Democratic Republic,

Poland and Hungary. The Soviet Union, however leads in the scope, intensity and number of applications. In some instances it was only a matter of laboratory verification of the physical principle and the possibility of application, but in other instances the methods are widely applied in the research in the field of hydrodynamics and aerodynamics and even for technological research serving directly the improvement of the construction of machines and equipment. Among the more remarkable works, as far as the methods used, I should mention here the presently used holographic interferometry in Czechoslovakia and in the Soviet Union and the application of holography in connection with schlieren system. Not less interesting are works aiming at the utilization of numerical methods for interferogram evaluation by means of computers, which is the case of Czechoslovakia and the Soviet Union. The method of small bubbles in flow visualization is very useful in ventilation, as used in Czechoslovakia. Among the earlier methods of interest I should mention the schlieren system with colour image, used in Czechoslovakia and German Democratic Republic, and the method of electric discharge which was studied in German Democratic Republic, and also the method of flow birefringence, used in Czechoslovakia. As far as application is concerned, I should mention the wide-spread use of interferometers for the study of flow in turbine cascades, the tuft method in the investigation of the flow around airplanes and automobiles and surface finishing by means of kaolin coatings as well as the optical study of cavitation effects on the water turbine blades. In most cases the interferometers were made in the Soviet Union and Czechoslovakia. German Democratic Republic was making schlieren instruments.

Before the conclusion, after having more or less fulfilled the task which was put before me by the topic of my paper, I would like to make a suggestion which, in case of your favourable reception, could serve as a recommendation for the competent international organization. The suggestion concerns instructional films and slides illustrating visualization. Most of the participants of this symposium are workers of institutes of higher learning. Most of us lecture in the field of hydrodynamics and aerodynamics. I am sure you will all agree with me that the use of films and slides depicting various examples of flows by means of visualization is extremely instructive and it considerably increases the quality of the teaching process. After all, this fact has been confirmed even by professors who do not deal directly with the study of the methods of visualization. They appreciate the importance of such films for the intensification of the teaching of specialists. Many such instruction films or slides have been produced in various countries. As far as I know, there exist more than 200 such films or loops in the United States. But for many countries it is impossible to get hold of them. That is why I suggest to work out a recommendation for the UNESCO, for instance, or for some other international institution or organization, to authorise a special commission, for example under the management of our host Institute, to make a selection of such films, loops and slides and to organize distribution of these materials under the patronage of the UNESCO. Many countries who need to quickly educate a larger number of specialists (this concerns, for instance, even the developing countries) would no doubt be very grateful for such a service.

Towards the end of my duty, I would like to express a wish: A wish that this symposium be successful; that we all have excellent papers, excellent listeners and excellent discussion, that we all feel happy, that we start a number of friendly contacts and that we all return home with a new inspiration both for our professional activities and for our personal lives.

Literal Referances

1) Řezníček, R. : Visualisace prouděni. Praha, Academia 1972.

2) Řezníček, R., Blahovec, J. : Použiti metod Zviditelňování prouděn̆í v zemědělství. Zemědělská technika, 21 (XLVIII), 1975, 2, pp. 235-243.

3) Belozerov, A.F., Berezkin, A.N., Razumovskaya, N.M., Spornik, A.F. : Aerobalistic Gas Flow Investigation Using Holographic Device to Schlieren System. In High Speed Photography, Tenth International Congress, Nice, 1972, pp. 401-406.

4) Belorezov, A.F., Spornik, N.M. : Polucheniye tenievykh snimkov pri vastanovleniyi volnovo fronta s gologramy v byelom sviete. Optico - mekhanicheskaya promishlenost, 1971, 3, pp. 9-11.

5) Thiele, W. : Schlierengeräte und ihre Anwendungen. VEB Carl Zeiss Jena Nachrichten, 1958, 2, pp. 75-100.

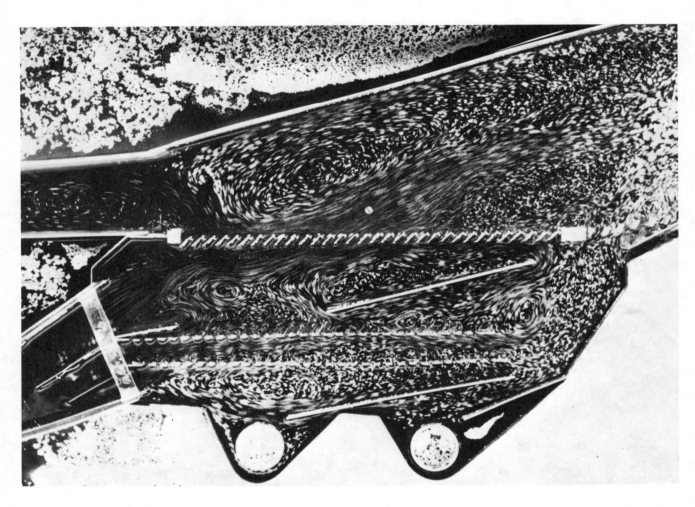

Fig. 1. Flow visualization by means of particles on liquid surface in the study
of the properties of more complex aerotechnical equipment. (Lit.1)

(a)

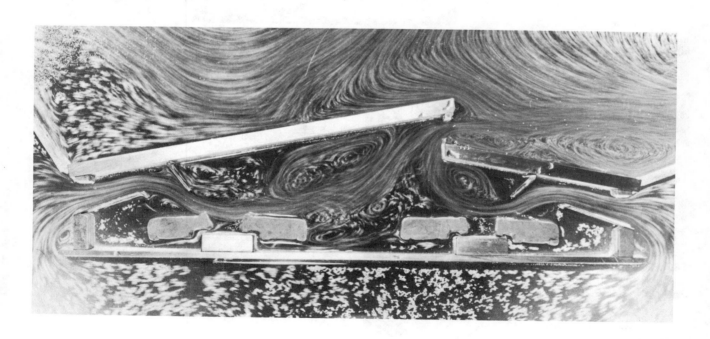

(b)

Fig. 2. Flow visualization of a thin layer of liquid in the study of air -
         conditioning problems (a), (b). (Lit. 1 and 2)

25

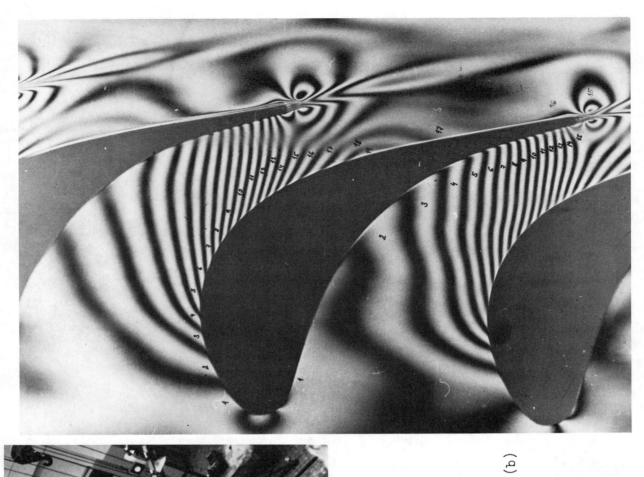

(b)

(a)

Fig. 3. Mach-Zehnder interferometer (a) used in
Czechoslovakia, and the interferogram
(b) obtained by means of this instrument.
(Lit. 1)

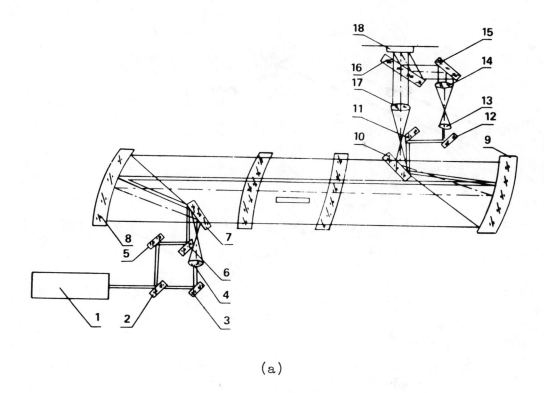

(a)

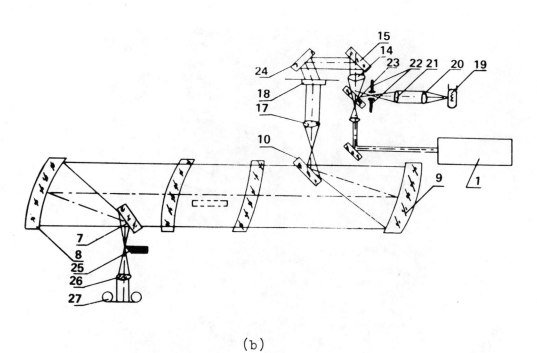

(b)

Fig. 4. Scheme of the Soviet holographic device added to a conventional
Schlieren system IAB-451, (a) obtaining and (b) reconstruction of
holograms. (Lit. 3 and 4)

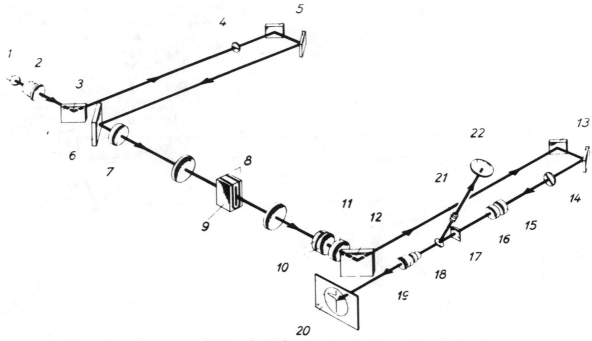

Fig. 5. Scheme of the Schlieren system produced in German
         Democratic Republic. (Lit. 5)

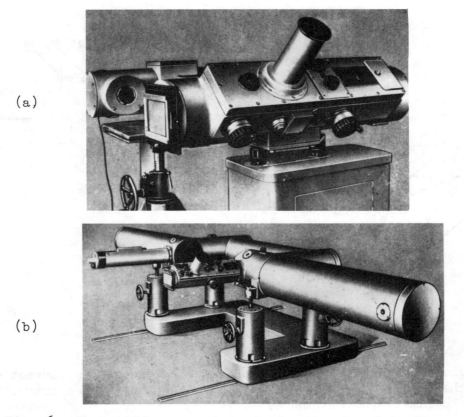

(a)

(b)

Fig. 6. Two schlieren systerm devices produced in German
         Democratic Republic. Schlieren - Aufnahme - gerät
         80 (a) and 300 (b), Carl - Zeiss, Jena. (Lit. 5)

FLOW VISUALIZATION RESEARCH IN WESTERN EUROPE

W. MERZKIRCH*

Selected examples of experimental work on flow visualization are described. Existing methods have been improved, e.g. with respect to sensitivity or quantitative evaluation, and applied to new boundary conditions and practical problems.

Introduction

The main role of flow visualization in experimental fluid mechanics is to allow one to observe and survey a large portion of a flow field, and thereby to facilitate the understanding of the physics of a flow problem. A successful example in this sense is the new concept for the structure of turbulent boundary layers which has been developed as a result of visual studies performed in such flow. Here, flow visualization, primarily, is a qualitative research tool. A great number of methods is known, however, which can also produce quantitative data of a flow field, e.g. the values of velocity, density or shear stress. There is an increasing interest in this field, since most of these methods can be considered as non-intrusive techniques which do not disturb the flow field.

Since the techniques of flow visualization are employed as tools in so many research fields, it is difficult to obtain a complete survey of this field, although "Flow Visualization" has become a key-word for the classification of technical literature. Individual sectors or aspects of the visualization techniques have been reported in special meetings during the last years (Ref. 1,2). It should also be mentioned that the methods of flow visualization can be regarded as part of a newly established field which has been named "Photonics" (Ref. 3,4). Within this framework one includes all the methods which produce, transport and record information by means of photons.

In a review of flow visualization research performed during the past years one must come to the conclusion that no basic, new methods have been developed. The major activities in this field consist of

- improving the ability of existing methods to produce quantitative flow data,
- increasing the sensitivity of the methods, and

---

* Professor of Fluid Mechanics, Institut für Thermo- und Fluiddynamik, Ruhr-Universität, 4630 Bochum 1, Germany

- applying the techniques to practical and more complicated situations.

A number of selected examples of such research is presented in this paper. The selection of examples is rather arbitrary: The original intension of this paper was to provide information in addition to the work which was presented in the form of individual papers at the International Symposium on Flow Visualization in Tokyo (1977). Therefore, this survey neither is complete nor representative for the whole research work performed on this subject in Western Europe. No reference is made to other individual papers of this Symposium and to the work performed at the reporting research institutions.

Visualization Using Smoke and Dye

The production of dye lines in water or smoke lines in air is used for the basic investigation of boundary layers or shear flows as well as for the solution of technical flow problems. This kind of visualization generally yields qualitative results and, therefore, is applied often in addition to another experimental technique. The visualized flow pattern then permits a better interpretation of the results recorded e.g. with a measuring probe. This applies in particular if the measured flow pattern is unsteady. A typical example of such combination and synchronization is described by Bandyopadhyay (Ref. 5). With the aid of smoke flow visualization one was able to recognize a certain structure of the velocity fluctuations in a turbulent boundary layer recorded with a hot-wireanemometer (Fig. 1). Structures of mean regular motions in the near wake behind a circular cylinder have been visualized by Etzold and Fiedler (Ref. 6). A smoke-wire consisting of a smoking woolen thread has been used in these investigations.

Hydrogen Bubbles

Among the electrolytic visualization techniques the hydrogen-bubble method has found the widest application, e.g. to study pipe flows (Ref. 7), stability problems in boundary layers (Ref. 8) or vortex configurations (Ref. 9). These problems are determined by the three-dimensional nature of the flow. In order to resolve the spatial velocity field, Bippes (Ref. 10) has developed a photogrammetric method which uses two separate cameras for recording the bubble pattern under different viewing directions. An evaluation procedure adapted from stereo-photography is described in Ref. 10; the light refraction at the water-air interface is included in the evaluation in order to correctly determine the local position of a individual bubble in the water flow. Fif. 2 shows the pattern of Taylor-Görtler vortices in the flow along a longitudinally curved wall; Bippes (Ref. 10) has succeeded to measure the three velocity components u,v,w of this three-dimensional flow field.

Spark-Tracing Technique

An apparatus for the spark-tracing technique is commercially available. The method can be used for the mapping of velocity fields in air flows of moderate to high speed. The application of the spark-tracing technique should be considered when smoke is not appropriate for being used. The models to be tested have to be made of a transparent material which additionally provides enough isolation for the applied voltages.

Bernotat and Umhauer (Ref. 11) have visualized the two-phase(gas-solid) flow in an air classifier. It is demonstrated that low concentration rates of the solid particles do not appreciably affect the air flow pattern, while high concentration rates cause a strongly disturbed flow field. It was thus possible to explain the loss of efficiency of the classifier when the concentration of the solid phase exceeds a certain value.

Velocity measurements in rotating turbomachines are associated with severe technical complications. Fister (Ref. 12, 13) has found a solution to visualize the flow between rotating turbine-blades with the spark-tracing technique. The key to the problem is a rotating Dove prism which yields constant orientation of the flow image, independently of the instantaneous position of the rotating blades (Fig. 3). The Dove prism has been employed also to perform schlieren measurements in a rotating turbine system.

Interferometric Measurement of Skin Friction

In numerous applications liquid oil films have been used for the visualization of surface air flow pattern. Such investigations are mainly of qualitative nature, indicating the direction of surface streamlines, separation lines, etc.

Tanner (Ref. 14-16) has shown that this oil film can also provide quantitative results. The oil placed on the surface of a rigid body in an air flow moves under the action of skin friction. Tanner gives the following relationship for the instantaneous film thickness $\Delta(x,t)$ at a position x from the (upstream)leading edge of the oil film (see also Fig. 4):

$$\Delta(x,t) = \frac{\mu}{t \cdot \tau^{1/2}} \int_{0}^{x} \frac{dx}{\tau^{1/2}}$$

t is the time passed since the flow has been started, μ the oil viscosity. The skin friction (wall shear stress) τ can be determined if one measures the film thickness. This measurement can be performed with an interferometric method (Fizeau interferometer). The interferogram records the skin friction distribution over a large portion of the surface. A comparison of this method with Preston tube measurements has shown good agreement (Ref. 17). Tanner also reports on the application of this method to the study of separated flow phenomena in a supersonic boundary layer (Ref. 18).

High Resolution Interferometry

The shift of an interference fringe as caused by an inhomogeneous phase object (compressible flow field) is a quantity which can be measured easily. Flow visualization by means of optical interferometers therefore is a quantitative test method. Weak phase objects produce small changes in optical phase or optical path-length and therefore small shifts of an interference fringe. The limit below which a quantitative evaluation does not appear possible usually is assumed as $\lambda/10$ for optical path-length changes or a shift of 1/10 of a fringe width. This limit in sensitivity of conventional interferometers does not allow for measuring the weak density changes occuring e.g. in low-density high-speed flows or acoustical problems.

Smeets (Ref. 19/20) has shown that the sensitivity of interferometers can be increased by at least one order of magnitude. i.e. up to the resolution of changes in optical path-length of $\lambda/100$ (λ = wave length of light). Two operations are necessary for this purpose: Firstly, two complementary interferograms have to be taken and superimposed, one with the test object in place and the second without the object to be tested. Complementary interferograms can be obtained e.g. in the two exit beams of a Mach-Zehnder interferometer. The exact superposition of the two complementary interferograms compensates for all the phase changes introduced by imperfections of the optical system. The second operation required is to take interferograms of infinite fringe width at a position where the intensity profile of the (extremely wide) interference fringe is linear. The pattern resulting from a weak phase object is a variation of the illumination in the field of view, the change in illumination then being directly proportional to the change in optical phase or path-length. This method has been applied to the study of noise generation in air jets (Fig. 15) (Ref. 21/22), and it should be mentioned that a modification of this very sensitive technique has been used by Smeets to investigate the density fluctuations in turbulent boundary layers (Ref. 23).

The quantitative evaluation of these high resolution interferograms is based on the linear relationship between optical phase shift and illumination intensity, and it requires to measure the relative photographic density on a recorded flow picture. This can be done with densitometric techniques. A photographic process to visualize curves of equal density in such flow photographs of gradually changing transparency has been described by Schweiger, Wanders, Wiegand and Becker (Ref. 24,25). This equidensitometric method allows a direct contour-mapping of schlieren photographs, electron beam pictures of low-density gas flows, or of the above mentioned interferograms taken of weak phase objects. For a better discrimination the curves of equal density can be varied in color. One obtains a fringe-like pattern, the distance between two fringes being associated to a change in optical path length of much less than one wavelength.

Speckle Interferometry

The interaction of light with the rough surface of a body or with an inhomogeneous transparent medium produces speckles. The speckle pattern carries information on the kind of inhomogeneity of either the surface or the transparent medium. This information must be decoded, and a number of decoding procedures have been developed in order to analyse the fine structure of such surfaces (Ref. 26). Superposing two speckle photographs of the same object being slightly displaced between the two exposures, results in an interference pattern which yields information on the correlation of the roughness elements (Ref. 27). Tribillon (Ref. 28) has shown that this technique can also be applied to measure correlation parameters of a turbulent field with fluctuating index of refraction.

Flames and Plasma Flows

Preferred objects for optical testing are flows with combustion (Figs. 6,7) and plasma flows, due to the big changes in optical phase achieved in these flow fields. A great portion of the basic work on optical flame research (particularly by interferometric means) has been performed in the laboratory of F.J. Weinberg at Imperial College in London (Ref. 29). Applied problems of plasma flows (e.g. air jets through high-current electric arcs) have been investigated in the research center of BBC in Baden, Switzerland (Ref. 30, 31).

The evaluation of interferograms taken of flames and plasmas suffers from the fact that the flowing gas is a mixture with unknown concentration of the individual components. A possible solution to this problem would be to take several interferograms of the same object at different wavelengths. In practice this has been done only with two wavelengths for the plasma of a monoatomic gas where the two unknown components are the atom gas and the electron gas. Since the refractivity of the electron gas is proportional to λ^2, the sensitivity of this two-wavelengths interferometry increases with increasing difference of the two selected wavelengths. Hugenschmidt and Vollrath (Ref. 32) have shown that this difference $\Delta\lambda$ can be extended beyond the visual range by taking one of the two interferograms with infrared illumination at $\lambda = 10.6\ \mu$ (CO_2-laser). This infrared interferogram is recorded on a layer of cholesteric liquid crystals.

Spectroscopy is still a major tool for optical flame or plasma research. The usual spectroscopic methods should be considered to be beyond the field of visualization. An interesting, purely visualizing technique, however, is the application of multispectral color photography which may yield information on the structure of such flow fields (Ref. 33).

Interferometry of Three-dimensional Flow Fields and Refractive Errors

Since the optical information is integrated along the light path through the flow field, interferometric techniques are most often applied to two-dimensional flows. The investigation of an arbitrary three-dimensional test field requires taking more than one interferogram of the object, each at a different viewing angle. If the field has some kind of symmetry, the number of interferograms necessary for evaluation is reduced. In the case of an axisymmetric field, with the axis being normal to the viewing direction, the number of necessary photographs is reduced to one. A number of such evaluation procedures has been reviewed in Ref. 34.

Also included in Ref. 34 is a review of the refractive errors which arise when the light does not propagate along a straight path through the test object. Such strongly refractive flow fields are plasma flows, flames, flows with strong heat transfer rates, and, as a recent application of optical visualization methods, flows in liquids having a density stratification. The strong deflection of light in a stratified fluid is shown in Fig. 8 for the case of a laser beam which enters horizontally into a water-salt solution having a vertical gradient of the salt concentration. The beam, seen here from the side due to light scattering, is bended towards the direction of higher salt concentration or higher density. The deflection must be taken into account if one wants to perform interferometric measurements of the motion of such fluid.

References

(1) C. Véret and W. Merzkirch, Optical Interferometry in Experimental Gasdynamics, Report on Euromech Colloquium 55, Appl. Opt. 14 (1975), 801

(2) Applications of Non-Instrusive Instrumentation in Fluid Flow Research, AGARD Conference Proceedings No. 193, Paris, 1976

(3) M. Balkanski and P. Lallemand (Editors), Photonics, Gauthier-Villars, Paris 1975

(4) Proceedings of the 12th International Congress on High-Speed Photography (Photonics), published 1977 by Society of Photo-Optical Instrumentation Engineers, Palos Verdes, Calif.

(5) P. Bandyopadhyay, Combined Smoke-Flow Visualization and Hot-Wire Anemometry in Turbulent Boundary Layers, Paper presented at the Symposium on Turbulence, Techn. Univ. Berlin, 1.-5. Aug. 1977, Proceedings to be published by Springer.

(6) F. Etzold and H. Fiedler, The Near-Wake Structure of a Cantilevered Cylinder in a Cross-Flow, Z. Flugwiss. 24 (1976), 77.

(7) M. Clamen and P. Minton, An Experimental Investigation on Flow in an Oscillating Pipe, J. Fluid Mech. 81 (1977), 421

(8) H. Bippes, Experiment 11e Untersuchung sekundärer Instabilitäten in der instabilen laminaren Grenzschicht einer parallel angeströmten konkaven Wand, Z. Ang. Math. Mech. 53 (1973), T. 92.

(9) H. Bippes, Experimente zur Entwicklung der freien Wirbel hinter einem Rechteckflügel, Acta Mech. 26 (1977), 223.

(10) H. Bippes, Eine photogrammetrische Methode zur Messung dreidimensionaler Geschwindigkeitsfelder in einer mit Wasserstoffbläschen sichtbar gemachten Strömung, DLR-FB 74-37 (1974)

(11) S. Bernotat and H. Umhauer, Application of Spark Tracing Method to Flow Measurements in an Air Classifier, Optoelectronics 5 (1973), 107

(12) W. Fister, Sichtbarmachung der Strömungen in Radialverdichterstufen, besonders der Relativströmung in rotierenden Laufrädern, durch Funkenblitze, Brennstoff-Wärme-Kraft 18 (1966), 425

(13) W. Fister, Photographische Erfassung der Strömungsverhältnisse in Turbomaschinen, in: Photographie und Film in Industrie und Technik (O. Helwich, Editor), Verlag Helwich, Darmstadt (1966), 53

(14) L.H. Tanner and L.G. Blows, A Study of the Motion of Oil Films on Surfaces in Air Flow, with Application to the Measurement of Skin Friction, J. Phys. E: Sci. Instrum. 9 (1976), 194

(15) L.H. Tanner and V.G. Kulkarni, The Viscosity Balance Method of Skin Friction Measurement: Further Developments Including Applications to Three-dimensional Flow, J. Phys. E: Sci. Instrum. 9 (1976), 1114

(16) L.H. Tanner, A Skin Friction Meter, Using the Viscosity Balance Principle, Suitable for Use with Flat or Curved Metal Surfaces, J. Phys. E: Sci. Instrum. 10 (1977), 278

(17) L.H. Tanner, Comparison of the Viscosity Balance Meter with Preston Tubes, J. Phys. E: Sci. Instrum. 10 (1977), 627

(18) L.H. Tanner, R.G. Peckham, C.R. Burrows, The Use of Interferometry as an Aid to Studying Flapper-Valve Flow Characteristics, J. Mech. Eng. Sci. 18 (1976), 229

(19) G. Smeets and A. George,Doppelbelichtungs-Interferometrie, Deutsch-Französisches Forschungsinstitut St. Louis, ISL Bericht 39/72 (1972)

(20) G. Smeets and A. George,Visualisierungsverfahren mit dem Differential-interferometer, Deutsch-Französisches Forschungsinstitut St. Louis, ISL Bericht 38/74 (1974)

(21) G. Smeets, Dichteprofile des Unterschallfreistrahls (D = 80 mm), gemessen mit dem Laser-Differentialinterferometer, Deutsch-Französisches Forschungsinst. St. Louis, ISL N 620/75 (1975)

(22) H. Oertel, Jet Noise Research by Means of Shock Tubes, Proceedings 10th Int. Shock Tube Symposium (G. Kamimoto, editor), Kyoto (1975), 488

(23) G. Smeets, Laser Interferometer for High Sensitivity Measurements of Transient Phase Objects, IEEE-Trans., Vol. AES-8 (1972), 186

(24) G. Schweiger, K. Wanders, H. Wiegand, Die Anwendung der Äquidensitometrie zur Diagnose von Zweiphasenströmungen und Strömungen geringer Dichte, DLR-Mitt. 73-02 (1973)

(25) G. Schweiger, K. Wanders, M. Becker, Influence of Electron-Beam Blunt-Body Interactions on Density Measurements in Transition Flow, in: Rarefied Gas Dynamics (K. Karamcheti, editor), Academic Press (1974), 295

(26) J. Bulabois, M.E. Guillaume, J.Ch. Viénot, Applications of the Speckle Pattern Techniques to the Visualization of Modulation Transfer Functions and Quantita-tive Study of Vibrations of Mechanical Structures, Appl. Opt. 12 (1973), 1686

(27) G. Tribillon, Sur quelques méthodes de speckle en métrologie, Optica Acta 24 (1977), 877

(28) G. Tribillon, Determination of R.M.S. σ_z in Some Kind of Turbulence by a Speckle Interferometric Method with Two Wavelengths, Paper presented at Euromech 55, Bochum (1974)

(29) F.J. Weinberg, Optics of Flames, Butterworths, London (1963)

(30) B. Ineichen, U. Kogelschatz, R. Dändliker, Schlieren Diagnostics and Inter-ferometry of an Arc Discharge Using Pulsed Holography, Appl. Opt. 12 (1973), 2554

(31) U. Kogelschatz, Application of a Simple Differential Interferometer to High-
      Current Arc Discharges, Appl. Opt. 13 (1974), 1749

(32) M. Hugenschmidt, K. Vollrath, Interferometry of Transient Phenomena Using
      Lasers Emitting at Different Wavelengths in UV- and Infrared Spectral Range.
      Deutsch-Französisches Forschungsinstitut St. Louis, Bericht ISL 36/74 (1974)

(33) G. Stoffers, Anwendung der Multispektralfotografie zur Sichtbarmachung von
      Flammenstrukturen, DLR-Mitt. 74-23 (1974)

(34) W. Merzkirch, Current Problems of Optical Interferometry Used in Experimental
      Gas Dynamics, AGARD Conference Proceedings No. 193, Paris (1976), 24-1

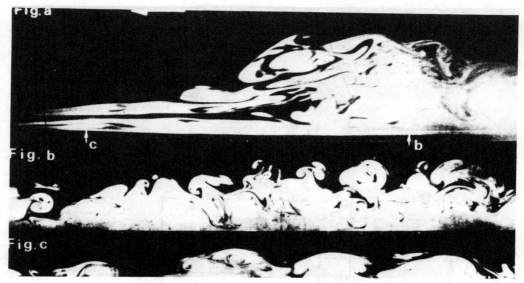

Fig. 1 Smoke tunnel visualization of turbulent boundary layer with abrupt negative
pressure gradient in down-stream direction:
a) side-view; b and c) view of cross-sections at positions b and c (respectively) with the illuminated light sheet being normal to the main flow
direction. By courtesy of Dr. Bandyopadhyay (Ref. 5).

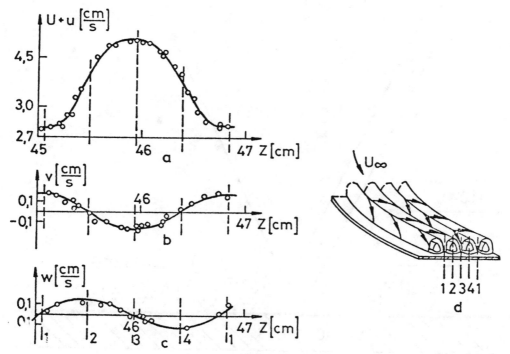

Fig. 2 Results obtained by the stereo-photographic recording of hydrogen-bubble
measurements in a three-dimensional flow field (Taylor-Görtler vortices in
the flow along a cambered wall): a) Velocity component in main flow direction along the wall, b) Velocity component normal to the wall, c) Velocity
component parallel to the wall and normal to the main flow (from H. Bippes,
Ref. 8, 10).

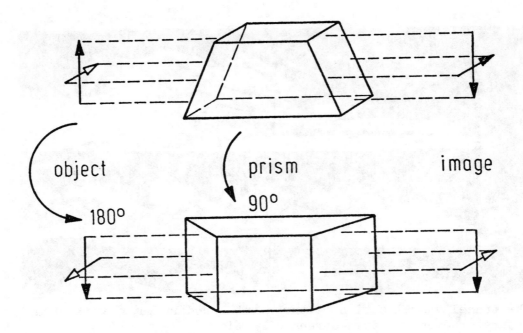

Fig. 3 An image of the object is formed through the Dove prism. Rotation of the
        prism with one half of the rotational speed of the object yields an image of
        constant orientation (according to Fister, Ref. 12, 13).

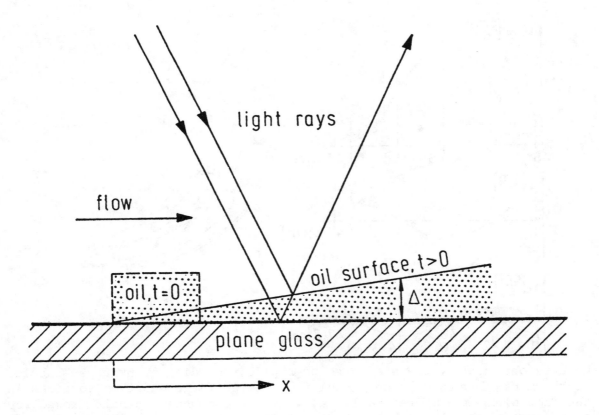

Fig. 4 Interferometric measurement of oil film thickness Δ in the air flow over
        a plane wall: Determination of skin friction according to Tanner (Ref.14-17)

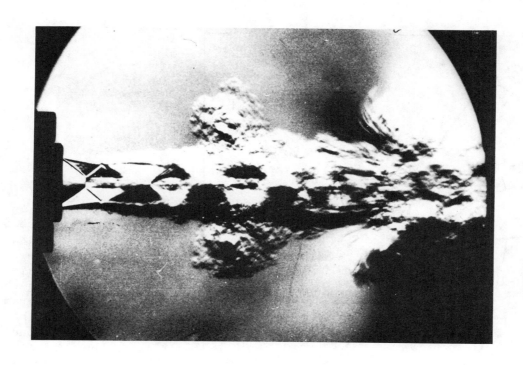

Fig. 5 Schlieren interferogram
of a super-sonic jet
exhausting from the
open end of a shock
tube. The photographic
density of this highly
sensitive interferogram
is directly proportional
to the gradient of the
gas density in the flow
field. By courtesy of
Dr. Smeets (Ref. 20).

Fig. 6 Schlieren interferogram
of a laminar propane
flame and the raising
burned gases. The hori-
zontal bar on the right
side indicates the
height of the flame
above the burner.

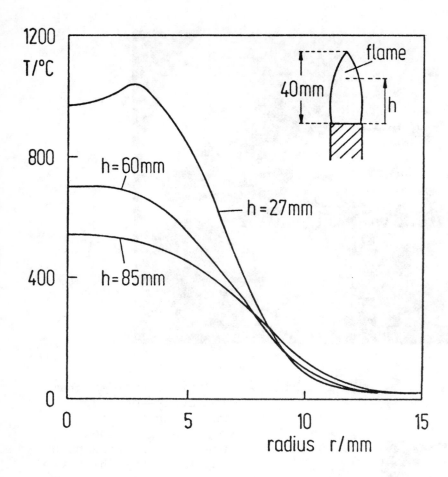

Fig. 7 Temperature distribution in the flame
and in the burned
gases obtained by
evaluating the
interferogram of
Fig. 6 with a sim-
plified model for
the concentration
of gas species.

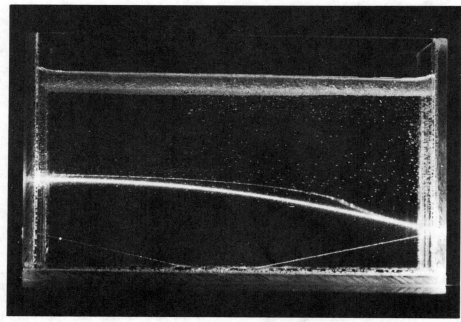

Fig. 8 Side view of laser beam entering horizontally a vessel filled with a
stratified water-salt solution. The laser beam is curved due to strong
refraction. From the opposite (transparent) wall a portion of the laser
beam is reflected back into the fluid. Width of the vessel is about 50
cm.

FLOW VISUALIZATION STUDIES IN THE UNITED STATES AND CANADA

WEN-JEI YANG*

Introduction

There are many researchers among various institutions in the U.S. and Canada who have been and/or are engaged in flow visualization studies. The range of topics are very broad and new innovative flow visualization techniques have been constantly introduced. Applications of the techniques originally intended for engineering and experimental physics have been broadened to various fields such as medicine, biology, architecture, and others. Hence, there is little doubt that this survey must remain incomplete. However, readers are urged to consult two information sources on flow visualization: bibliographies published by the U.S. National Technical Information Service (NTIS): and the Lockheed Information Systems. From the latter source, one can obtain relevant citations and exhaustive bibliographies on flow visualization through a computerized bibliographic search service. Two important and major data bases on the subject of flow visualization are "COMPENDEX" (referring to engineering index data base) and "NTIS" (NTIS data base).

One realizes that the art, science, and technology of flow visualization have always been central to the study of problems in transport phenomena. This is particularly true in case of fluid mechanics including hydraulics, hydrodynamics, aerodynamics, and plasma dynamics. The techniques can be applied to the studies in both simulation facilities and on the actual hardware. Since pioneering application in engineering sciences and experimental physics, new flow-visualization techniques have been constantly introduced to improve the efficiency and efficacy of these studies in transport phenomena.

This survey emphasizes recent studies on flow visualization in the U.S. and Canada. The visualization techniques are classified into three categories: (1) addition of foreign materials, (2) optical methods, and (3) their combinations. Some new setups and principles are described.

* Professor of Mechanical Engineering, The University of Michigan, Ann Arbor, Michigan 48109, U.S.A.

PART I   ADDITION OF FOREIGN MATERIALS

1.   Direct Injection Methods

   1-A.   Smoke

   Introducing a thin jet of smoke into a stream of air is the most popular
method of visualizing flow patterns in air. The smoke is generally produced by
(i) exposing hygroscopic salt (certain substances containing bromide or chloride,
e.g., titanium tetrachloride) to air, (ii) incomplete combustion of organic substance such as wood, straw or cigar, or (iii) smoke generated by the vaporization of
mineral oil.

   In a visualization study of the flow in an axial flow inducer for rocket
pumps or other types of spacecraft pumping machinery, Lakshminarayana (1972) used
the smoke generated by kerosene for the study of both absolute flow and relative
flow. The smoke produced by a solution of titanium tetrachloride and carbon tetrachloride was used to visualize the relative flow near the blade surfaces. Cheng
et al (1977) at The University of Alberta visualized the formation of vortex cells
in convective flows in curved rectangular channels with smoke.

   For applications to cooling problems encountered in gas turbines, compressors,
and other rotating devices, Yu et al (1973) at the University of Minnesota investigated the heat transfer and fluid flow characteristics of a cylinder enclosure having
both rotating and stationary walls in the presence of coolant (air) throughflow.
High rates of sustained smoke generation is required because of the highly disturbed
nature of the flow, the relatively large physical dimensions of the apparatus, and
the long-time viewing needed to observe the complex flow pattern. This requirement,
along with limitations with respect to flammability and toxicity, motivates the
development of a new smoke generator (Yu et al, 1972).

   Mueller and Goddard (1977) at the University of Notre Dame studied the boundary
layer transition on an ogive nose cylinder and transition in the free shear layer
after leading edge separation from an airfoil, the periodicity of the vortex formation in the laminar wake behind thin flat plates and airfoils, the supersonic flow
over a wedge, and the laminar wake behind a glunt body. The flows were visualized
by using cigarette smoke, silk threads and glowing iron particles.

   Model studies of directed vertical sterile air flows were conducted by
Buchberg (1974) in order to develop design criteria for the achievement of effective
forward and reverse isolation of hospital patients with respect to airborne contamination. CO was used as a gaseous tracer to simulate airborne contamination in
flow pattern visualization for the model evaluation.

   1-B.   Small particles and droplets

   Flow visualization in aerodynamics is often achieved by the injection of smoke
or a fog of kerosene into a wind tunnel or other experimental apparatus.

   Atomized DOP - Griffin and Votaw (1973) at the U.S. Naval Research Laboratory
developed the aerosol generation and injection system Fig. 1, that produces a polydisperse liquid aerosol of DOP (d_i(2-ethylhexyl)-phthalate. Compressed air is fed
through the atomizer nozzles to produce a high velocity air stream through four
holes which lie at least 25 mm below the surface of the DOP liquid in the nozzle
head. The exiting streams of air atomize the nearby liquid by means of high shearing

action and produce an aerosol of small particles. The supply of liquid DOP into the high shear region is maintained by vertical feed holes in a collar mounted on the body of the nozzle pipe. The atmoized DOP particles are next passed into a jet impactor that is designed to reduce the size range of dispersed particles in the aerosol. They are then passed vertically through the slit and undergoes stagnation flow on a small horizontal plate. Many of the large particles are stripped from the mixture, and the particle size spectrum is reduced. The average size of aerosol particles is 0.69 micron. The system has the advantages of no moving parts and no need for ignition, combustion, and cooling of the tracer ingredients. The particles produced are of controlled size and flow rate, and the system is continuously operable with the only external requirement being a filtered and regulated supply of compressed air. Whereas the DOP aerosol is especially useful for the visualization of oscillatory flows (Griffin and Votaw, 1972; Griffin, Romberg and Votaw, 1973; Griffin and Ramberg, 1976), the small particle size ranges obtainable also make possible the study of turbulent fluid motion. Figure 2 is one photograph of a van Karman vortex street formed in the wake of a vibrating cable.

Water fog - Bisplinghoff et al (1976) at the MIT Aerophysics Laboratory developed a liquid nitrogen-steam fog generating system for subsonic flow visualization.

CO_2 fog - Erickson used CO_2 fog to visualize the flow field surrounding a single injection hole in his experimental heat transfer study on adiabatic wall film effectiveness and augmented heat transfer coefficients due to blowing for one hole and a single row of holes at various injection angles and center-to-center spacing. The fog diffused so rapidly due to the high turbulent mixing in the injection region that only large scale turbulent motion near the hole was visible.

Liquid droplets in stratified flows - A good laboratory model of an ocean thermocline or of atmospheric inversion layer is a system consisting of two superimposed layers of liquid of different densities, with a thin stationary layer at the interface. Hurdis and Pao (1976) studied the propagation of disturbances within these two geophysical systems induced by the motion of a vertical flat plate through visualizing the fluid motion in the gradient layer by injecting droplets of a mixture of carbon tetrachloride, mineral oil, and oil red dye into the upper layer with a syringe and hypodermic needle. The density of the mixture was such that the droplets were neutrally buoyant at the middle of the gradient layer.

Suspended particles - For stereoscopic (3-dimensional) visual studies of complex turbulence shear flows, Brodkey et al (Nychas et al, 1973; Praturi et al, 1976; Brodkey, 1977) of Ohio State University employed small suspended tracer particles (pliolite of 44 to 47 μm in diameter) in water. Saintsbury and Sampath (1974) evaluated variable combustor geometry for reducing gas turbine emissions. Water models of PT-6 combustor models were tested in a water tunnel to visualize the flow patterns in the simulated combustor using Polystyrene pellets as the tracer. The location and the shape of the baffle were decided from the studies.

Natural convection applications - The buoyancy driven flow induced by natural convection is characterized by low velocities. For a study of natural convection between a body and its finite enclosure, cigar smoke may be used to visualize the flow patterns when air is the working fluid in the gap (Yin et al, 1973). Powe et al (1973) at the Montana State University developed a technique to visualize the very slow motion of water in confined spaces by adding drops of "Ajax" liquid detergent. They (1975) have also studied natural convection flow patterns between a heated body and its cooled spherical enclosure of complicated geometries including vertically eccentric spheres and concentric vertical cylinders with hemispherical end cape. When 20 cs silicone oil (Pr = 150-300) and 350 cs silicone oil (Pr = 2000-5000) were used as test fluids, tracer particles were obtained by spraying fluorescent point over the free surface of the oil. This paint would separate into small particles which would diffuse throughout the fluid and remain in suspension indefinitely.

1-C. Dye

Dye streaks in water are similar to smoke streaks in air for flow visualization. However, the visibility of colored dye streaks is substantially higher and its diffusion rate is usually lower than that of smoke. In view of the above advantages, Liang et al (1976) at the University of Michigan utilized a combined dye filament and hydrogen bubble technique to determine the mechanism of heat transfer augmentation in air-cooled heat exchanger fins resulting from surface perforations. Test models consisting of either single or multiple plates and slots were tested in a towing tank. While the flow behavior in the regions close to the plate surfaces and the wake flow inside the slots was visualized by the traces of food color dyes of different colors, the characteristics of the bulk flow was displayed by the hydrogen bubbles. The study covered all three flow regimes. A similar study using the same technique in a water flume was conducted at Colorado State University (Loehrke et al, 1976).

Dye was employed by Pater et al (1974) at Arizona State University to define the flow regimes (laminar and turbulent) for flow between closely spaced corotating disks for both radially inward and radially outward through flows. Applications include a multiple disk turbine (when the flow is supplied to the outer periphery of the disks, the pair of disks constitute an element of a multiple disk turbine) and a multiple disk pump (when the flow is supplied at an inner radius of the disk). White and Dee (1973) have attempted to visualize the injection molding of three different plastics into a series of molds under varying molding conditions.

To visualize the steady laminar flow pattern induced inside a cylindrical tank by a spinning bottom, Alonso (1974) at the University of Mississippi used a blue aniline dye mixed with methyl alcohol (to adjust the dye density to that of the tap water). A lubricant (yellow oil) was used by Bein et al (1975) as a dye in visualizing a flow field in the narrow gaps between the sides of the rotor and the side plates of the housing of rotary machines such as vane compressors and Wankel engines. The lubricant was injected into the gap to grease a seal to seal the clearances between high and low pressure regions in the cavities surrounding the rotor. The separation line between the lubricant and the leaking fluid (dark blue oil) is sharp on photographs.

Brodkey (1977) suggested the possibility of using combined bye injection and particle marking for stereoscopic visualization of complex turbulence shear flows. Przirembel et al (1977) at Rutgers University observed the flow phenomena during fluid resonance in a tube with the aid of dye injection techniques. The jet flow and the flow in the vicinity of the resonance tube inlet was visualized through color schlieren and shadowgraph techniques.

2.   Tufts and Surface Tracing Methods

2-A. Tufts

Foster and Haji-Sheikh (1975) investigated the boundary layer and heat transfer phenomena in the region of separated flow immediately downstream of flush, normal injection slots. Air was injected through these slots into a turbulent primary flow. Tuft studies confirm the existence of a region of reversed flow immediately downstream of the injection location. Very thin nylon tufts mounted on a perfectly flexible universal hinge was used to evaluate the qualitative nature of backflows and radial flows at the inlet and exit (absolute flow) of an axial flow inducer (Lakshminarayana, 1972). Bangham (1973) employed surface tufting techniques to define flow angularity and degree of turbulence outside the boundary layer of an elliptical cone. The tufts (flags) were attached to the end of a pon mounted normal to the model surface. The data obtained during the supersonic and subsonic wind-tunnel tests using both surface oil and tufts are comparable.

2-B. Wall tracer (surface flows) methods

In studying the air flow pattern around the existing Calgary Health Science Center, a 1-to-1200 scale model was tested in an open jet wind tunnel. The prevailing wind directions were simulated by rotating the model in the airstream. The model was painted black and the oil flow visualization technique was used to show the wind patterns on the ground.

3. Electrical Control Methods

3-A. Smoke wire

Smoke visualization in wind tunnels is subject to one or more of the following limitations: maximum useful velocities and turbulent levels of the flow; means of local introduction and ease of repositioning of the smoke streaks; generation of flow disturbances by the technique; contamination of tunnels especially in recirculating systems; and density and quality of the smoke. Nagib (1977) at the Illinois Institute of Technology employed a smoke-wire technique which can produce controlled sheets of smoke streaklines for visualizing various complex flows such as flows over single and multiple bluff bodies, flows in thick turbulent boundary layers, laminar and turbulent wakes, and laminar and turbulent stagnation flows. The objective is to arrive at a method in air which would be comparable in its quality and simplicity to the hydrogen-bubble technique in water.

3-B. Soap and hydrogen bubbles

Sage Action Inc. (1971) manufactures a bubble generator for producing neutrally-buoyant helium-filled bubbles. It consists of a head (Fig. 3) which is the device that actually forms the bubbles and a console containing micrometering valves which regulate the flow of helium, bubble solution, and air to the head. Neutrally buoyant helium-filled bubbles about 1 mm in diameter, form on the tip of the concentric tubes and are blown off the tip by a continuous blast of air flowing through the shroud passage. The bubble solution flows through the annular passage and is formed into a bubble inflated with the helium passing through the inner concentric tube. The desired size and neutral buoyancy are achieved by proper adjustment of the air, the bubble solution, and the helium flow rates. As many as 300 bubbles per second can be generated in this device.

Colladay and Russell (1976-a, -b) of the NASA-Lewis Research Center injected these bubbles into a turbulent boundary layer through discrete holes in the test surface of an ambient air-wind tunnel, Fig. 4. The paths traced by the bubbles map streakline patterns of the injected film air mixing with the mainstream. Unlike fog or smoke, which diffuses rapidly, the bubble streaklines are clearly identifiable as continuous thread-like streaks which can be traced through the film injected region.

At the University of Michigan (Liang et al, 1976), hydrogen bubbles were generated in a towing tank from a fine platinum wire (0.025 mm in. dia.) through application of a pulsed direct current. Consecutive rows (time lines) of the hydrogen bubbles show the turbulent structure of the flow in the passage through the test models consisting of multiple plates and slots (see Section 1-C).

4. Chemical Methods

4-A. Electrolytic dye production

In 1966, Baker described a technique which is applicable in aqueous solutions and uses a pH indicator for visualizing three-dimensional flow fields. The method consists of placing two electrodes in a solution of thymol blue (thymolsulphone-phthalein). A d.c. voltage is impressed between the electrodes, one of which is a fine wire placed in the flow field. The resulting current flow induces a proton transfer near the wire, resulting in the color change there. If the voltage is pulsed, a small cylinder of colored solution will form around the wire, then move away from the wire with the flow.

Marple et al (1974) applied the method to visualize the flow field in an inertial impactor such as the aerosol impactor used for the inertial separation of airborne particles.

4-B. Photochemical dye production

The electrolytic dye production method suffers from the presence of the cathode wire in the flow. Hummel at the University of Toronto is a pioneer in developing this flash photolysis technique (Popovich and Hummel, 1967-a). A very dilute solution of 2- (2, 4-dinitro-benzyl)-pyridine in 95 percent ethylalcohol was employed. When the light from a flash tube is focussed onto a point in the fluid, a photochemical reaction is initiated which yields a spot of blue dye within a few microseconds. A tracer can thus be introduced at the desired position and time without disturbing the flow. The method was employed to study the viscous sub-layer in turbulent pipe flow (Popovich and Hummel, 1967-a and -b). Later (Iribarne et al, 1969; Smith and Hummel, 1973), a pulsed ruby laser was used to re-place the flash tube. The photochemical dye can be produced along the whole path of the laser beam through the fluid. Velocity profile may be obtained, without disturbing the flow, by directing the beam normal to the mean flow direction. The technique has been used to investigate a laminar jet (Iribarne et al, 1972) and normal and tangential velocities in the boundary layer and out into the free stream in laminar flow around spheres (Seeley et al, 1975).

5. Cavitation

Polymer additives - Using a camera specially desinged for photography of water jets, Hoty and Taylor (1977) of the U.S. Navy studied the effect of the drag-reducing additive (at a concentration of 25 ppm), called polyacrylamide, on under-water jet cavitation. The cavitation inception index is greatly decreased in the presence of polymer additives. Whereas the bubble appearance in pure water resembles ragged groups of small, sharp and rough bubbles, the cavitation bubbles in polymer solution are larger, rounded and of completely normal appearance.

PART II. OPTICAL METHODS

Optical methods are non-disturbing and most suitable for unsteady flows in compressible fluid media. The methods are based on the idea that variations in the density of the fluid medium (ρ) imply variations in the local index of light refraction (n). The Gladstone-Dale equation describes

$$n = 1 + K\rho$$

K is the Gladstone-Dale constant which is a property of the fluid medium.

Both shadowgraph (responding to the second derivatives of the density) and schlieren (responding to the density gradient) methods utilize the light deflection in a fluid medium. On the other hand, the associated phase alteration is the basic effect for the visualization of flows with interferometers. The light sources for optical visualization methods includes standard light sources such as a mercury vapor lamp and the laser-light source. Laser light is highly monochromatic and coherent and, in certain types, can produce light pulses of extremely short duration and high-energy concentration. It is especially suitable for interferometry. Recently, laser light has been used as a light source in holographic flow visualization methods.

1.   Shadowgraph

In studying the location of transition from laminar to turbulent flow in the boundary layer of high-speed test models, Bock (1974) of the ARO Inc. used the focused shadowgraph systems with schlieren quality optics, Fig. 5. Specially designed spark light sources provide illumination for shadowgraphs of relatively low velocity models (on the order of 833 m/s); at higher velocities ($\sim$ 5000 m/s) a pulsed ruby laser (G-switched) is used as the shadowgraph system light source. Use of the laser light source has also allowed a combination front light/focused shadowgraph system, Fig. 6. Typical photographs are shown in Figs. 7. The model in Fig. 7-a had a smooth surface, whereas a portion of the model in Fig.7-b had been roughened with grooves. Figure 7-c would have been blurred at this model velocity had a spark light source been used. The surface detail of the model in Fig. 7-d is provided by the simultaneous front-light photography.

2.   Schlieren Methods

Dye injection and oil film techniques are difficult to use in water at high velocities. Arakeri et al (1973) at the California Institute of Technology used the optical setup, Fig. 8, in a high speed water tunnel to determine the incipient condition for boundary layer cavitation on two axisymmetric bodies by visual observation. Photographs of the thermal boundary layer itself (with the test body heated or cooled to create the density gradients) were taken with a schlieren system. Both bodies were found to exhibit a laminar boundary layer separation, while cavitation inception was observed to occur within the region of separated flow. Boundary layer transition was observed in Arakeri (1974).

Hannah (1975) at the U.S. Naval Ordnance Laboratory developed a technique to measure the boundary layer phenomena in two-dimensional and axisymmetric flow fields with the aid of a schlieren technique. A schlieren visualization study was performed on thermal stratification in Freon 113 and water with localized heat sources by Lovrich et al (1974.). Page and Przirembel (1977) at Rutgers University developed several techniques for visualizing regions of separated flows in supersonic and subsonic flows. The techniques include color schlieren, shadowgraph and Wollaston prism schlieren interferometry.

3.   Interferometry

Various surveys on this optical method have appeared in the past, for example (Marzkirch, 1973 and 1974). Holographic interferometry is presented separately in the succeeding section.

Figure 9 illustrates the optical set-up which was used to study an unsteady, underexpanded axisymmetric jet from a 2.54 cm diameter tube at supersonic velocity (Oertel, 1973): jet tube location, multiple pulse laser, Mach-Zehnder interferometry, and high speed camera. A bleachable absorber is used to generate multiple laser pulses so that instantaneous images of the laser-illuminated event are swept onto a fixed film drum with a mirror rotating at about 75000 RPM for a film writing speed of about 1.63 cm/µs. From sequence (mutiple frame) laser interferograms made by blacklighting a Mach-Zehnder optical interferometer, the density distribution in the entire flow field may be inferred. Spark shadow photographs give qualitative support to the finite fringe interferograms, as shown in Figs. 10 and 11.

Farhadieh and Tankin at Northwestern University employed a Mach-Zehnder interferometer to study two-dimensional Benard convection cells in distilled water and sea water in the region where density is a linear function of temperature. A dye particle (DuPont Victoria Green) was dropped into the test section to indicate the nature of the convective pattern. This dye particle sank to the bottom and dissolved, yielding streaklines which were green and clearly visible on the interferograms. An M-Z interferometer was also used to visualize the diffusion of heat and mass in a porous tube generated flow field. A pulsed Xenon laser (λ = 0.5126 µm) as the illumination source was used to obtain interferograms of the mixing layer that develops at the interface between two dissimilar gas flows having different temperatures or consisting of two different species and generated by two adjacent grids of porous tubes.

4.   Holography

Flow visualization holography may be grouped into two: hologram (single exposure holography), and holographic interferometry (double exposure holography). The hologram (the interference patterns resulting from the sum of a reference beam and an object beam) can be reconstructed (reilluminated with the reference beam) to make 3-dimensional photograph, shadowgraph, schlieren and velocimeter. It is the holographic interferometry that has been expanded more than any other by the introduction of holography and has been applied to flow visualization more often than any other.

TRW Systems Group were perhaps the first to make extensive applications of holography in flow diagnostics, including the study of projectiles in flight, rocket engine combustion chambers, coal and petroleum fired furnaces, and some basic studies in ignition and flame propagation. The usual procedure of producing two beams, one for the scene and one for the reference beam, is used in their holography arrangement, Fig. 12. However, the holocamera is unique in that the individual rays of light are brought together again at the hologram plane, even though diffuse light is used to illuminate the scene. This method, combined with pathlength matching of the two beams, greatly relaxes coherence requirements of the laser. It also provides efficient use of the light.

Arnold Research Organization (ARO), Inc. has applied holography in a wide variety of aerodynamic studies including nozzles, jets, spraying systems, rocket and jet engine exhausts, wind tunnels, diesel engines, combustion chambers, arc heaters, erosion and ablation chambers. Figure 13 shows a converted schlieren system with 50 cm diameter primary mirrors. It provides holograms from which schlieren, shadowgraph, interferometry and other spatial filtering recordings can be made. A comparison of a flow field made with a conventional white light schlieren system with that made from a hologram is shown in Figs. 14(a) and (b). Diffraction effects are more predominant in the laser illuminated case. An obvious difference is observed in turbulent regions where the extremely short duration laser pulse stops the flow to the extent that fine structure can be seen.

Double plate hologram interferometry has been applied to the study of boundary layers, JP-4 sprays, combustion boron particle clouds, wind tunnels, turbulent boundary layer separation in supersonic flow, and shock-boundary layer interactions, at the U.S.A.F. Aerospace Research Laboratories. Figure 15 shows quantitative density distribution in a 4 mm thick boundary layer. At Pratt-Whitney, holography has been applied to the study of flow fields in nozzles, turbine blades, and wind tunnels using a system shown in Fig. 16. In high energy laser development, the reaction of hydrogen and fluorine in chemical lasers requires that these chemicals be injected and uniformly mixed at low pressure. The U.S. Army Missile Command has applied holographic interferometry in the study of mixing in chemical lasers, yielding design information which is vital to the injection technique. Science Applications, Inc. has applied holography to study materials response in erosive and ablative environments. Both ruby and frequency doubled YAG lasers have been used in these studies. Spring and Ragsdale (1974) at the U.S. Naval Ordnance Laboratory used holography in studying supersonic flow in air inlets. The shape and location of the shock wave were obtained from the photographs of the holographic interferograms by a graphical procedure.

Burner (1973) at the NASA Langley Center derived an expression to determine the average gas density across the test section by measuring fringe shifts from the reference position at a known density. Collins et al (Jagota and Collins, 1972; Kosakoski and Collins, 1973) at the U.S. Naval Postgraduate School developed the computer programs to reduce fringe shift data obtained from the interferograms to density. The three-dimensional density fields thus obtained for transonic corner flows agree well with those obtained from an analytical solution of the governing equations.

Debler and Vest (1977) at The University of Michigan applied holographic interferometry to the study of two-dimensional motions of salted-stratified water in tests that spanned several decades of Reynolds numbers. The isopycnic contours are coincident with the contours of constant refractive index and the streamlines. The interferometric fringe pattern thus provides a mapping of the streamlines.

PART III. COMBINED COMPATIBLE-NONINTERFERING TECHNIQUES

Sedney et al (1973) at the U.S. Ballistic Research Laboratories exploited several combined compatible-noninterfering techniques for studying complex three-dimensional flow fields without the use of sophisticated instrumentation: (i) a combined optical (shadowgraph and schlieren) and surface indicator (oil flow) technique; (ii) a versatile vapor screen technique using an inexpensive CWlaser; and (iii) a reversible dye technique using the color reaction of chemical pH indicators to give erasable surface flow patterns. The first method has several advantages over the standard one: (i) improved sensitivity and resolution of the surface patterns, (ii) need of a very small amount of oil; (iii) monitoring of development in the flow pattern possible; (iv) recording of some shock positions; and (v) short down time between experiments. The main advantages of the second method is the ease of installation and the ability to move and rotate the screen. It can be easily combined with the first and third methods. The third method has several advantages over the surface indicators: (i) entrained particulate matter is not used in the indicator; (ii) no indicator accumulation in the regions of flow separation or re-attachment; (iii) the patterns are formed after the flow is established; and (iv) the patterns are erasable.

Withjack et al (1974) visualized the Couette instability of a linearly stratified salt-water solution between rotating cylinders by means of shadowgraph and dye-trace methods. Although the shadowgraph reveals the growth of instabilities, it fails to show the physical nature of the associated wave form. Conventional methods such as dye injection and mixing aluminum powder or pearl essence into the test fluid, are prohibited owing to the linear density stratification. In their method, methynol blue dye was diffused into pearl essence. The ingredient was painted onto the inner cylinder surface while the stratified fluid was filling the annulus to minimize exposure to air, which dries the mixture and results in flaking when the cylinder is started to rotate. The shadowgraph and dye traces are photographed using a camera or a movie camera.

Dewey (1971) at the University of Victoria used smoke to visualize flow fields behind a hemispherical blast wave (resulting from a TNT explosion) using a modified shadowgraph technique. Linear smoke (tobacco) streams have been injected in a conventional diaphragm shock tube (Dewey and Witten, 1975) to study one-dimensional flows with schlieren photography. Rectangular arrays of isolated smoke tracers have been employed to study two-dimensional flows such as those produced by shock reflections (Dewey, 1973; Dewey and Walker, 1975-a, and -b). A double pass schlieren system in which a plane mirror is incorporated as a wall of the shock tube is developed for the injection of the smoke tracers, Fig. 17. Ammonium chloride is used as the particle tracer which is formed by bubbling air through concentrated ammonium hydroxide, hydrochloric acid and then water. The optical system uses a pulsed ruby laser as the light source. The system appears to be as accurate as other shock wave monitoring techniques such as pressure gauges and it is more versatile than interferometry. The system is economical, and relatively low quality optical components can be used. The introduction of the particle tracers does not appear to have had any measurable effect on the flow properties. A summary of these techniques is presented in Dewey and McMillin (1977).

REFERENCES

Abbreviations:

ICIASF = International Congress on Instrumentation in Aerospace Simulation
Facilities; AGARD = Advisory Group for Aerospace Research and Development,
NATO; SMPTE = Society of Motion Picture and Television Engineers;
ISFV = International Symposium on Flow Visualization.

Alonso, C.V. (1975). Steady laminar flow in a stationary tank with a spinning
bottom. J. Appl. Mech. 42, 771-776.

Arakeri, V.H. and Acosta, A.J. (1973). Viscous effects in the inception of
cavitation on axisymmetric bodies. J. Fluids Eng. 95, 519-527.

Arakeri, V.H. (1974). A note on the transition observations on an axisymmetric
body and some related fluctuating wall pressure measurements. ASME Paper No.
74-WA/FE-1 presented at the ASME Winter Annual Meeting, New York, Nov. 17-22.

Aivdor, J.M., Kemp, N.H. and Knight, C.J. (1976). Experimental and theoretical
investigation of flow generated by an array of porous tubes. AIAA J. 14, 1534-1540.

Avidor, J.M. and Delichatsios, M. (1977). Flow-field visualization of heat and mass
diffusion in a porous tube generated flow. ISFV Paper No. 39, Oct. 11-15, Tokyo.

Baker, D.J. (1966). A technique for the precise measurement of small fluid
velocities. J. Fluid Mech. 26, 573-575.

Bangham, M.L. (1973). Visualization of the flow field near an elliptical cone body
surface at angle of attack. ICIASF '73 Record, pp. 172-174, IEEE, New York.

Bien, M., Shavit, A. and Solan, A. (1975). Non-axisymmetric flow in the narrow gap
between a rotating and a stationary disk. ASME Paper No. 75-WA/FE-14 presented at
the ASME Winter Annual Meeting, San Francisco, Calif.

Bisplinghoff, R.L., Coffin, J.B. and Haldeman, C.W. (1976). Water fog generation
system for subsonic flow visualization. AIAA J. 14, 1133-1135.

Bock, O.H. (1974). Focused shadowgraph visualization of boundary-layer transition
in aeroballistic range studies. Opt. Eng. 13, 143-146.

Brodkey, R.S. (1977). Stereoscopic visual studies of complex turbulence shear
flows. ISFV, Paper No. 7, Oct. 11-15, Tokyo.

Buchberg, H. (1974). Model studies of directed sterile air flow for hospital
isolation. Ann. Biomed. Eng. 2, 106-122.

Burner, A.W. (1973). A holographic interferometer system for measuring density
profiles in high-velocity flows. ICIASF '73 Record, 140-145.

Cheng, K.C., Nakayama, J. and Akiyama, M. (1977). Effects of finite and infinite
aspect ratios on flow patterns in curved rectangular channels.
Proc. ISFV, Paper No. 17, Oct. 11-15, Tokyo.

Colladay, R.S. and Russell, L.M. (1976-a). Streakline flow visualization of discrete hole film cooling with holes inclined 30° to the surface. NASA TN D-8175.

Colladay, R.S. and Russell, L.M. (1976-b). Streakline flow visualization of discrete hole film cooling for gas turbine applications. J. Heat Transfer 98, 245-250.

Debler, W.R. and Vest, C.M. (1977). Visualization of a stratified flow by holographic interferometry. Proc. Roy. Soc. London A, in print.

Dewey, J.M. (1971). The properties of a blast wave obtained from an analysis of the particle trajectories. Proc. Roy. Soc. London A. 324, 275-299.

Dewey, J.M. (1973). The analysis of the particle trajectories in nusteady shock flows. ICIASF '73 Record, 119-124.

Dewey, J.M. and Walker, D.K. (1975-a). A multiply pulsed double-pass laser schlieren system for recording the movement of shocks and particle tracers within a shock tube. J. Appl. Phys. 46, 3454-3458.

Dewey, J.M. and Walker, D.K. (1975-b). High-speed photography of particle tracers behind non-planar shocks. Proc. 11th Int. Congr. on High Speed Photogra., Chapman & Hall, 386-392.

Dewey, J.M. and Whitten, B.T. (1975). Calibration of a shock tube flow by analysis of the particle trajectories. Phys. of Fluids 18, 437-445.

Dewey, J.M. and McMillin, D.J. (1977). Visualization of shock and blast wave flows. ISFV Paper No. 33, Oct. 11-15, Tokyo.

Erikson, V.L. (1971). Film cooling effectiveness and heat transfer with injection through holes. NASA CR-72991 (Report HTL-TR-102), University of Minnesota).

Farhadieh, R. and Tankin, R.S. (1974). Interferometric study of two-dimensional Benard convection cells. J. Fluid Mech. 66, 739-752.

Foster, R.C. and Haji-Sheikh, A. (1975). An experimental investigation of boundary layer and heat transfer in the region of separated flow downstream of normal injection slots. J. Heat Transfer 97, 260-266.

Griffin, O.M. and Votaw, C.W. (1972). The vortex street in the wake of a vibrating cylinder. J. Fluid Mech. 55, 31-48.

Griffin, O.M. and Votaw, C.W. (1973). The use of aerosols for the visualization of flow phenomena. Int. J. Heat Mass Transfer 16, 217-219.

Griffin, O.M., Ramberg, S.E. and Votaw, C.W. (1973). The generation of liquid aerosols for the visualization of oscillatory flows. ICIASF '73 Record, pp. 133-139, New York.

Griffin, O.M. and Ramberg, S.E. (1976). Vortex shedding from a cylinder vibrating in line with an incident uniform flow. J. Fluid Mech. 75, 257-271.

Griffin, O.M. and Ramberg, S.E. (1977). Wind tunnel flow visualization with liquid aerosol particle tracers. Proc. ISFV, Paper No. 1, Oct. 11-15, Tokyo.

Hale, R.W., Tan, P., Stowell, R.C. and Ordway, D.E. (1o71). Development of an integrated system for flow visualization in air using neutrally-buoyant bubbles. Sage Action Inc. Report 7107.

Hannah, B. (1975). Qualitative schlieren measurements of boundary layer phenomena. Proc. 11th Int. Congr. on High Speed Photogr., Chapman & Hall, 539-545.

Havener, A.G. and Radley, R.J., Jr. (1974). Turbulent boundary-layer flow separation measurements using holographic interferometry. AIAA J. 12, 1071-1075.

Hoyt, J.W. and Taylor, J.J. (1977). Visualization of jet cavitation in water and polymer solutions. Proc. ISFV, Paper No. 47, Oct. 11-15, Tokyo.

Hurdis, D.A. and Pao H.-P. (1976). Observations of wave motion and upstream influence in a strarified fluid. J. Appl. Mech. 43, 222-226.

Iribarne, A.P., Hummel, R.L., Smith, J.W. and Frantisak, F. (1969). Transition and turbulent flow parameters in a smooth pipe by direct flow visualization. Chem. Eng. Prog. Symp. Ser. 65, no. 91, 60-70.

Iribarne, A., Frantisak, F., Hummel, R.L. and Smith, J.W. (1972). An experimental study of instabilities and other flow properties of a laminar pipe jet. AIChE J. 18, 689-698.

Jagota, R.C. and Collins, D.J. (1972). Finite fringe holographic interferometry applied to a right circular cone at angle of attack. J. Appl. Mech. 39, 897-903.

Krasinski, J.S., Yahalom. R. and van Hardeveld, T. (1973). Oil visualization wind tunnel study of the Calgary Health Science Center. Proc. 4th Canadian Congr. of Appl. Mech., Montreal, Que., May 28-June 1, 617-618.

Kosakoski, R.A. and Collins, D.J. (1974). Application of holographic interferometry to density field determination in transonic corner flow. AIAA J. 12, 767-770.

Lakshminarayana, B. (1972). Visualization study of flow in axial flow inducer. J. Basic Eng. 97, 777-787.

Liang, C.Y., Lee, C.P. and Yang, Wen-Jei (1976). Visualization of fluid flow past perforated surfaces. Proc. Japan 4th Symp. on Flow Visualization, ISAS Univ. of Tokyo, July 15-16, 69-73.

Loehrke, R.I., Roadman, R.E. and Read, G.W. (1976). Low Reynolds number flow in plate wakes. ASME Paper No. 76-WA/HT-30 presented at the ASME Winter Annual Meeting, Dec. 5, New York.

Lovrich, T.M., Schwartz, S.H. and Holmes, L.A. (1974). Flow visualization of thermal stratification with localized heat sources. J. Spacecraft 11, 664-669.

Marple, V.A., Liu, B.Y.H. and Whitby, K.T. (1974). On the flow field of inertial impactors. J. Fluids Eng. 96, 394-400.

Mathews, B.J., Wuerker, R.F., Chambers, H.F., Jr. and Hojnacki, J. (1974). Holography of JP-4 droplets and combusting boron particles. in Instrumentation for Airbreathing Propulsion (eds. A. Fuhs and M. Kingery), Progress in Astronautics and Aeronautics, Vol. 34, 297-313.

McCutchen, C.W. (1977). Flow visualization by stereo shadowgraphs of stratified fluid. ISFV Paper No. 31, Oct. 11-15, Tokyo.

Merzkirch, W. (1973). Optical interferometry in gas dynamics. A review of recent progress. ICIASF '73 Record, 125-132.

Merzkirch, W. (1974). Flow visualization. Academic Press, Chap. 3.

Mueller, T.J. and Goddard, V.P. (1977). Smoke visualization of subsonic and super-sonic flows. ISFV, Paper No. 4, Oct. 11-15, Tokyo.

Nagib, H.M. (1977). Visualization of turbulent and complex flows using controlled sheets of smoke streaklines. Proc. ISFV, Paper No. 29, Oct. 11-15, Tokyo.

Nychas, S.G., Hershey, H.C. and Brodkey, R.S. (1973). A visual study of turbulent shear flow. J. Fluid Mech. 61, 513-540.

Oertel, F.H., Jr. (1973). Laser interferometry of unsteady, underexpanded jets. ICIASF '73 Record, 146-154.

Page, R.H. and Przirembel, C.E.G. (1977). Techniques for visualization of sepa-rated flows. ISFV Paper No. 37, Oct. 11-15, Tokyo.

Pater, L.L., Crowther, E. and Rice, W. (1974). Flow regime definition for flow between corotating disks. J. Fluids Eng. 96, 29-34.

Popovich, A.T. and Hummel, R.L. (1967-b). A new method for non-disturbing turbulent flow measurements very close to a wall. Chem. Eng. Sci. 22, 21-25.

Popovich, A.T. and Hummel, R.L. (1967-b). Experimental study of the viscous sub-layer in turbulent pipe flow. AIChE J. 13, 854-860.

Powe, R.E., Yin, S.H., Scanlan, J.A. and Bishop, E.H. (1973). A technique for visualization of the very slow motion of water in enclosed spaces. J. Heat Transfer 95, 408-409.

Powe, R.E., Baughman, R.C., Scanlan, J.A. and Teng, J.T. (1975). Free convection flow patterns between a body and its spherical enclosure. J. Heat Transfer 97, 296-298.

Praturi, A.K., Hershey, H.C. and Brodkey, R.S. (1976). Stereoscopic photography of shear flow turbulence. 4th Bienn. Symp. on Turbulence in Liquids, Paper 35, Univ. of Mo. at Rolla.

Przirembel, C.E.G. and Page, R.H. (1977). Visual studies of resonance tube phe-nomena. ISFV, Paper No. 33, Oct. 11-15, Tokyo.

Radley, R.J., Jr. and Havener, A.G. (1973). Application of dual hologram interfero-metry to wind-tunnel testing. AIAA J. 11, 1332-1333.

Saintsbury, J.A. and Sampath, P. (1974). Variable combustor geometry for reducing gas turbine emissions. in Fluid Mechanics of Combustion (eds. J.L. Dussourd, R.P. Lohman and E.M. Uram), ASME, 217-232.

Sedney, R., Kitchens, C.W., Jr. and Bush, C.C. (1973). The marriage of optical tracer and surface indicator techniques in flow visualization - a survey. ICIASF '73 Record, 155-171.

Seeley, L.E., Hummel, R.L. and Smith. J.W. (1975). Experimental velocity profiles in laminar flow around spheres at intermediate Reynolds numbers. J. Fluid Mech. 68, 591-608.

Smith, J.W. and Hummel, R.L. (1973). Studies of fluid flow by photography using non-disturbing light-sensitive indicator. J. SMPTE 82, 278-281.

Spring, W.C., III and Ragsdale, W.C. (1974). Investigation of the use of holography in studying supersonic flow in inlets. in Instrumentation for Airbreathing Propulsion (eds. A. Fuhs and M. Kingery), Progress in Astronautics and Aeronautics, Vol. 34, 95-106.

Trolinger, J.D. (1974). Laser instrumentation for flow field diagnostics. AGARDograph No. 186, NATO.

Trolinger, J.D., Belz, R.A. and O'Hare, J.E. (1974). Holography of nozzles, jets, and spraying systems. in Instrumentation for Airbreathing Propulsion (eds. A. Fuhs and M. Kingery), Progress in Astronautics and Aeronautics, Vol. 34, 249-261.

Trolinger, J.D., Bentley, H.T., Lennert, A.E. and Sowls, R.E. (1974). Application of electro-optical techniques in diesel engine research. SAE Paper 740125.

Trolinger, J.D. (1975). Flow visualization holography. Opt. Eng. 14, 470-481.

Trolinger, J.D. (1976). Holographic interferometry as a diagnostic tool for reactive flows. Combustion Sci. and Tech. 13, 229-244.

White, J.L. and Dee, H.B. (1973). Flow visualization of injection molding. Soc. Plastics Engrs., Eng. Property and Structure Div., Div. Tech. Conf., Downingtown, Pa., Oct. 9-10, 193-196.

Withjack, E.M. and Chen, C.F. (1974). Experimental study of Couette instability of stratified fluid. J. Fluid Mech. 66, 725-734.

Yin, S.H., Powe, R.E., Scanlan, J.A. and Bishop, E.H. (1973). Natural convection flow patterns in spherical annuli. Int. J. Heat Mass Transfer 16, 1785-1795.

Yu, J.P., Sparrow, E.M. and Eckert, E.R.G. (1972). A smoke generator for use in fluid flow visualization. Int. J. Heat Mass Transfer 15, 557-558.

Yu, J.P., Sparrow, E.M. and Eckert, E.R.G. (1973). Experiments on a shrouded, parallel system with rotation and coolant through flow. Int. J. Heat Mass Transfer 16, 311-328.

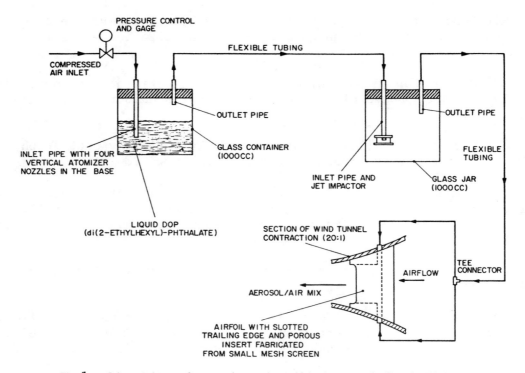

Fig 1 — Schematic layout of an aerosol generation and injection system for flow visualization.

FIG. 2 Flow visualization of vortex shedding from a vibrating six-stranded cable. The Reynolds number based on cable outer diameter and free stream speed is 220. The vibration frequency is 110 per cent of the Strouhal frequency for this Reynolds number and the amplitude is 30 per cent of a diameter.

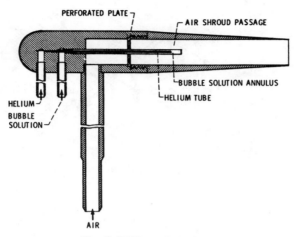

Fig. 3 Bubble generator head

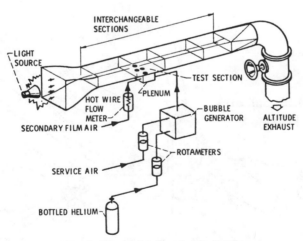

Fig. 4 Film cooling flow visualization rig

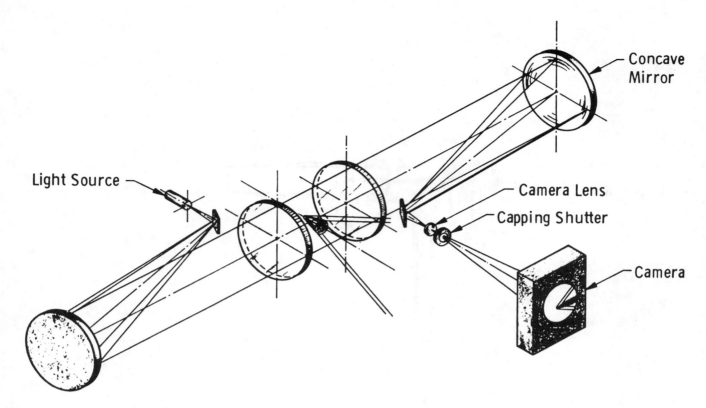

Fig. 5 *Focused shadowgraph.*

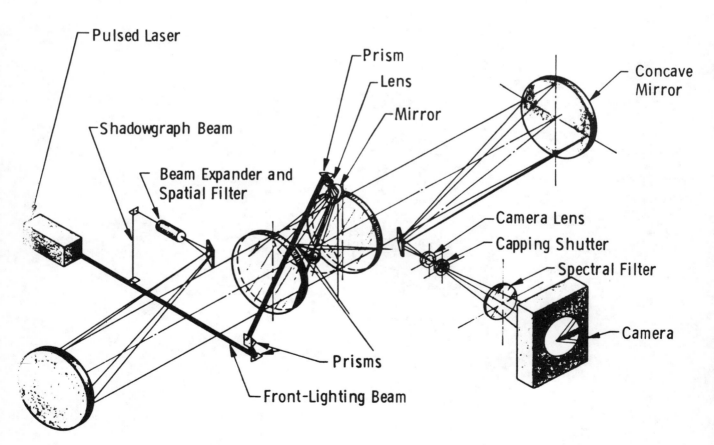

Fig. 6 *Laser front-lighted focused shadowgraph.*

58

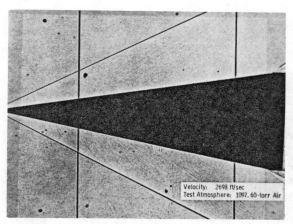

Fig 7 a *Focused shadowgram — spark light source.*

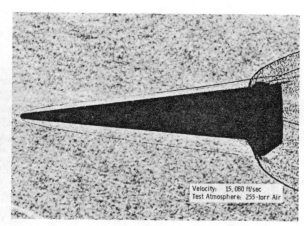

Fig. 7 c *Focused shadowgram — laser light source.*

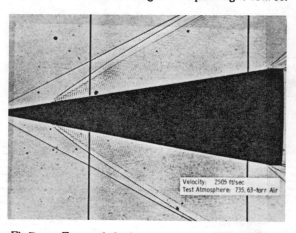

Fig 7 b *Focused shadowgram — spark light source.*

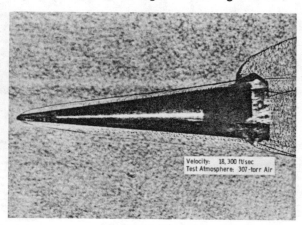

Fig 7 d *Laser front-light/focused shadowgraph photograph.*

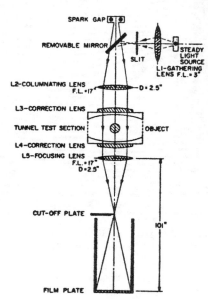

Fig 8 **A schematic diagram of the Schlieren set-up in the high speed water tunnel**

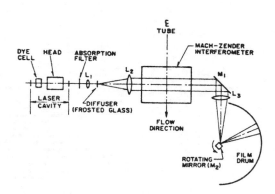

FIG 9 - Plan View of the Optical Set-up: Jet Tube Location, Multiple Pulse Laser, Mach-Zehnder Interferometer, and High Speed Camera.

FIG 10 - Sequence Spark Shadowgrams for Separate Runs in Air. Flow Conditions are Comparable to Those for Figure 9.

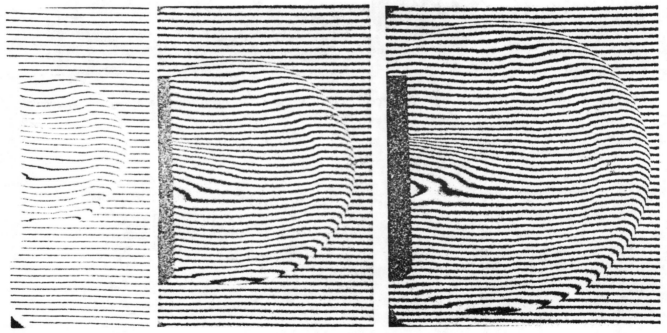

FIG 11 - Sequence of Multiple - Pulse Laser Interferograms for a Single Run in Air.
λ = 6943 Å, $M_s \sim 4.5$; $p_e/p_\infty \sim 75$; $M_e > 1$; Writing Speed $\sim$ 0.64 in/μs; Elapsed Times After Exit: 11 μs, 26 μs, and 40 μs, Respectively

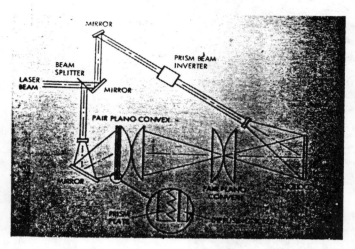

Fig. 12 TRW Systems focused diffusing screen holography arrangement.

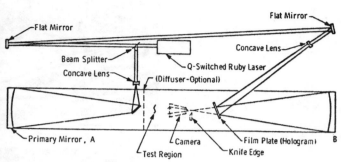

Fig 13 Flow visualization holography system at the Arnold Engineering Development Center (converted from a conventional 50 cm diameter Schlieren system).

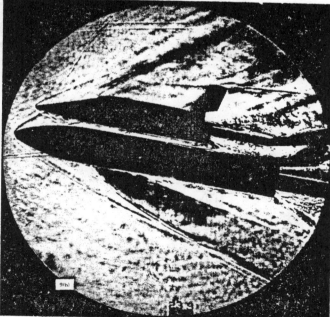

Fig 14 Comparison of conventional and holographic Schlieren. (a) White light Schlieren system; arc pulsewidth was 100 microseconds. (b) Holographic Schlieren; laser pulsewidth 20 nanoseconds.

Fig. 15 Double plate hologram interferometry applied to the study of boundary layers (contributed by Capt. George Havener).

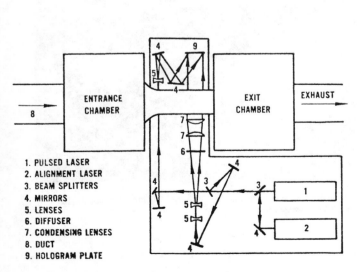

1. PULSED LASER
2. ALIGNMENT LASER
3. BEAM SPLITTERS
4. MIRRORS
5. LENSES
6. DIFFUSER
7. CONDENSING LENSES
8. DUCT
9. HOLOGRAM PLATE

Fig. 16 Holographic flow visualization system of Pratt Whitney.

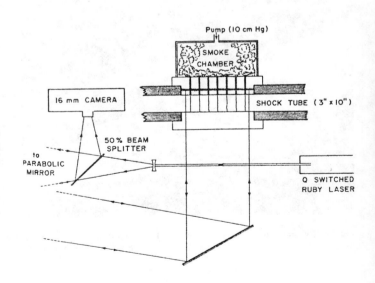

Fig. 17 Double-pass multiply-pulsed laser schlieren and smoke injection custom.

DIRECT INJECTION METHODS

WIND TUNNEL FLOW VISUALIZATION WITH LIQUID PARTICLE AEROSOLS

O. M. GRIFFIN* and S. E. RAMBERG*

A liquid aerosol generation system has been developed recently at the Naval Research Laboratory (NRL) and it is capable of producing sub-micron sized flow tracers of di (2-ethylhexyl)-phthalate, or DOP, particles which are non-toxic, noncorrosive and otherwise safe and convenient in normal usage. This paper describes the aerosol generation and injection systems presently in use at NRL, and briefly discusses the light scattering qualities of the aerosols so produced. Several typical examples are given of the results from experiments conducted recently in two low-speed wind tunnels. These examples include photographs of the vortex wakes behind vibrating bluff cylinders and the flow fields near circular cylinders of finite length at yawed incidence to a uniform flow.

Introduction

Flow visualization by some means is often the first step in the study of complex fluid dynamic phenomena. The visualization of flows in aerodynamics is often accomplished by the introduction of smoke generated from wood, straw or other combustible substance into the flow, but these and other materials in use (1,2) frequently pose problems of handling, cleanliness and safety. These factors suggest the need for a flow tracer that is nontoxic, noncorrosive and otherwise safe and convenient in normal use. Such a system has recently been developed (3,4) and utilizes di (2-ethylhexyl)-phthalate, or DOP, from which a polydisperse liquid aerosol of fine droplets is generated at room temperature with only an external requirement of clean compressed air. This DOP aerosol is nontoxic, chemically inert and noncorrosive under normal laboratory conditions, and also has a high flash point. The particles produced in the aerosol generator are typically of mean diameters less than 1 μm and are especially useful for studies of oscillatory and turbulent air flows. DOP and similar monodisperse aerosols of minute particles also have been used as seedings to enhance the light scattering properties of gases in which flow measurements are made by means of laser doppler anemometry, LDA (5), and laser streak velocimetry, LSV (6).

The purposes of this paper are to describe the aerosol generation and injection systems currently in use at the Naval Research Laboratory (NRL), to briefly discuss the light scattering qualities of the aerosol and to give typical examples of the results which have been obtained with these systems.

*Ocean Technology Division, Naval Research Laboratory, Washington, DC. 20375, USA.

Experimental Facilities and Equipment

The flow experiments reported in this paper were performed in open jet tunnels equipped with 75 mm × 75 mm and 150 mm × 150 mm exits and 20:1 area contraction ratios. These relatively small and convenient facilities are capable of speeds up to about 8 meters/second. The tunnels, together with their associated measuring devices and data acquisition systems, have been described in detail elsewhere (7,8,9,10). There are provisions for mounting cylindrical models in several vibration-isolated shaker & yoke assemblies at the exit sections of the tunnels, and for positioning various bluff model shapes at different angles of yaw to the incident flow.

Some measurements of the light scattering properties of the DOP aerosol and of smoke were also part of the experimental program. The mixtures from the aerosol and smoke generators were fed into a cylindrical glass container which was shielded except for 15° slits through which the incident and scattered light beams were passed. One of the slits was moveable so that the variation of the scattered light intensity could be measured as a function of the angle between the incident and scattered beams. The incident beam was generated by a dc light source, and the beam intensities were measured with a Gamma Scientific Log-Linear Photometer.

The Generation of Aerosols for Flow Visualization

The aerosol generation and injection system employed most recently is outlined in Figure 1. The injector sketched in the figure is similar to the aerosol and smoke injectors that were previously developed at NRL (3,4). Compressed air is filtered and then introduced to a pressure valve and gage that meters the flow to the inlet pipe and to the atomizer nozzles. The atomizer nozzles in the present system consist simply of four 0.25 mm diameter holes drilled in the base cap of the inlet pipe. The liquid ingredient for the aerosol is di (2-ethylhexyl)-phthalate (11), which is chemically inert, nontoxic and noncorrosive. This version of the atomizer nozzle was particularly suitable for use in the low-speed flow visualization experiments described in this paper because of the low outlet pressure and flow rate through the system. Additional atomizer nozzles, similar in design but capable of larger aerosol generation rates, have been developed at NRL and are described elsewhere (3,12). One of the larger nozzles was employed in the light scattering experiments described in a later section.

The atomized particles of DOP are routed to a jet impactor that removes most of the largest particles in the aerosol. This particular impactor consists of a brass tube with a slit jet orifice at one end. The aerosol is accelerated through the slit and impinges on a small flat plate, where most of the largest particles are stripped from the mixture. In principle, the slit and plate arrangement can be adjusted to obtain the desired particle size distribution. The design and construction of the atomizer nozzle and jet impactor are quite simple and economical. Experiments undertaken at NRL by Echols & Young (12) showed that the mean size of aerosol particles was 0.69 μm after passing through a system similar to that of the present study with an inlet air pressure of 1.73×10^2 kPa. The particle size distribution was such that most particles were less than 1.3 μm in diameter after passing through the jet impactor under these conditions. After the jet impactor, the aerosol is ready for introduction into a wind tunnel or other experimental apparatus.

In order to introduce the aerosol into the larger wind tunnel, a slender airfoil with a slotted trailing edge was positioned vertically in the tunnel contraction section as shown in Figure 2. The airfoil is hollow and contains a porous filler of 60 mesh wind tunnel screening to achieve a uniform injection rate of aerosol along the vertical length of the airfoil's trailing edge. A velocity profile containing a vertical sheet of aerosol with a uniformity of ± 2 percent in mean velocity has been achieved at the exit jet of the tunnel by proper adjustment of the tunnel speed and the aerosol bleed rate. Equally good results have been obtained with a similarly-configured airfoil (3) mounted in the smaller wind tunnel.

The generation and injection systems for the DOP aerosol have the advantages of no moving parts and no need for heating, cooling or combustion of the tracer material. It is possible to generate aerosols with adjustable size spectra and mass concentrations by appropriately choosing the dimensions of the atomizer nozzle and the jet impactor; average particle sizes between 0.6 and 0.8 μm can be obtained routinely. In recent studies of laser doppler anemometry (5, 13, 14), it is noted that an upper limit on particle size of about 1 μm should be imposed to insure that the motions of suspended tracer particles accurately follow a given turbulent velocity field. An aerosol generator of the type designed at NRL by Echols & Young (12) was recently employed by Simpson, Strickland & Barr (14) to seed a wind tunnel flowfield in order to measure turbulent boundary layer separation with a laser-doppler anemometer.

Much the same particle size limitations apply to laser streak velocimetry, as Sparks & Ezekiel (6) have indicated. Chemically inert aerosols that satisfy these mean particle size criteria, at least in the mean, can be generated and injected into a narrow plane of a wind tunnel with systems such as the one described in this paper. These aerosols approach a condition of monodispersity and are suitable for general purpose applications.

Light Scattering Qualities of the DOP Aerosol

The variation of the incident light scattered by the injected tracers during flow visualization plays an important role in the optimization of system parameters such as light placement and intensity, camera positions and camera settings. Some measurements were made to investigate the effects of the angle between the incident beam and scattered light for a DOP aerosol and a laboratory-generated tobacco smoke. Some experiments were also performed to investigate the relative effectiveness of aerosols and tobacco smoke as flow indicators, using a tobacco smoke generator employed by Koopmann (15) during some earlier flow visualization experiments at NRL. The atomizer nozzle used to generate the aerosols for the light scattering experiments has been described previously by Griffin & Votaw (3). This nozzle is capable of producing larger flow rates than the flow visualization nozzle, but still produces polydisperse aerosols with a mean diameter less than 1 μm after the mixture has been passed through a jet impactor.

The angular dependence of the scattered light intensity $I(\theta)$ is plotted in Figure 3 for the DOP aerosol and the tobacco smoke. The intensity $I(\theta)$ is normalized by the value measured at an angle of 90° between the incident and scattered light beams. These measurements indicate the increased light scattering efficiency that is obtained with forward scatter ($\theta > 90°$) of the incident light; the scattered light reaches a minimum value between $\theta = 75°$ and 90° in both cases.

A comparison of the effectiveness of a DOP aerosol relative to tobacco smoke is plotted in Figure 4 as a function of the angle between the incident and scattered light beams. The mixture flow rates from the two generators were held constant in order to obtain this comparison, and the light scattering quality of the aerosol is consistently better throughout the angular range investigated by as much as a factor of two in many instances.

Photographic Set-Up and Procedures

The aerosol generation systems whose characteristics are outlined in the preceding sections have been employed in photographic studies of a number of flow fields. The most recent arrangement (16) that was employed to photograph the vortex wakes behind vibrating cylinders in the small (75 mm × 75 mm exit) wind tunnel is depicted in Figure 5. A uniform velocity profile containing a vertical sheet of the DOP aerosol exited from the tunnel contraction section, and flowed about the cylinder mounted in the test section. Two strobe lights were used to illuminate the flow field and were positioned to take advantage of the good forward light scattering characteristics of the aerosol. Both light sources were placed on the opposite side of the aerosol sheet from the camera at angles of approximately 110° and 150° from the camera axis. By driving the strobe system from the same signal generator that was used to vibrate the cylinder, the flashing rate of the light system was precisely synchronized with the frequency of the cylinder's in-line motion. For the conditions investigated, this frequency was also synchronized with the shedding frequency of the vortices (16).

The camera used in all of the experiments was a 35 mm Nikon F which was fitted with a Micro-Nikkor Auto 55 mm, f 3.5 lens. This camera enabled flow field photographs to be taken as close as 150-200 mm from the plane of the DOP when the camera lens was inserted through a small opening in the side of the test section. The photographs were taken by simultaneously single-flashing the two strobes in a blacked-out room while the camera shutter was held open.

A similar arrangement also has been employed in order to photograph the wakes of bluff bodies in the 150 mm × 150 mm exit wind tunnel. In that configuration both large and small airfoils have been used to inject the DOP aerosol into the wind tunnel. The spanwise character of the vortex shedding from a vibrating cable (10) was photographed at various positions along the cable by varying the position of the small airfoil in the wind tunnel contraction chamber. The wakes of finite-length cylinders at yawed incidence were photographed by fixing the large airfoil in the tunnel contraction section as shown in Figure 2.

Flow Visualization with the Aerosol System

Two series of flow visualization experiments have recently been completed in the wind tunnels at NRL. The first was a study of the effects of in-line oscillations on the von Karman vortex streets formed behind circular cylinders placed in a uniform stream, and the second was the observation of the flow around finite-length cylinders at yawed incidence. These two studies are discussed briefly here to indicate the flow visualization results that can be obtained with a liquid aerosol of DOP particles.

The phenomena of periodic flow separation and the formation of vortices that accompany the flow past a bluff body have been observed for many years, but many questions relating to the mechanisms of vortex formation still are unanswered. If one of the natural frequencies of the body placed in an incident flow is near the Strouhal frequency of vortex shedding from a stationary cylinder, then self-excited vibrations can occur if the damping of the system is sufficiently low. A striking characteristic of this fluid-structural interaction is that of synchronization, or "locking-on" between the vortex and vibration frequencies. This means the cylinder and wake have the same characteristic frequency and that the natural Strouhal frequency, characteristic of vortex shedding from a stationary cylinder, is suppressed. Synchronization takes place when the amplitude of cylinder vibrations reaches a critical threshold and is accompanied by increased uniformity in the phase of the vortex shedding along the span of the cylinder (8,15).

This frequency capture or locking-on occurs not only for crossflow, vortex-excited oscillations near the Strouhal frequency, but also when a cylinder vibrates in line with the flow at a frequency near twice the Strouhal value (16). The low Reynolds number photographs shown in Figures 6 and 7 are similar in wake geometry to the flow patterns observed when model and full-scale offshore structures undergo vortex-excited oscillations. In Figure 6 two vortices are shed from the cylinder during each cycle of the cylinder's oscillation and a complex wake pattern of pairs of vortices is formed down-stream. The frequency of the vortex shedding was locked onto the vibration frequency at slightly less than twice the Strouhal frequency under these conditions.

Under certain conditions only a single vortex is formed during a cycle of the cylinder's in-line oscillation, as shown in Figure 7. Here again the frequency of the vortex street is locked onto the cylinder vibration, but the shedding frequency for vortices of like sign is half the vibration frequency. The wake pattern in the case of Figure 7 is virtually the same as has been found when a cylinder is vibrating laterally to the incident flow (7,8).

As an additional example, the complex secondary vortex formation that takes place at large lateral vibration amplitudes is shown in Figure 8. In this photograph two counter-clockwise vortices and a single clockwise vortex are shed during a cycle of the cylinder vibration (8).

The flow about a finite-length cylinder is necessarily influenced by the flow at the ends of the body. These end effects are often detrimental and considerable effort is usually given to eliminate or to at least minimize their influence. For a finite-length cylinder at yawed incidence, the end conditions can profoundly affect the flow field over the entire span even for large aspect ratios. The spanwise velocity component on a yawed body is responsible for extending the influence of the end condition and at the same time prevents the use of usual methods, i.e. end plates, for eliminating end effects. Flow visualization techniques are particularly helpful in resolving this dilemma. The photographs in Figures 9 and 10 give two examples of the wakes behind finite-length circular cylinders at yawed incidence. The cylinders are mounted in the plane of the aerosol sheet and the flow is from left to right in each photograph. The yaw angle in Figure 9 is about 30° and in this configuration the wake is still partly of the Karman vortex street type. However, the flow region nearest the cylinder's free end has begun the transition to a different form of trailing vortex system. In Figure 10, at a larger yaw angle near 45°, the transition to a steady trailing vortex system is complete and the vortices are no longer being shed periodically in time. They emanate alternately along the span in a steady pattern. The steady vortex wake in Figure 10 is similar to the shedding pattern observed during other investigations related to yawed missile wakes (17).

Concluding Remarks

A method for the generation of polydisperse liquid aerosols has been applied to the visualization of complex fluid motions. Particles with controllable size ranges are produced in the generator system, and mean particle diameters of less than 1 μm are readily obtained.

The design and construction of typical aerosol generator and wind tunnel injection systems have been described. More detailed information relating to the assembly of systems capable of greater aerosol generating capacities is available in the cited references. The system described in this paper has no moving parts and requires no heating, cooling or combustion of the tracer material. The aerosol of di (2-ethylhexyl)-phthalate (DOP) is nontoxic, noncorrosive and nonhygroscopic.

Light scattering measurements have been made in order to compare the scattering capabilities of the DOP aerosol with those of tobacco smoke. For the flow rates at which the measurements were made, the scattered light intensity from the aerosol is nearly double the scattered light intensity of smoke when the angle between the incident and scattered beams is in the range between 45° and 150°.

Examples are given of typical flow visualization results obtained in two configurations using a DOP aerosol as the tracer. Photographs of the vortex street wakes formed behind vibrating cylinders and the wake patterns behind finite-length cylinders at yawed incidence give ample evidence of the successful operation of the aerosol generator.

Acknowledgment

This paper has been prepared as part of the overall research program of the Naval Research Laboratory (NRL) and is published by permission. The authors wish to acknowledge the many contributions made by C.W. Votaw of NRL during the early stages of this work.

References

1. J.P. Yu, E.M. Sparrow & E.R.G. Eckert "A Smoke Generator For Use in Fluid Flow Visualization" Int. J. Heat Mass Transfer, Vol. 15, 557-559 (1972).
2. R.L. Maltby & R.L.A. Keating "Smoke Techniques for Use in Low Speed Wind Tunnels" in AGARDograph 70, Flow Visualization in Tunnels Using Indicators, 87-109 (1969).
3. O.M. Griffin & C.W. Votaw "The Use of Aerosols for the Visualization of Flow Phenomena" Int. J. Heat Mass Transfer, Vol. 16, 217-219 (1973).
4. O.M. Griffin, S.E. Ramberg, C.W. Votaw & M.D. Kelleher "The Generation of Liquid Aerosols for the Visualization of Oscillatory Flows" Proc. Int. Cong. Instrumentation in Aerospace Simulation Facilities (IEEE), 133-139 (1973).
5. F. Durst, A. Melling & J.H. Whitelaw "Laser anemonetry: a report on EUROMECH 36" J. Fluid Mech., Vol. 56, 143-160 (1972).
6. G.W. Sparks, Jr. & S. Ezekiel "Laser Streak Velocimetry for Two-Dimensional Flows in Gases" AIAA J., 110-113 (1977).
7. O.M. Griffin & C.W. Votaw "The vortex street wakes of vibrating cylinders" J. Fluid Mech., Vol. 55, 31-48 (1972).
8. O.M. Griffin & S.E. Ramberg "The vortex street in the wake of a vibrating cylinder" J. Fluid Mech., Vol. 66, 553-578 (1974).
9. S.E. Ramberg & O.M. Griffin "Velocity Correlation and Vortex Spacing in the Wake of a Vibrating Cable" Trans. ASME, J. Fluids Eng., Vol. 98, 10-16 (1976).
10. S.E. Ramberg & O.M. Griffin "The Effects of Vortex Coherence, Spacing and Circulation on the Flow-Induced Forces on Vibrating Cables and Bluff Structures" Naval Research Laboratory Report 7945 (1976).
11. Union Carbide Corporation, Chemical and Plastics Physical Properties Tables, 10-11 (1969).
12. W.H. Echols & J.A. Young "Studies of Portable Air-Operated Aerosol Generators" Naval Research Laboratory Report 5929 (1963).
13. M.K. Mazumder & K.J. Kirsch "Flow Tracing Fidelity of Scattering Aerosols in Laser Doppler Velocimetry" Applied Optics, Vol. 14, 894-901 (1975).
14. R.L. Simpson, J.H. Strickland & P.W. Barr "Features of a separating turbulent boundary layer in the vicinity of separation" J. Fluid Mech., Vol. 79, 553-594 (1977).
15. G.H. Koopmann "The vortex wakes of vibrating cylinders at low Reynolds numbers" J. Fluid Mech., Vol. 28, 501-512 (1967).
16. O.M. Griffin & S.E. Ramberg "Vortex shedding from a cylinder vibrating in line with an incident uniform Flow" J. Fluid Mech., Vol. 75, 257-271 (1976).
17. K.D. Thompson & D.F. Morrison "The spacing, position and strength of vortices in the wake of slender cylindrical bodies at large incidence" J. Fluid Mech., Vol. 50, 751-783 (1971).

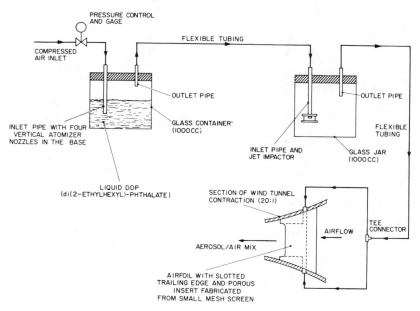

Figure 1 Schematic layout of an aerosol generation
and injection system for flow visualization.

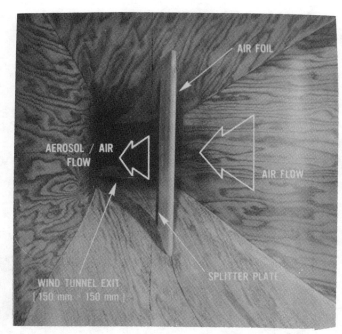

Figure 2 A slender airfoil employed at
NRL for the introduction of aerosol
tracer particles into a 150 mm × 150 mm
exit wind tunnel.

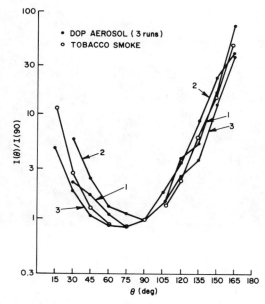

Figure 3 The measured intensity
I(θ) of scattered light as a
function of the angle θ between the incident and scattered light beams. DOP aerosol
(three runs)————●————;
Tobacco smoke (one run)
————O————.

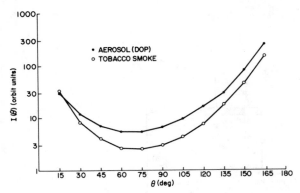

Figure 4 A comparison of the scattered light intensities $I(\theta)$ of a DOP aerosol and tobacco smoke as a function of the angle θ between the incident and scattered beams. Flow rate (DOP and smoke mixture) = 7.2 (10^2) cc/sec. DOP aerosol ————●————; Tobacco smoke ———— 0 ————.

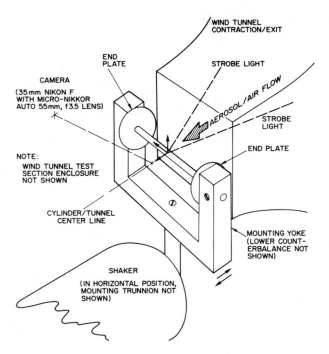

Figure 5 A line diagram of the experimental set-up employed to photograph the wakes of cylinders vibrating in line with an incident uniform flow.

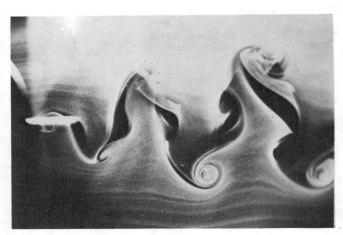

Figure 6 The vortex wake of a circular cylinder vibrating in line with an incident uniform flow at a Reynolds number of 190. Two vortices are shed during each cycle of the cylinder's vibration.

Figure 7 The vortex wake of a circular cylinder vibrating in line with an incident uniform flow at a Reynolds number of 190. A single vortex is shed during each cycle of the cylinder's vibration.

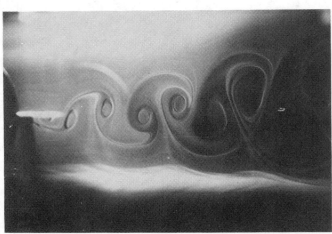

Figure 8 Secondary vortex formation in the wake of a cylinder vibrating transverse to an incident uniform flow at a Reynolds number of 190.

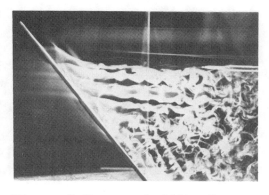

Figure 9 Vortex shedding from a stationary, yawed cylinder at a Reynolds number of 300. The yaw angle is about 30° from the vertical.

Figure 10 Vortex shedding from a stationary yawed cylinder at a Reynolds number of 300. The yaw angle is about 45° from the vertical.

FLOW VISUALIZATION TECHNIQUES IN AN AUTOMOTIVE FULL-SCALE WIND TUNNEL

TETSUO IMAIZUMI,* SHINRI MUTO,* and YASUSHI YOSHIDA*

The use of smoke in flow visualization techniques is suitable for observing the flow around a vehicle. In Japan Automobile Research Institute (JARI), a hand-held probe type smoke generator was used for flow visualization, and the flow patterns around a passenger car and a heavy duty truck were observed in the full-scale wind tunnel with it.

1. Introduction

Flow visualization techniques, which include the tuft method, oil film method and smoke technique, play an important role in understanding the relationship between the body configuration of a vehicle and its aerodynamic characteristics, and they are widely used in small-scale and full-scale wind tunnels.

Flow visualization tests often used in full-scale wind tunnels are the tuft and oil film methods, which contribute to observation of the flow pattern at the surface of the vehicle. Smoke technique,which contributes to the observation of the external airflow around the vehicle or internal airflow through the vehicle, has been attempted sometimes in a full-scale wind tunnel. The smoke technique used at the Japan Automobile Research Institute (JARI) uses a hand-held probe type smoke generator which has been studied (Ref. 3, 4, 5).

The smoke generator usually used for the smoke technique has the following weak points.

(1) Smoke tends to change into liquid in the smoke duct.

(2) The generator is more complicated and larger.

The authors have developed a simple and efficient generator that can improve the above-mentioned weak points. The smoke generator, one of the hand-held probe type, has the producing part (heater) included in the ejecting part (nozzle). Therefore, a mineral oil is vaporized in the producing part (in the air stream) so that smoke may be introduced directly into the air stream without liquefaction.

This report describes some results in flow visualization tests around a vehicle in a full-scale wind tunnel by using the smoke technique with the oil smoke generator.

---

* Japan Automobile Research Institute, Inc. (JARI);
   Yatabe-cho, Tsukuba-gun, Ibaraki, Japan

2. JARI Full-Scale Wind Tunnel

Flow visualization tests and aerodynamic force measurements are made in the JARI full-scale wind tunnel. This wind tunnel is of open circuit type (Eiffel type), with two test sections upstream and downstream of the fan, as seen in Fig. 1. The main dimensions and general specification of the wind tunnel are shown in table 1.

3. Smoke generator and its performance

This smoke generator was developed by applying the principle that the smoke is generated by the vaporization of a mineral oil. It is used in a full-scale wind tunnel. The whole view is shown in Fig. 2, in which the generator is roughly divided into two parts, one is the hand-held probe, namely, the nozzle (2 mm dia. and 1 m length) in which the heater is installed, and the other is the part to feed the oil into the nozzle through the vinyl tube from the oil container. In order to generate the desired smoke, it is necessary to control heater temperature and the oil flow rate.

Liquid paraffine, light oil and kerosene are generally used for the smoke oil. It was found from the preliminary experiments that the critical temperatures at which liquid paraffine, light oil and kerosene change completely to mist are approximately 400, 260 and 190°C respectively. Based on the above data kerosene is found to be suitable for the smoke oil, because it changes into mist completely at the lowest temperature. The performance of this smoke generator is described below.

Applicable wind velocity range : 10 - 45 m/s (see Fig. 3)
Oil consumption averages*   : 480 ml/h

4. Experimental results and discussion

4.1 Flow visualization tests on a passenger car

Generally speaking the aerodynamic front lift depends mainly upon the configurations of the front body and the inner state of the engine room of the test vehicle. The above described affairs may be understood qualitatively from the flow visualization tests. It was attempted in these tests to study the relationship between the aerodynamic force and flow pattern on the test vehicle in the first test section of the JARI full-scale wind tunnel with the smoke generator. The configurations of the test vehicle are: (1) basic configuration (fastback type test vehicle); (2) front spoiler added; (3) grille seals; and (4) front spoiler added and grille seals.

Aerodynamic forces on the test vehicle are measured with the six-component balance in the first test section of the wind tunnel, and flow visualization tests are carried out with two hand-held probe type smoke generators. The two smoke generators are located of 2 m upstream of the vehicle as shown in Fig. 5. One of the two streaklines flows near the top of the bonnet, and the other attaches to the bumper of the test vehicle, respectively. Fig. 4 shows the effect of the configuration on the front lift. Fig. 6 shows flow patterns around such vehicles as described in (1), (2), (3) and (4). As seen from Fig. 6, the streakline which flows near the upper part of the bonnet, is lifted upward in order of (1), (3), (2) and (4), while the streakline which attaches to the bumper, in the case of (2) is lifted upward the bumper, and then enters the grille. The streakline in (3) flows beneath the underbody. The streakline in (4) is observed to flow upward rapidly in front of the bumper.

---

* This indicates the amount of the oil consumption when it changes to the mist completely at the outlet of the nozzle.

4.2 Flow visualization tests on a heavy duty truck

Flow visualization tests on a heavy duty truck of cab-over type were carried out in the second test section of the wind tunnel. The flow pattern around the cab of the truck at a uniform flow velocity of approximately 15 m/s is shown as follows.

(A) Flow in the proximity of the outside surface of the cab.

Fig. 8(a) shows the existence of the reverse flow with an anticlockwise direction, running from rear to front in the upper region of the side of the cab. From this result, the above-mentioned reverse flow causes the side window to become covered with dust or muddy water splashed by the tires of a truck. In order to maintain clear vision through the side window, it may be necessary to weaken the reverse flow as much as possible, with reference to the flow condition of the side of the cab. To apply this idea an A-post vane should be attached on the front corner of the cab as shown in Fig. 8(b). The vane can induce a high velocity flow in the reverse flow region, and this almost all runs in the same direction as the main flow.

(B) Flow in the gap between the cab and the rear body.

The flow visualization tests are performed under constant engine load. Fig. 9 (a) shows the flow in the gap running upward, and also the phenomenon of the flow running into the rear inlet of the engine room. On account of the above flow condition, the flow in the gap is thought to seriously influence the efficiency of engine cooling and the deposit of dirt on the rear window of the cab. Fig. 9(b) shows the flow pattern with a device under the front part of the cab as shown in Fig. 7. The flow pattern in this case where the flow runs out of the engine room, may be considerably different from those without a device.

5. Conclusions

It was found from the test that the hand-held probe type smoke generator is applicable to the flow visualization tests in the full-scale wind tunnel to clarify such problems as;

1. Understanding of the relationship between the body configuration of the vehicle and its aerodynamic characteristics;
2. Reduction of the dirtiest areas on the vehicle's surface, particularly when driving in the rain on dirty or even muddy ground and in the wake of other vehicles;
3. Reduction of aerodynamic noise;
4. Development of various kinds of aerodynamic attachments to be added to the vehicle;

The authors studied fragmentarily aerodynamic characteristics, but hope to do more precise measurement of the flow pattern around a vehicle in order to compare the data in flow visualization tests with those in aerodynamic force measurement tests.

References

(1) W.Merzkirich, Flow Visualization, (1974), 250, Academic Press.
(2) H.Hoshino, Nissan Diesel Technical Review, (in Japanese), No. 39, (1976), 6.
(3) M.E.Olson, The Society of Automotive Engineers, Feb. Paper No. 760188, (1976), 10.
(4) H.E.Pringham and W.D.Bowman, The Society of Automotive Engineers, Jan. Paper No. 730236, (1973), 12.
(5) W.T.Mason,Jr., The Society of Automotive Engineers, Aug. Paper No. 750705, (1975), 21.
(6) F.K.Schenkel, The Society of Automotive Engineers, Feb. Paper No. 770389, (1977), 11.

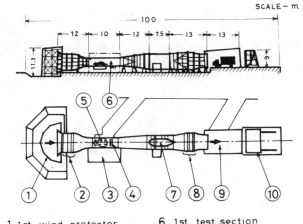

Table 1-The main dimensions and general specification, JARI Full-Scale Wind Tunnel

1) Dimension of test sections
   1 st   3m(H) x 4m(W) x 10m(L)
   2 nd   5m(H) x 6m(W) x 13m(L)
2) Contraction ratio
   1 st   4.06
   2 nd   1.63
3) Dimension of fan
   Fan diameter   5.5 m
   Number of fan blades   12
   Driving motor power   1200 kw
   Motor and fan maximum rotation
   speed   350 rpm
4) Maximum air speed
   1 st   57 m/s
   2 nd   23 m/s
5) Transverse velocity distribution inside
   the 80 % of cross sectional area in the
   test sections
   1 st   ±1.0 % of the mean velocity
   2 nd   ±8.0 % of the mean velocity

1. 1st wind protector
2. honeycomb and screens
3. 1st control room
4. chassis dynamometer
5. aerodynamic balance
6. 1st test section
7. fan and motor
8. honeycomb and screens
9. 2nd test section
10. 2nd wind protector

Fig. 1-JARI Full-Scale Wind Tunnel

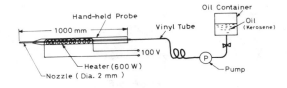

Fig. 2-Oil smoke generator

wind velocity: U=15 m/s

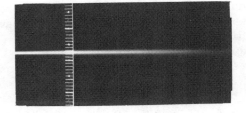

U=45 m/s

Fig. 3-Applicable wind velocity range of streakline

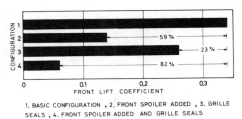

1. BASIC CONFIGURATION , 2. FRONT SPOILER ADDED , 3. GRILLE SEALS , 4. FRONT SPOILER ADDED AND GRILLE SEALS

Fig. 4-Variation of front lift coefficient with configuration of the test vehicle

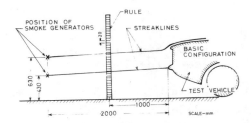

Fig. 5-Arrangement of smoke generator and test vehicle in test section

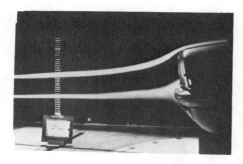
(1) Basic configuration

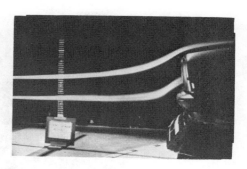
(2) Front spoiler added

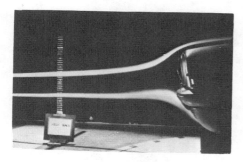
(3) Grille seals

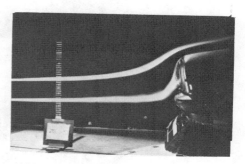
(4) Front spoiler added and grille seals

Fig. 6-Streaklines around the test vehicle (U=15 m/s)

Fig. 7-Front view of the truck, with a device added (under the front part of the cab)

(a) Basic configuration

(b) A-post vane added

Fig. 8-Flow pattern on the side of the cab (U=15 m/s)

(a) Basic configuration

(b) A device added (under the front
    part of the cab)

Fig. 9—Flow patterns in the gap between the cab
and the rear body (U=15 m/s)

INTERACTION OF TWO VORTEX RINGS

YUKO OSHIMA* and SABURO ASAKA**

Yuko OSHIMA*and Saburo ASAKA**

Studies on interaction of two vortex rings due to the mutual induced velocity were carried out in air using the smoke of NH4Cl.   Various cases carried out were as follows. [A] Two rings proceed   along parallel axes. [B] Head on collision along a common axis. [C] Collision with an arbitrary angle. [D] Two rings one after another in a common axis. Various cross-linking phenomena of the vortex filaments were observed in each case [A,B,C] and the game of leap-frogging of the rings were realized.

Introduction

It has been passed over 100 years since Reynolds (Ref.1) presented a paper concerning the experimental studies on a vortex ring.   Since then, a number of papers were presented on the vortex rings.   Some of new attempts in the vortex rings were listed in the references. (Refs.2-6)   Half of the studies were devoted to the theories and the other to the experiments.   The fluid was considered as an invicid one in these theories for long time, but the effect of viscosity cannot be separated in the experiments of the real fluids.   So there were always some discrepancies between them.   Especially in the interaction of vortices, some interesting phenomena were occurred due to the viscosity, that is, the cross-linking of vortex filaments and merging of vortices.

On the other hand, turbulence is considered as an assembly of vortices with various sizes and intensities.   To clarify a mechanism of turbulence or of complex fluid flow, it is one approach to investigate the behavior of vortex rings. Because vortex ring is the simplest way to realize the vortex motion in laboratory with good reproducibility and it make a loop of vortex filament by itself without the effect of the boundary.

Furthermore, vortices have various roles in nature, for example, in a flapping of the wing of insects and birds, and in a locomotion of fish.   In these motion, circulation is occurred by vortices and the momentum is shed to the fluid.

In the following sections, we report the investigation of the interactions of two vortex rings at various conditions, such as [A] Two rings proceed along parallel axes, [B] Head on collision along a common axis, [C] Collision with an arbitrary angle, [D] Two rings one after another in a common axis.   The schematic configuration of these four interactions are shown in Fig.1.

 * Research Associate of Physics, Ochanomizu University. Otsuka, Bunkyo-ku, Tokyo.
** Hornored Professor, Ochanomizu University, Otsuka, Bunkyo-ku, Tokyo.

Experimental Apparatus and Method

Experiments were carried out in closed container to avoid the disturbance from the sorroundings. Schematic diagram of the apparatus is shown in Fig.2 in the case of [A]. A loud speaker system was used to generate a vortex ring which was drived electronically controlled by d.c. current. When an amount of air passed through an orifice attached in front of the speaker, vortex ring was generated by a shear motion at a sharp edge of the orifice, and it proceeded perpendicularly to its plane. The relation between the initial speed of the ring and the d.c.current was calibrated for each system. Two such speaker system were used in the case of [B] and [C]. In case [D], delay device was added to realize the second ring. These vortex rings were visualized using a smoke of NH_4Cl. Liquid of HCl and HNO_3 were kept in funnels with cock separately, and a few drops of them were shed drop by drop into individual plates respectively, which were placed in a box. (Fig.3) The smoke of NH_4Cl produced in the box was flowed into the speakers through hoses by the excess pressure of the air pump which was connected at the other side of the box. The merit of this apparatus is not to be produced the excess smoke which is harmful for other apparatus and a health.

Using the intense beam of slide projector, the vortex rings were visible in dark sorroundings. Meridian cross section of the rings was illuminated by the narrow beam of slit light. Photographs were taken by motor-driven camera at each instant after the some duration from the vortex ejection using the electronical delay device. Also 16mm cine-camera and video recording system were used to record the continuous recording.

Results of Experiments

As there are many results, here we like to list the brief results in each case with figures and photographs. Details of the individual case commit to each report in respective references.

[A] Two rings proceed along parallel axes. (Refs.7 and 8)
Two rings came nearer due to the mutual interaction of the other ring; and the cross-linking of the vortex filaments was occurred. Depending on the initial speed of the rings, cross-linkings of the filaments were occurred from one to three times as follows; (1) Two rings merged into one distorted ring after one time of cross-linking at lower speed. The ring changed its shape oscillately in three dimensions (forward and backward, vertical and horizontal). (2) At higher initial speed, merged ring splitted into two rings again. After the first cross-linking, the second cross-linking succeeded at the portion most separated initially, and one distorted ring splitted into two rings, each of which consisted of the respective halves of the original vortex rings. They proceeded perpendicularly to their original direction plane. A series of photographs are shown in Fig.4. (3) At still higher speed, two rings merged, splitted and merged again into one ring through three times of cross-linkings. The measured inclination of two rings against the original plane is shown in Fig.5 in non-dimensional time variation at various initial speed. From this figure, the inclination of the vortex rings due to the interaction depends on the non-dimensional time and it does not on the initial velocity.

[B] Head on collision along a common axis. (Refs. 9 and 10)
When two rings came nearer, the diameters of the rings expanded due to the mutual induced velocity of the other ring. The trajectories of the rings were agreed with that of the numerical simulation as shown in Fig.6. These trajectories were different with that of the case of colliding against the plate, because the

boundary layer was developed due to the viscosity of the fluid in the latter case. At higher initial speed of the rings, some instabilities were observed along the circumferences of the rings, and the two rings were divided into a number of small vortex rings like a chain of necklace before they decay as shown in Fig.7.

[C] Collision with an arbitrary angle. (Ref.11)

This is the combination of the previous two cases [A] and [B]. Depending on the initial speed of the rings and the colliding angle, interaction was divided into six patterns. Figure 8 is the diagram of the interaction in polar coordinate. In determining the boundary of the domains, simple reasoning using the vortex filament approximation was carried out and got the adequate agreements. In Fig.9, some examples of the photographs are shown.

[D] Two rings one after another in a common axis. (Refs.4,12 and 13)

This interaction is popular in the text books of the fluid dynamics (Refs.14 and 15), but the game of leap-frogging could not realize as is reasoned in the inviced flow theory. In most case, it is observed that the two rings consolidated into one ring in water. The game of leap-frogging could be observed in air at preferable condition, that is, the diameter of the orifice D = 7.0 -8.5cm and the time interval between the two rings ΔT = 0.15 - 0.25sec, and the condition seemed to be tight. It seems to us that this phenomena depend on the Reynolds number and the non-dimensional time interval between two rings $U_0 \Delta T/D$, and also depends on the concentration of the vorticity into the vortex filaments, that is, formation of the vortex rings. Figure 10 is a series of photographs of the game of leap-frogging.

Concluding Remarks

In the several kinds of interaction of two vortex rings, some new observations were found. These are the cross-linking of vortex filaments in various types, scale up phenomena in vortices consolidation into one, and the jump cascade into a number of small vortices from large scale. And also the game of leap-frogging could be realized.

References

1) O.Reynolds; Nature 14 (1876) 477.
2) F.W.Dyson; Phil.Trans. A 184 (1893) 1014.
3) C.H.Krutzsch; Annalen Phys. 35 (1939) 497.
4) T.Maxworthy; Jour.Fluid Mech. 51 (1972) 15.
5) P.G.Saffman; Studies Appl.Math. 49 (1970) 371.
6) T.Kambe and Y.Oshima; Jour.Phys.Soc.Japan 38 (1975) 271.
7) Y.Oshima and S.Asaka; Natural Science Rep. Ochanomizu Univ. 26 (1975) 31.
8) Y.Oshima and S.Asaka; Jour.Phys.Soc.Japan 42 (1977) 708.
9) Y.Oshima; Natural Science Rep. Ochanomizu Univ. 24 (1973) 61.
10) Y.Oshima; to be published in Jour.Phys. Soc.Japan.
11) Y.Oshima; to be published.
12) Y.Oshima, T.Kambe and S.Asaka; Jour.Phys.Soc.Japan 38 (1975) 1159.
13) H.Yamada and T.Matsui; Proceeding of Symposium on Turbulence 1977 [in Japanese].
14) A.Sommerfeld; Mechanik der deformierbaren Medien (Akademische Verlag.Leipzig, 1957) Kap.4.
15) G.K.Batchelor; An Introduction to Fluid Dynamics (Camblidge University Press, 1967) Ch.7.

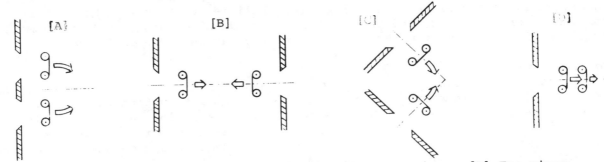

Fig.1   Schematic diagram of interaction of two vortex rings. [A] Two rings
    proceed along parallel axes. [B] Head on collision. [C] Collision with arbitrary
angle. [D] One after another in a common axis.

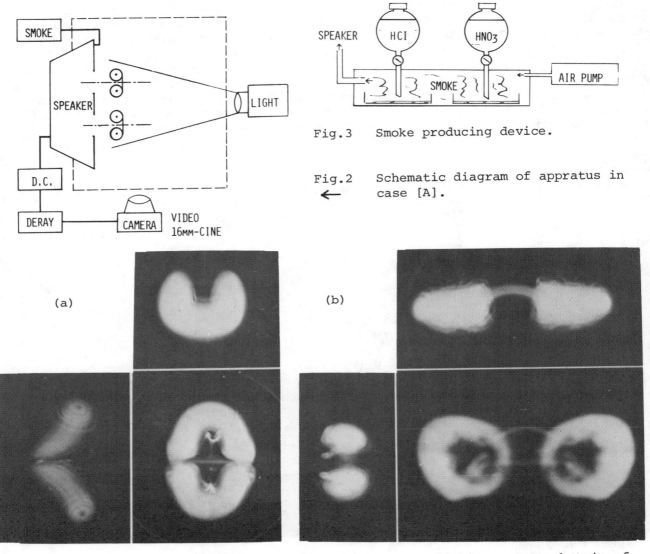

Fig.3   Smoke producing device.

Fig.2   Schematic diagram of appratus in
 ←   case [A].

Fig.4   A series of photographs of the two times cross-linking.   In each trio of
    photographs, the left one is the side view, the upper one is the top view and
    the lower one is the front view, respectively, at one moment.   The ring proceed
from the left to the right in side view, and from the up to the down in top view,
respectively. U_O = 25cm/sec, Re = 360. (a) t = 0.22sec, (b) 0.73sec.

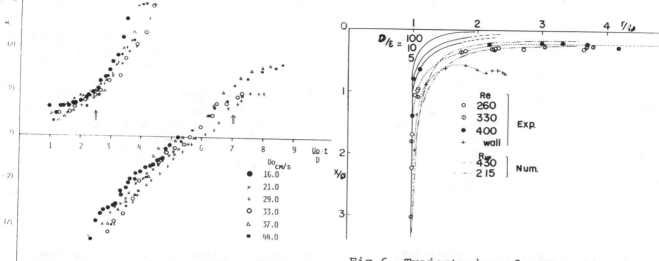

Fig.5 Inclination of two rings θ° vs non-dimensional time Uot/D, Arrow signs show the first and the second cross-linking. Points at right-lower half are measured from the direction in 90° rotation around the axis in comparison with the points in left-upper half.

Fig.6 Trajectories of vortex ring in head on collision. The result agreeds with that of the numerical simulation. Solid line shows the result of invicid flow theory, and + signs show the trajectories against the wall.

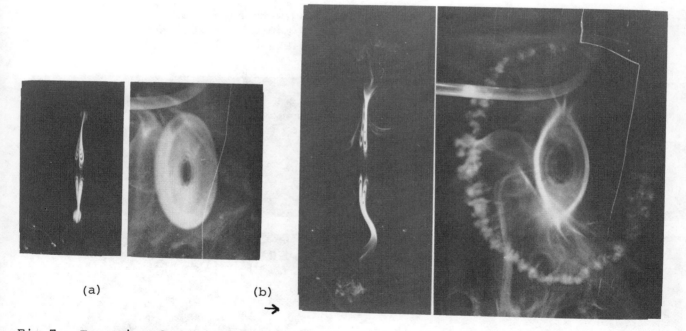

(a)   (b)

Fig.7 Two pairs of photographs showing the jump cascade into a number of small vortex rings. In each pair, the left one is the cross-section of meridian plane, and the right one is the whole view looking from 30° angle from the colliding plane. U_0 = 58cm/sec, (a) t = 0.30sec, (b) 0.48sec.

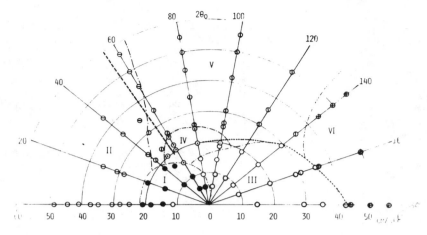

Fig.8   Relation between
         the initial speed of the
         rings and the colliding
         angle. Broken line and
         dotted line show the
         boundary predicted by the
         vortex filament approxi-
         mation.

(a) 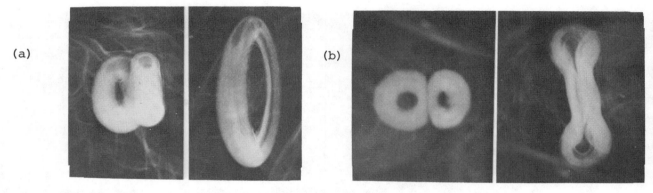 (b)

Fig.9   Two pairs of photographs colliding with an angle.
(a) $2\theta_O = 50°$, $U_O = 36cm/sec$.   (b) $2\theta_O = 140°$, $U_O = 26cm/sec$.
     $t = 0.77sec$, $t = 1.24sec$. $t = 0.98sec$, $t = 1.34sec$.

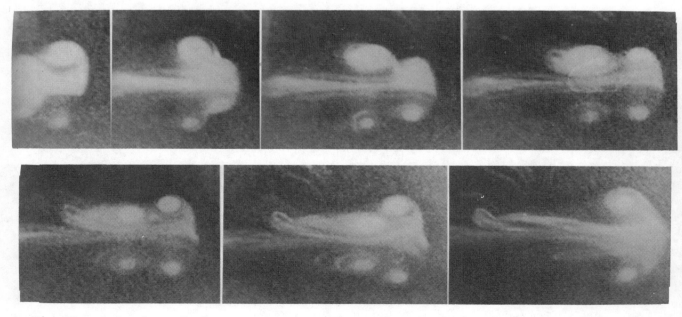

Fig.10   A series of photographs of the game of leap-frogging.   T = 0.175sec,
    upper ▵ T = 0.4sec,   0.6sec,   0.7sec,   0.8sec,
    lower   0.9sec,   1.0sec,   1.2sec.

86

SMOKE VISUALIZATION OF SUBSONIC AND SUPERSONIC FLOWS

T. J. MUELLER* and V. P. GODDARD**

The smoke flow visualization facilities and techniques developed by F. N. M. Brown for low speed flows and extended by V. P. Goddard for high speeds are presented. Although a large number of flow problems have been studied at the University of Notre Dame using these techniques, the present paper describes research on transition in attached and separated shear layers and laminar wakes at subsonic speeds and the flow over a wedge and the laminar wake behind a blunt body at supersonic speeds. Agreement between the smoke flow data and other techniques for obtaining the same results is good for all the problems studied.

Introduction

The use of smoke for flow visualization at low subsonic speeds may be traced to Dr. Ludwig Mach of Vienna in 1893 (Ref.1). Mach observed and photographed the flows by using silk threads, cigarette smoke and glowing iron particles. In France about 1900 to 1901, E. J. Marey (Ref.2), extending his earlier experience with the visualization of liquid flows, and cognizant of the work of L. Mach, produced excellent wood smoke photographs. Although many other small low speed smoke tunnels followed, Professor F. N. M. Brown at the University of Notre Dame was the first to develop a large three dimensional research smoke tunnel capable of speeds up to 67 meters/second in 1950. This facility evolved from his experience with two and three dimensional smoke tunnels which began in 1935. In fact, most of the progress and refinement of smoke visualization techniques are credited to Brown (Ref.3). Brown used an indraft wind tunnel with 12 screens followed by and inlet with a contraction ratio of 24 to 1 for the largest test section of 0.61 by 0.61 by 1.83 meters. The smoke was introduced upstream of the first screen from a rake which can be moved and positioned to place smoke lines anywhere in the test section. He was also the first to develop a practical smoke generator using kerosene and to take three-dimensional, stereo, and high speed pictures of smoke flows. These wind tunnels, smoke generators, and photographic techniques have been extensively copied. Extending the techniques of Brown, V. P. Goddard was able to produce smoke streamlines in supersonic flows in 1959 (Ref.4). An indraft supersonic wind tunnel with seven screens and an inlet contraction in area of 100:1 to the nozzle throat was used by Goddard. A modified schlieren system permits the simultaneous photographing of both the smoke and shock-wave patterns (Refs. 4 and 5). When nitrous oxide is used in place of smoke, the streamlines become visible along with the shock pattern in any ordinary schlieren system. The use of smoke visualization at supersonic speeds appears to be unique at Notre Dame.

*Professor of Aerospace and Mechanical Engineering, University of Notre Dame, Notre Dame, Indiana 46556, USA.

**Associate Professor of Aerospace and Mechanical Engineering, University of Notre Dame, Notre Dame, Indiana 46556, USA.

Subsonic Flow Visualization

Professor Brown spent the period from 1957 to 1969 developing the flow visualization techniques needed to study the boundary layer transition process. He was able to present a physical description of transition by means of data obtained from 20 micro-second still photographs and 16mm motion pictures up to 4000 frames/sec. His studies focused on natural transition of the boundary layer on a tangent ogive nose cylindrical body (Ref.6). Brown characterized the transition process at zero pressure gradient as four states as shown in Fig.1. In the first state the sinusoidal or two-dimensional wave motion appears as rings around the body. The frequency of these waves is a function of the freestream velocity to the 3/2's power. These waves then begin to distort three dimensionally and a thatching pattern is evident in the second state. In the third state, the thatching pattern breaks down into the turbulent boundary layer. The fourth state is a laminar flat which is drawn behind the turbulent layer which accelerates downstream yielding an intermittent process. The laminar layer (fourth state) is eventually overtaken by a new turbulent section. The intermittency of the repetition of these states is about 1/5 the frequency of the wave state. Brown also investigated the influence of an adverse pressure gradient on the transition process, see Fig. 2. As the adverse pressure gradient was increased the transition process was intensified. The laminar flat or fourth state shortens and eventually disappears with increasing adverse pressure gradient. Among other observable changes, the thatched pattern takes on a furculate aspect. In a later investigation Knapp, Roache and Mueller (Ref. 7) correlated the flow visualization data with hot-wire anemometer measurements. They also investigated the influence of sound on the transition process. When the sound frequency was near the natural formation frequencies of the two-dimensional waves the sound caused the transition region to become more clearly visible and to move upstream. The frequency of the two-dimensional wave becomes locked into the sound frequency. Transition process is no longer intermittent and the sound causes the vortex trusses to align themselves in rows rather than in a thatched pattern. These techniques are presently being used to study the problem of the separation bubble near the leading edge of airfoils including transition in the free shear layer. Figure 3 clearly shows the separation near the leading edge of a NACA 66_3018 airfoil and the first stage of transition in the free shear layer.

Brown and Goddard (Ref.8) investigated the effect of sound on the laminar wake behind airfoils and flat plates with sharp leading and trailing edges. In the visual study of the flow past an airfoil or flat plate the three basic characteristics of the wake flow with or without sound are; the frequency of vortex pair formation, the geometry of the vortices (i.e. the spacing ratio, h/λ, where h is the vertical distance between rows and λ is the horizontal distance between vortices in the same row) and V_W the wake speed as compared to V_{FS} the free stream velocity. Since the wake becomes visible with the introduction of smoke, the frequency of vortex formation can be determined using standard stroboscopic techniques. The geometry of the wake can be determined by linear measurements made from photographic prints on dimensionably stable photographic paper. Knowing both the free stream velocity and the frequency of vortex formation the wake speed can be determined. The introduction of sound had a direct influence upon the frequency of vortex formation. The wake frequency follows the sound frequency for a limited range both above and below the natural frequency of formation. The vortex spacing ratio varied with the sound frequency. However, the wake speed was invarient with change in sound frequency. The effects of sound on the wake flow behind a typical flat plate are summarized in Fig. 4 where results for two different free stream speeds are shown. The sound control limits, upper and lower, obtained from experiment are indicated in Fig. 4. Theoretical curves of the spacing ratio variation with change in sound frequency are also shown. These curves were generated by making use of von Karman's work where he showed that the drag of an object can be determined from wake characteristics. Since the drag is unchanged when sound is introduced and knowing the wake characteristics without sound, the theoretical curve can be generated from the change in wake characteristics with sound. Figures 5 and 6 are smoke photographs of the flat plate wakes used in Fig. 4 and show the difference in spacing ratio when sound is introduced.

Supersonic Flow Visualization

Figure 7 shows the supersonic flow past a wedge. This photograph was made using a modified schlieren system in order to be able to obtain simultaneous schlieren and smoke lines. This photograph shows the two fundamental ways in which supersonic flow tends to follow parallel surfaces; flow through a shock wave and expansion around a corner. In the flow through the shock wave the abrupt deflection of the streamline so as to flow parallel to the front surface of the wedge can be observed. The expansion at the shoulder of the wedge so as to follow parallel to the main body of the wedge is clearly seen. The Mach number of the flow can be determined from the photograph by measuring the shock angle and the streamline deflection. In a similar way streamlines can be followed throughout the flow field and so map the entire flow field. In order to study wake flows, a lucite wedge-shaped plug with a rounded leading edge was inserted slightly upstream of the nozzle throat. This configuration resembles a plug nozzle referred to as the expansion-deflection nozzle, or may be thought of as representing a strut, a flame holder, a Scramjet fuel injector, etc. A simultaneous smokeline and opaque-stop schlieren photograph of the wake flow using a laser light source is presented in Fig. 8. Smoke is not visible in a schlieren system since it has approximately the same index of refraction as air. By measuring the local deflection of the smokelines passing through the recompression shock wave and wave angles, the Mach number immediately ahead of and behind the recompression shock can be determined. Another series of experiments were run with an actual expansion-deflection nozzle (Ref.5). A comparison of the Mach numbers obtained from smokeline/shock wave pattern and from total and static pressure measurements is shown in Fig. 9. This correlation is for the Mach number immediately downstream of the recompression shock versus the distance measured along the recompression shock from the nozzle centerline. These high speed flow visualization techniques are currently being used to study a transonic cascade problem.

References

(1) L. Mach, Uber die Sichtbarmachung von Luftstromlinien, Zeitschrift fur Luftschiffahrt and Physik der Atmosphare, Vol.15,Heft 6 (1896), plates I-III, 129-139.

(2) E. J. Marey, Changements de direction et de vitesse d'un courant d'air qui rencontre des corps de formes diverses. Compt.rend.,132, (1901), 1291-1296.

(3) W. Merykirch, Flow Visualization, (1974) Academic Press.

(4) V.P.Goddard, J.A. McLaughlin and F.N.M.Brown, Visual Supersonic Flow Patterns by Means of Smoke Lines. J.Aero/Space Sci.Vol.26, No.11, (1959), 761-762.

(5) T.J.Mueller, C.R.Hall,Jr., and W.P.Sule, Supersonic Wake Flow Visualization, AIAA Journal, Vol.7, No. 11, (1969), 2151-2153.

(6) F.N.M.Brown, The Physical Model of Boundary Layer Transition, Reprinted from the Proceedings of the Ninth Midwestern Mechanics Conference Publication, Univ. of Wisconsin (1965), 421-429.

(7) C.F.Knapp, P.J.Roache, and T.J.Mueller, A Combined Visual and Hot-Wire Anemometer Investigation of Boundary Layer Transition, UNDAS-TR-866CK, Univ. of Notre Dame, (1966), 1-97.

(8) F.N.M.Brown and V.P.Goddard, The Effect of Sound on the Separated Laminar Boundary Layer, Final Report NSF Grant G11712 (1963), Notre Dame, IN. 1-67.

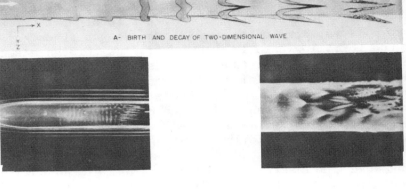

Fig. 1. Natural Transition on an Ogive Nose Cylinder for Zero Pressure Gradient.

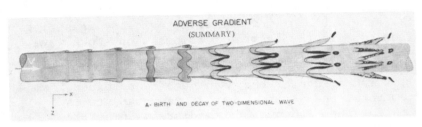

Fig. 2. Natural Transition on an Ogive Nose Cylinder for an Adverse Pressure Gradient of 0.4316 Q/ Meter.

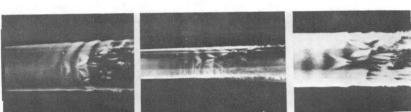

Fig. 3. Smoke Flow Photographs of a NACA 66_3-018 Airfoil at 19.24° Angle of Attack and Reynolds Number/Meter of 6.33×10^5.

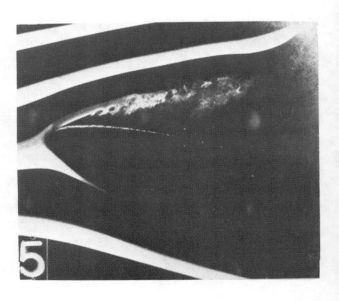

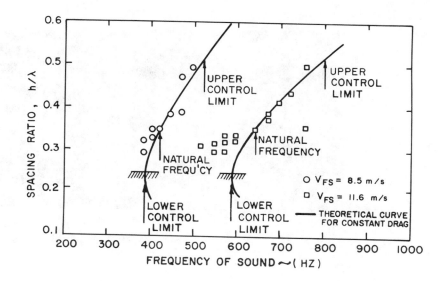

Fig. 4. Variation of Spacing Ration with Sound Frequency for a Flat Plate.

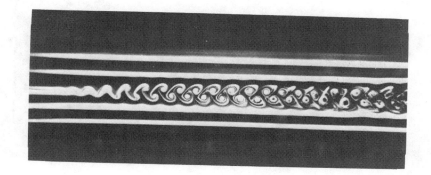

Fig. 5. Flat Plate of Figure 4, V_{FS} = 8.5 m/s and Sound Frequency of 0 Hz.

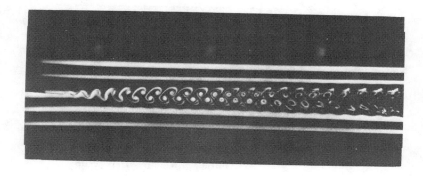

Fig. 6. Flat Plate of Figure 4, V_{FS} = 8.5 m/s and Sound Frequency of 470 Hz.

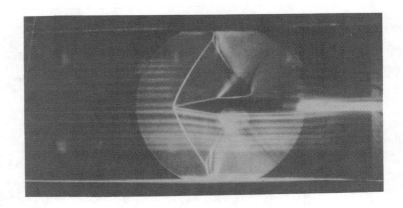

Fig. 7. Supersonic Flow Past a 5° Half-Angle Wedge, M = 1.38.

Fig. 8. Simultaneous Smokeline and Opaque-Stop Schlieren Photograph of Wake Flow with Laser Light Source.

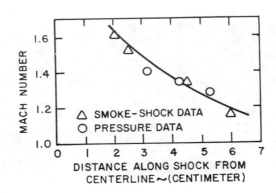

Fig. 9. Correlation of Smokeline-Shock Data and Pressure Data for Mach Number Immediately Downstream of Recompression Shock for E-D Nozzle.

FLOW UP A STEP: AS AN ILLUSTRATION OF THE PHENOMENON OF BOUNDARY LAYER SEPARATION AND RE-ATTACHMENT

S. L. GAI*

It is shown that the study of salient features of boundary layer separation in low speed flow is facilitated by certain well-defined flow configurations. The particular example of the low speed flow up a step is considered and some experimental data on the geometry of separation in front of the step is presented.

1. Introduction

Generally separation is defined as the phenomenon of flow break-away from the surface because the boundary layer is unable to withstand the adverse pressure gradient impressed on it by the external flow. The separation point on the surface in two dimensional flow is defined as the point where the gradient $(\partial u/\partial z)_s = 0$, that is, incipient back flow (see for example, Ref. 1, 2, 3). Downstream of separation the flow moves away from the wall so markedly that normal velocities are large and the Prandtl boundary layer equations are no longer valid.

It may, however, be more instructive sometimes to describe the boundary layer phenomenon and separation via the vorticity concept as has been done by Lighthill (Ref. 4). An interesting explanation of separation along these lines is given by London (Ref. 5), who defines separation as due to the vorticity confined in the boundary layer fluid penetrating into the main stream. This, by conservation of circulation implies mixing of the fluid in the neighbourhood of the boundary with the external flow, resulting in flow breakaway from the surface.

2. Theoretical Considerations

Lighthill (Ref. 4; see also Ref. 6,7) defines the boundary layer as a vortex layer of thickness 'δ', whose influence is felt entirely

---

*Assistant Professor, Aeronautical Engineering Department, Indian Institute of Technology, BOMBAY 400 076, INDIA.

within itself and which, when thin enough, alters the potential flow into one around a thickened body. Also, since the tangential vorticity is not significantly affected by convection along the surface, streamlines near the surface are closely parallel to it.

However, when there is separation, streamlines no longer remain parallel to the surface and the small diffusion distance is far exceeded by the effect of convection in separating the vorticity layer from the surface. In fact, the line of separation is the loci of points on a surface where Ω_s is zero and $\Omega'_s < 0$. At such points the streamlines come right away from the surface so that the flow direction is no longer parallel to the surface. According to Lighthill (Ref. 4), the separation or dividing streamline makes, with the surface, an angle given by

$$\tan \theta = 3\mu \, \Omega'_s / p'_s \qquad \qquad \dots (1)$$

A similar expression by Oswatitsch (Ref. 8) gives this angle as

$$\tan^{-1}\left[-3\, \tau'_s / p'_s \right]$$

which becomes identical to eqn.(1) when it is noted that $\Omega_s = -\dfrac{\partial u}{\partial z}$.

The separation angle θ, therefore, depends on the rate of fall of the wall shear stress and the pressure gradient imposed on the flow. It is then easy to see that for large adverse pressure gradients the DSL will be highly curved. This is unlike the supersonic flow separation wherein there is a direct connection between the flow angle and the pressure rise imposed on the flow. However, when the separation geometry is well-defined such as in the case of flow up a step, which is a prototype of rapidly separating flow, the DSL shape is governed by the step height and the separation length. It is then of interest to know how varying the step height will affect the flow separation ahead of the step.

3. Results and Discussion

The low speed flow up a **step** provides a flow geometry (Fig. 1) where the re-attachment point (C) is approximately fixed, while the point of separation (B) moves towards or away from the step face depending on the step height. The ratio of step height (h) to plate length (L), in turn, governs the **pressure field** impressed on the flow over the step.

Fig. 2 shows a typical flow visualisation result obtained of laminar flow over a step. The flow visualisation experiments using smoke were conducted in a low speed wind tunnel of cross-section 230 mm x 100 mm at a speed of approximately 4m/sec, giving a plate Reynolds number of about 5×10^4. The flat plate was 200 mm long and the step heights varied from 6.35 mm to 35 mm, giving (L/h) values ranging from 32 to 5.8. The laminar flow condition was established by hot wire measurements and Fig. 3 shows a typical profile measured about 200 mm from the leading edge of the plate in the absence of any step. The experimentally determined skin friction values also agreed with the zero pressure gradient flat plate theoretical values within about 15 per cent.

After thus establishing the laminar nature of flow over the plate, separation was provoked by the steps. It is noticed from the photographs that the dividing streamline is clearly defined and the separation region bounded by the plate surface, step face, and the dividing streamline encloses a strong standing vortex, whose sense is clockwise. The outer flow is also steady. It was possible to observe, during the experiments, the point of separation of flow with an accuracy of ± 2 mm. It was also observed that while the dividing streamline is fairly straight for smaller step heights, it was somewhat curved for the larger steps. Another feature of the flow is that the standing vortex is symmetrically disposed along the bisectrix dividing the corner area. This is in agreement with the theoretical result of Ringleb (Ref. 9).

Measurements of wall pressures on the plate were also made for various step heights. These were consistent with the flow visualisation results.

4. Geometry of the Separation Cavity

As noted in § 3, it was possible to identify the dividing streamline with reasonable certainty. Knowledge of the distance of the separation point together with the height of the step and the shape of the dividing streamline gives an indication of the geometry of the separation cavity. Fig. 4 shows the results obtained for different ratios of step height to boundary layer thickness (h/δ) where 'δ' is the reference boundary layer thickness; that is, the boundary layer thickness at the step location in the absence of the step. It is seen that the results exhibit two distinct trends. When the step height is small and comparable to the boundary layer thickness, there is an almost linear variation between the relative separation length and the relative step height. However, beyond $h/\delta > 4$, that is, when the step heights are large in comparison with the boundary layer, the variation in separation length with change in step height is not very significant.

It would thus appear that for small step heights when the dividing streamline is fairly straight, the relative lengths of separation and step height bear a nearly linear relationship. Secondly, when the step heights are quite large and the dividing streamline is highly curved, the separation length scales on step height alone (Ref. 10).

5. Conclusions

The flow up a step is quite frequently mentioned as the model of rapidly separating flow but surprisingly very little experimental data exists on this subject. The present experiments provide some data on the laminar flow up a step and the results obtained show that the separation distance ahead of step varies differently depending on whether the step height to boundary layer ratio is either very small or very large.

Nomenclature

x, z co-ordinate axes, along and normal to the surface

u, w velocity components tangential and normal to the surface

p pressure

U_∞ free stream velocity

θ separation angle

τ shear stress

μ coefficient of viscosity

ν kinematic viscosity

Ω vorticity

Primes denote differentiation with respect to x and the subscript 's' refers to conditions at the surface.

References

(1) S. Goldstein, Modern Developments in Fluid Dynamics, Vol.I, (1965), 124, Dover.

(2) H. Schlischting, Boundary Layer Theory, (1960), 113, McGraw-Hill.

(3) L. Prandtl and O.G. Tietjens, Applied Hydro and Aero Mechanics, (1957), 64, Dover.

(4) M.J. Lighthill, Laminar Boundary Layer - Ed. L. Rosenhead, (1963), 60, Oxford Univ.

(5) L.D. Landau and E.M. Liftschiz, Fluid Mechanics, (1959), 151, Pergammon.

(6) G.K. Batchelor, An Introduction to Fluid Dynamics, (1967), 325, Camb. Univ.

(7) H. Liepmann and A. Roshko, Elements of Gas Dynamics, (1957), 313, John Wiley.

(8) K. Oswatitsch, Boundary Layer Research - Ed. H. Gortler, (1958), 357, Springer-Verlag.

(9) F.O. Ringleb, Boundary Layer and Flow Control, Vol. I, Ed. G.V. Lachmann, (1961), 265, Pergammon.

(10) M.J. Lighthill, On Boundary Layers and Upstream Influence - A Comparison between Subsonic and Supersonic Flows, Proceedings of the Roy. Soc. (Lon), A 217, (1953), 344.

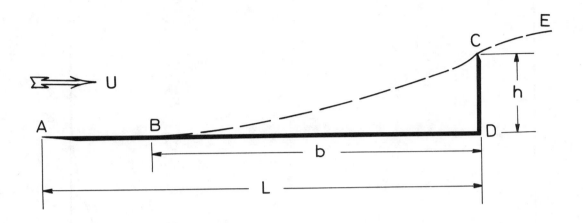

Fig.1   Schematic Diagram.

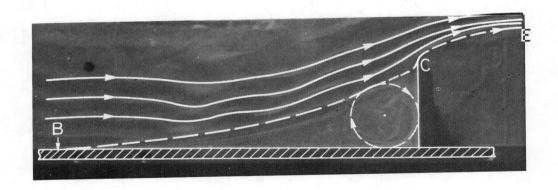

Fig.2   Typical Photograph of Flow up a Step.

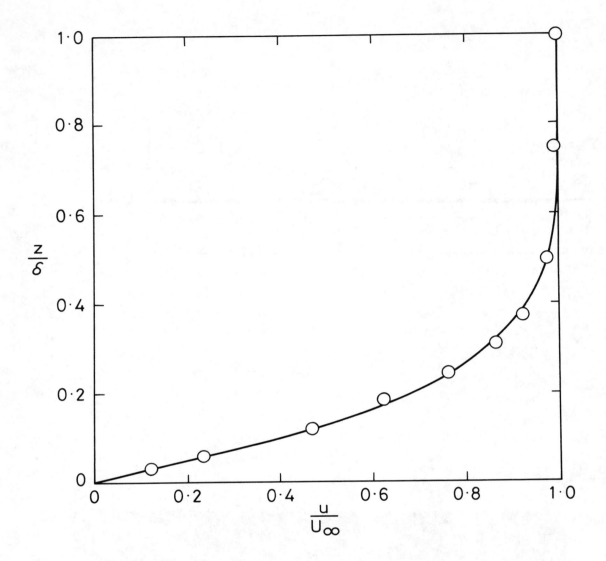

Fig.3   Typical Velocity Profile.

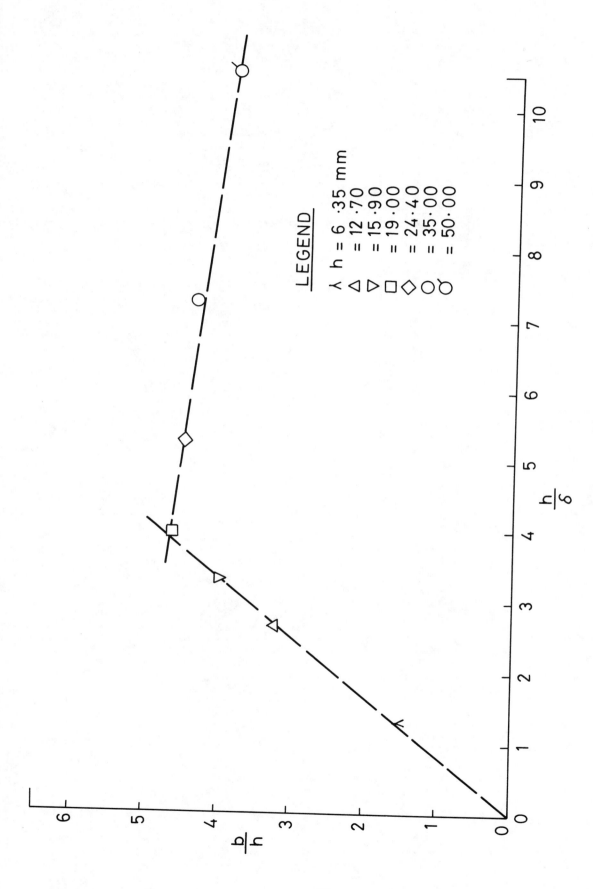

Fig. 4   Relative Separation Length Vs Relative Step height.

99

FLOW VISUALISATION AND VELOCITY MEASUREMENTS NEAR THE WALL OF A CIRCULAR PIPE

O. SCRIVENER, C. BERNER, and P. MUNTZER*

A special test section was designed for flow visualisation and velocity measurements near the wall of a pipe using
a strobophotographic method.
An analysis of the validity of the method together with
qualitative and quantitative results is presented both for
newtonian and non-newtonian flows.

Introduction

It was shown, for example by Smith et al (ref. 1), that Pitot tubes, hot wires or
hot films are not suitable for drag reducing fluid flow measurements.
It was also shown that the response of the Pitot tube is related to its diameter
due to the existence of an elastic behaviour of the molecules. Nevertheless some
results on the polymer solutions characteristics were obtained by this method : for
example, the existence of a length scale (see Fruman (ref. 2) and Piau (ref. 3)),
the presence and measurement of the diameter of macromolecule agglomerates (see
Kalashnikov (ref. 4)). Concerning the hot films and hot wires the response of the
probe is anomalous, due to the ionic character of several polymers, to the deposit
of macromolecules on the probe and to calibration difficulties due to the modification of the heat transfer coefficient.
At present, the only suitable methods to determine the velocity field in polymer
solutions are optical and, with reserve, electrochemical. In our measurements we used
both Laser Doppler Anemometry and photographic methods. Their common advantage is
that pertubations due to the presence of probes are avoided.

1. Experimental apparatus.

1.1. Flow continuations

The fluid was flowing in a circular cross-section pipe (fig. 1) to prevent
secondary flows. In order to obtain different wall shear stresses with the same
hydrodynamic conditions, two pipe diameters were used : 21 mm and 50 mm. The pipe was
supplied with tap water through a constant head   and a tranquillising tank.
To maintain the same flow conditions a second constant pressure tank is set at the
exit of the pipe. The flow rate was measured by filling a volumetric tank.

---

* Institut de Mécanique des Fluides - ERA CNRS 0594, Strasbourg, France

The distance between the entrance and the test section was more than 100 diameters in order to measure in the fully developped flow even for polymer solutions.

The pipe was built from smooth copper or glass tube-elements of 1 m carefully joined. Pressure taps were located along the pipe to measure the pressure drop and also to calculate the friction velocity. The highest Reynolds number obtained by this configuration was of the order of 50 000.

To prevent the mechanical degradation phenomena of the macromolecules a concentrated polymer solution (0,5 %) was continuously injected between the constant head and tranquillising tanks and mixed homogeneously with the primary tap water flow. The final concentration of the polymer solution (PEO 301) was between 5 and 100 ppm. The injection flow rate was controled by a rotameter. Two test sections were inserted in the pipe, the one for Laser Doppler Anemometry, the other specially designed for measurements near the wall by the strobophotographic method.

1.2. LDA measurements

The principles of Laser Doppler Anemometry are reported in the litterature (ref. 5). For our experiments we used an anemometer working in the Doppler Differential Mode with forward scattering. An optical glass tube was inserted in the pipe as test section.

It was necessary to seed the flow by mixing a few parts of silicone emulsion with the polymer solution. The required particle concentration was low and had no effect on the rheological characteristics of the solutions.

Due to the circular geometry of the pipe it was difficult to measure the velocities closer than 0.5 mm from the wall ; an increase of the signal to noise ratio was observed mainly due to light scattering by the walls.

The advantage of LDA compared to the photographic method is that an analogue signal is obtained the magnitude of which is proportionnal to the velocity. This allows a real time analysis of the velocity fluctuations.

It was necessary to apply corrections to the measurements to account for error produced by the finite dimensions of the optical probe and velocity gradients.

1.3. Test section for visualisation and photographic methods

A test section was designed (fig. 2) to allow the measurements of the velocities from 0.05 mm to 1 mm of the wall in a meridian plane of the pipe. It was built from a perspex bloc and two rectangular channels were drilled which emerged at each side of the tube (fig. 2). From the same figure it can be shown that the shape of the pipe remainded undisturbed in a region of 4 mm width. The pertubation due to the outlet of the channels in the pipe was found to be negligible by observing the streamlines with the dye method in laminar flow.

Closed outside by glass plates, one of these channels was used for observation and photography through a microscope, the second for the lighting of the wall region. Optical glass fibers were used to transmit the light from the source to the test section. Due to the small depth of field of the microscope (less than 0.2 mm) the observed region was located with good precision.

To locate the wall, a micrometer was introduced in the pipe and put in contact with the wall at the beginning of each trial.

The same method was used for an artificial rugged pipe. Ruggedness, made from steel wire arches of 0.1 mm of diameter, were glued on the wall of the pipe.

2. Experimental procedures

2.1. Visualisation technique

Dye was injected at the wall of the pipe through several injectors of 0.1 mm diameter located at different distances upstream $^{of}/_{and}$ in the observed region. The injection pressure was regulated to prevent perturbations formed by the dye jet emerging into the pipe.

The same technique was used to visualise the wake behind a ruggedness cemented at the wall of the observed region. Dye injectors are located before or behind the cylinder.

2.2. Strobophotographic method

The streamwise velocities were measured by taking pictures of particles (mixed with fluid) at two successive times. The instantaneous velocities were determined from the distance between two traces after enlargement of the pictures and the time between the flashes which is measured with a photocell and a time counter.

A double flash generator was used to light the particles as shown in fig. 3. Two flash tubes were controled by a rotating contact. The duration of the flashes were limited by two slits on a rotating disk.

The tracers were plastic particles of about 0.01 mm diameter. The equation of motion of the plastic particle may be written as follows :

$$(1 - (\rho)\,)\vec{\gamma} + K\vec{u} = 0$$

$$(\rho) = \rho'/\rho = \text{relative density of the particle}$$
$$\vec{\gamma} = \text{acceleration of the fluid particle}$$
$$K = \text{Stokes coefficient}$$

The validity of the assumption that the plastic particle follows the fluid is determined by the following parameters : (see ref. 6)

α = angle between the velocity vectors of the plastic and the fluid particle =
$$\frac{1 - (\rho)}{K} \frac{v}{R}$$

β = relative error on the particle velocity = $- \dfrac{1 - (\rho)}{K} \dfrac{dv}{ds}$

Under our experimental conditions the particles followed the flow with good accuracy.

3. Experimental results

3.1. Flow visualisation

3.1.1 Smooth pipe

In newtonian flow, by injection of dye (fluoresent) in the observed region at very low injection pressures (a few mm of water over the static pressure), the wall region was visualised for y^+ values less then 10. It was shown (fig. 4) that, even very close to the wall, there existed velocity fluctuations of very low frequency. The shape of the dye filet revealed the existence of streamwise vortex. An injection with a higher pressure also showed the existence of higher frequency vortex further from the wall (fig. 5). By injection of dye, upstream from the observed region, the existence of a bursting process was shown (fig. 6). This is also described by other authors. It was not possible to determine the frequency of the bursting process, but it was clearly shown that each burst (ejection) was followed by a rest time before the arrival of the smooth sweep. The structure of the flow observed in the sublayer was consistent with the model proposed by Offen et al (ref. 7)

We have not observed very significant differences in the case of non-newtonian flows, except for the thickness of the region of low frequency streamwise vortex which was increased.

3.1.2 Rough pipe

The existence of instabilities in the sublayer was confirmed by the visualisation of the flow behind the cylinder (fig. 7). From laminar to turbulent flow, different flow "domains" were found. The existence of a vortex in the wake of the cylinder was observed even for ruggedness the height of which was less than the sublayer thickness. From a quantitative study of the detachment point on the cylinders (see ref. 8) fluctuations of the flow were shown in the region of $y^+ < 5$ confirming the non-permanent character of the flow in the viscous sublayer.

3.2 Velocity measurements

3.2.1 Treatment of the results

Due to the velocity fluctuations, the dispersion of the values of the instantaneous streamwise velocity was large (see fig. 8, 9). Because the timing of the pictures was not recorded it was necessary to ensure that the error on the calculation of the mean velocity was negligible. At a given distance from the wall the measurement points obtained were not enough to calculate the mean value with good accuracy. Therefore we calculated the mean velocity in a slit of 0.1 mm thickness. This was allowed because the distribution of the velocity values was not far from Gaussian.

3.2.2 Newtonian flow

The dispersion of the measured instantaneous velocities and also the calculated mean reduced profile have confirmed the existence of instabilities even in the viscous sublayer. An universal transition profile between the wall and the turbulent core was determined (fig. 10).

3.2.3 Non-newtonian flow

The addition of polymers to the flow resulted to a reduction of the mean velocities near the wall (fig. 11). The polymer concentration affected the velocity reduction. The maximum velocity reduction occurred for the same concentration which corresponds to the maximum of drag reduction. These results obtained by the strobophotographic method near the wall produced a continuous fitting with those obtained by Laser Doppler Anemometry further from the wall (fig. 11). The velocity fluctuations in the viscous sublayer were also shown to be greater than in the newtonian case.

4. Conclusion

We have shown that flow visualisation, strobophotographic and Laser Doppler Anemometry are complementary methods to study the structure of turbulence in pipes. By means of these three methods we confirmed that important instabilities are located in the wall region of a newtonian flow. These instabilities were increased by the presence of polymers. A statistical analysis of the velocity fluctuations suggested that the drag reduction mechanism is located in the viscous sublayer. The origin of drag reduction may be due to the modification of the structure of turbulence due to the influence of the polymers on the production and dissipation terms of turbulence energy.

Références

(1) K.A. SMITH, E.W. MERRILL, H.S. MICKLEY, P.S. VIRK : Anomalous Pitot tube and hot film measurements in dilute polymer solutions.
Chem. Eng. Sci., 22, 619, (1967)

(2) D. FRUMAN, P. SULMONT, G. LOISEAU : Mesure des vitesses dans les fluides visco-élastiques au moyen de tubes de Pitot.
Journal de Mécanique, 8, 4, (1969)

(3) J.M. PIAU : Flow of dilute polymer solutions.
Comm. at 1st Int. Conf. Drag Red. Cambridge (1974)

(4) V.N. KALASHNIKOV, A.M. KUDIN : Size and volume concentration of aggregates in drag reducing polymer solution.
Natural Phys. Sci., 242, 92, (1973)

(5) F. DURST : Scattering phenomena and their application in Optical Anemometry.
J. of Appl. Math. and Phys., 24, 617, (1973)

(6) O. SCRIVENER : Contribution à l'étude de l'écoulement d'un fluide en conduite lisse. Thèse Strasbourg (1967)
Etude de l'influence des solutions de polymères sur la structure de l'écoulement turbulent dans une conduite lisse. Thèse de Doctorat d'Etat Strasbourg (1975)

(7) G.R. OFFEN, S.J. KLINE : A proposed model of bursting process in turbulent boundary layers.
J. F. M., 70,   2, 209, (1975)

(8) P. MUNTZER : Contribution à l'étude du frottement turbulent des fluides en conduite rugueuse. Thèse Strasbourg (1967)

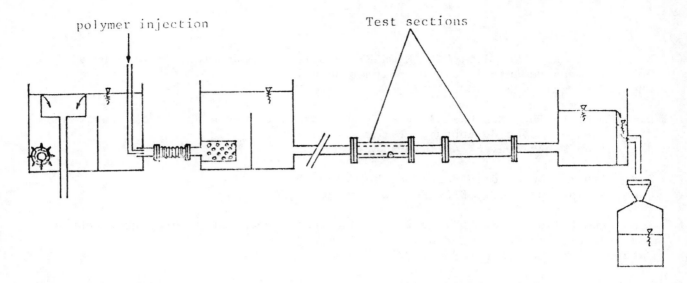

Fig. 1 : Experimental Apparatus

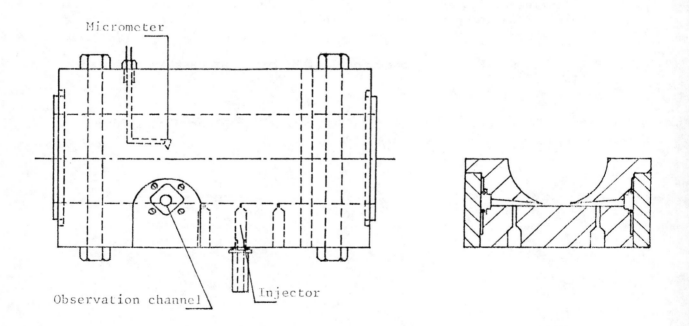

Front view

Fig. 2 : Visualisation and photographic test section

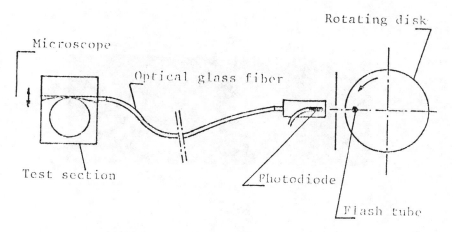

Fig. 3 : Lighting device

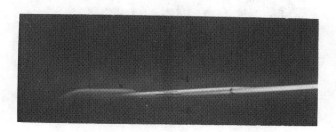

Fig. 4 : Flow pattern near the wall
Re = 30 000

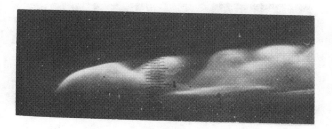

Fig. 5 : Flow pattern near the wall
Re = 30 000

Fig. 6 : Flow pattern near the wall
     Re = 30 000

Fig. 7 : Visualisation of the wake behind
     artificial ruggedness

Fig. 8 : Velocity measurements by
     strobophotographic method

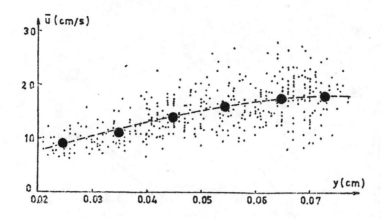

Fig. 9 : Local velocities .
         Mean velocities ●

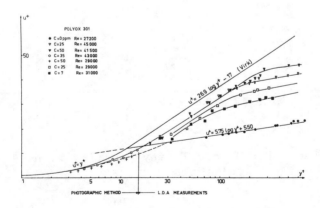

Fig. : 10

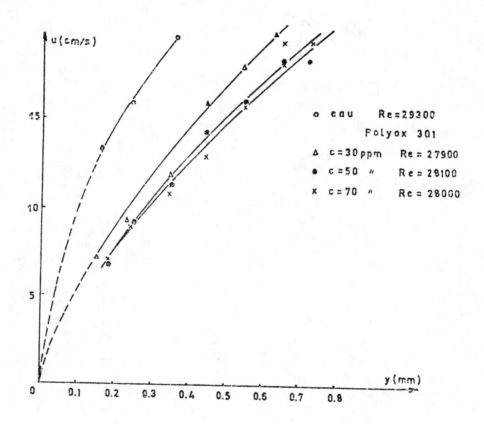

Fig. 11 : Mean velocity profiles with
polymer solutions

APPLICATION OF HYDRODYNAMIC VISUALIZATION TO THE STUDY OF LOW SPEED FLOW AROUND A DELTA WING AIRCRAFT

HENRI WERLÉ*

At the ONERA water tunnel, simple means and methods allowed a detailed study of the flow around a delta wing aircraft**.

In steady flow, visualizations reveal not only the flow structure and its evolution as a function of incidence and yaw with the fundamental phenomena that characterize them (separation, vortices, etc), but also the effects of a vertical jet (simulation of VTOL aircraft) or a transverse jet (control of upper surface flow separation).

These experiments were extended to the unsteady flow domain, in particular to the case of model either spinning or periodically oscillating. Lastly, such tests can also be used for an approach of more complex studies of applied aerodynamics, with air intakes, engine exhausts, ground effect, etc.

INTRODUCTION

For more than two decades, the ONERA water tunnel [1] devotes all its activity to visualizing flows at low speed, which made it possible to contribute to studies pertaining to a broad variety of domains : fundamental research, space, navy, industry, etc. [2]. But its main vocation remains essentially aeronautics, which explains the theme chosen for this paper.

Some results, both old and new, are presented of tests performed with the same type of aircraft model, either schematic or more complete and nearer to an actual aircraft. This paper is thus a survey of the fundamentel phenomena characterizing this type of flow and their evolution as a function of various parameters,   but it also approaches or illustrates many problems of applied aeronautics, such as flow control, spin, air intakes, exhaust jets, ground effect, etc.

EXPERIMENTAL TECHNIQUE

The ONERA water tunnel (fig. 1a) works by simple draining by gravity. In these conditions, the flow velocity does not exceed 0.5 m/s in the test section. This vertical section includes various mounting possibilities : half models at the wall, complete models mounted on lateral mast or rear sting, the supports serving for the passage of ducts for coloured liquids (visualization), for injection or suction (engine simulation, etc.) and for transmission of certain movements (rotation, oscillation, etc.).

Among the many methods of hydrodynamic visualization [3], we only used the two processes particularly well adapted to vertical facilities with open circuit, as the ONERA tunnel :

— liquid tracer method, in which opaque coloured streams of the same density and viscosity as water are emitted, usually from orifices bored through the surfaces of the model (fig. 1b),

— gaseous tracers, in this case air bubbles in suspension within the water and produced during the tank filling ; this process is associated with the method of the light plane that limits the illumination of the fluid to a thin layer. Set along the test section or the model axis, these layers contain the bubble trajectories and provide an image of the flow (fig. 1d) ; placed perpendicular to this axis, these layers are crossed by the air bubbles and visualize the pattern of the phenomena in a cross section (fig. 1c), which can be observed with a mirror installed downstream of the model.

TESTS PERFORMED WITH SCHEMATIC MODELS

The model schematizing an aircraft of Mirage III type comprises a cylindrical fuselage with conical nose and a delta-shaped ($\varphi_{LE} = 60°$), thin, sharp-edged low wing, and a swept fin.

---

* Division Head, Office National d'Etudes et de Recherches Aérospatiales (ONERA) - 92320 Châtillon (France)
** Some examples of visualizations have been put together in the ONERA film no. 888 [10], presented during the meeting.

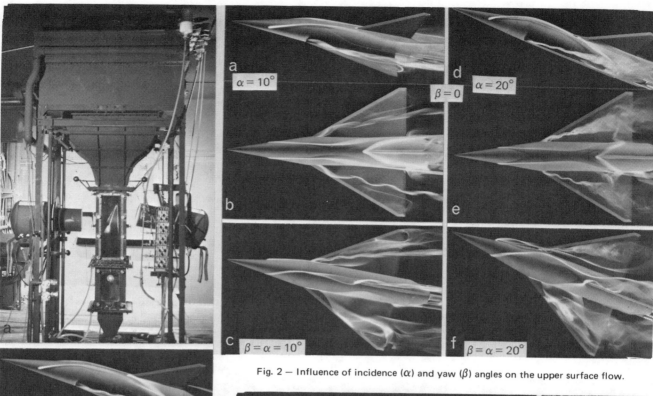

Fig. 2 — Influence of incidence (α) and yaw (β) angles on the upper surface flow.

Fig. 1 — General view of the water tunnel (a)
and visualization methods :
  - coloured emissions (b)
  - gaseous tracers seen in transverse (c)
    and longitudinal (d) sections.

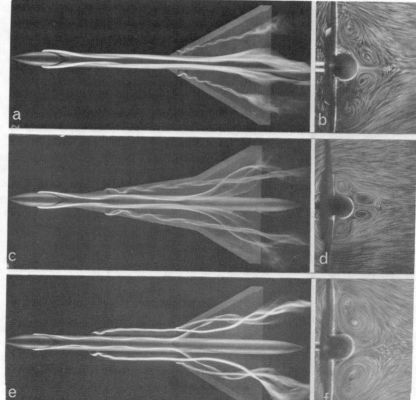

Fig. 3 — Comparison of various types of wing ($\alpha = 15°$) :
delta with long fuselage (a, b) ; double delta (c, d) ; delta with strakes (e, f).

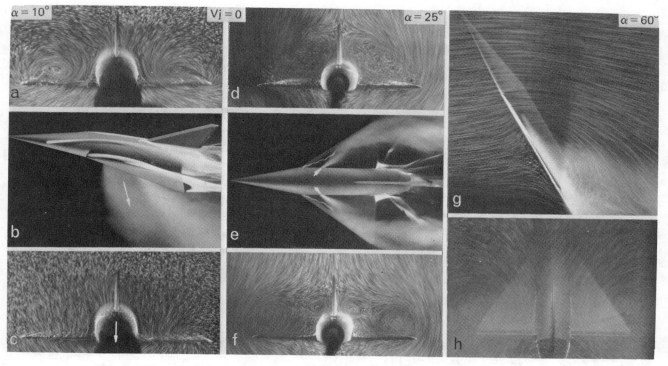

Fig. 4 — Effect of blowing on the flow around a delta wing aircraft - Simulation of the down blowing
jet on a VTOL aircraft (a-c) - Flow control by lateral jet (d-h).

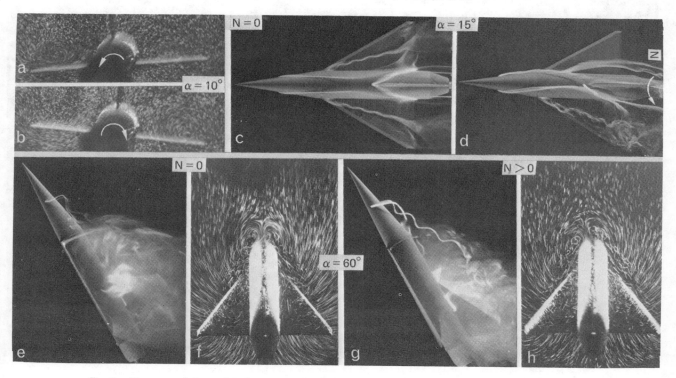

Fig. 5 — Flow around the model of a delta wing aircraft in unsteady regime. Periodic oscillations in roll (a, b).
Uniform rotation simulating a horizontal roll (c, d) or a spin (e-h).

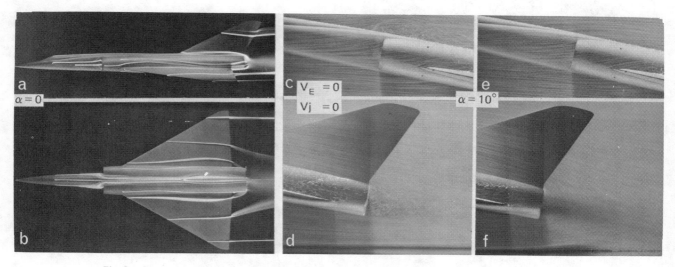

Fig. 6 — Flow around a model of Mirage IV aircraft without (c, d) or with (a, b, e, f) engine simulation.
Ground represented by a fixed floor (d, f).

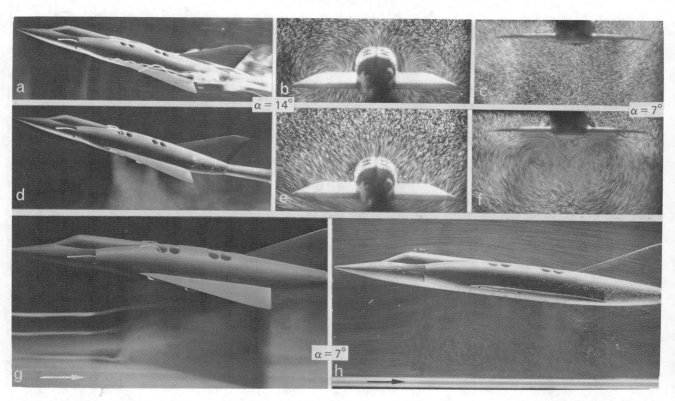

Fig. 7 — Flow round a model of VTOL aircraft without (a-c) or with (d-h) simulation of lift engines.
Ground represented by a moving belt (g, h).

In steady regime, the upper surface flow is characterized by the presence of two well-organized, symmetrical wing vortices for $\alpha = 10°$ and $\beta = 0$ (fig. 2a, b and 4a). When incidence α increases, fuselage vortices are formed while the wing vortices develop, then get disorganized : it is the breakdown phenomenon [4], observed at the trailing edge around $\alpha = 15°$ (fig. 1b, c and 5c), and around mid-chord at $\alpha = 20°$ (fig. 2d,e). Lastly, at very high angle of attack (fig. 4d, 1d, 5e, f) the flow over the wing eventually separates, and only the fuselage nose remains the seat of two organized vortices.

A setting in sideslip breaks the symmetry of the flow, and in particular creates a fin vortex, the displacement of fuselage and wing vortices, with down travel of the breakdown on the half wing whose effective sweep angle is increased by the yaw angle, and the inverse on the other half wing (fig. 2c, d).

Figure 3, 5c and 1c illustrate, for $\alpha = 15°$ ($\beta = 0$) the influence of the wing planform and the fuselage length on the number, trajectory and development of the various vortices forming on the upper surface. It is known that these vortices provide, on fast aircraft, the lift augmentation necessary at low speeds [5].

The effect of blowing on this type of flow can be observed on figure 4. The entry into service of a downblowing jet (simulation of a VTOL aircraft - fig. 4a, b, c) creates an induced downwards vertical component of the velocity and, consequently, entails the gradual resorption of the upper surface vortices observed without jet, a consequence of the decrease of the effective angle of attack of the wing. The flow control [6] by a jet issued from the apex and emitted above the upper surface along an axis close to that of the wing vortex is a process that remains efficient up to high angles of attack (fig. 4d-h) ; it makes it possible to reestablish on the upper surface an organized and stable vortex pattern.

A few visualizations observed in unsteady regime [7] are put together on figure 5 and reveal the influence, on the wing vortices, of a movement of periodic oscillations in roll (fig. 5a, b), and of uniform rotation (fig. 5c-h). At moderate angle of attack (continuous roll - fig. 5c, d) they reveal the accentuation of the vortex regime on the upper surface of the half wing with descending leading edge, and on the contrary its partial or complete resorption on the ascending leading edge side. At high incidence (spin movement, fig. 5e-h), the influence of rotation is limited to the vortices issued from the fuselage nose, while the flow over the wing is completely separated.

TESTS PERFORMED WITH COMPLETE MODELS

A first example (fig. 6) concerns a model of the twin-engined Mirage IV aircraft : the effects of a simulation of the lateral air intakes and of the rear exhaust jets can be detected both at zero incidence (fig. 6a, b) and at moderate incidence (compare fig. 6c, d with fig. 6e, f). We observe that the air intake blockage provokes a separation near the wing apex [8]. These tests show that it is sometimes mandatory to simulate the engine operation to ensure correct tests on a complete aircraft.

A last example is that of a complex model of VTOL aircraft, with simulation of the eight lifting engines set in the fuselage. The tests confirm the reduction of effective incidence under the effect of these eight jets (fig. 7a-f), as before with the single jet. They also provide the shapes and dimensions of the streamtubes entering the eight air intakes on top of the fuselage (fig. 7e, g, h), the vortex structure of the resulting jet under the fuselage (fig. 7f), and lastly the vortex phenomena related to the ground effect (fig. 7g, h) which, with this type of model, can only be simulated in realistic conditions by the moving belt method [9].

CONCLUSION

Even when the similarety conditions are poorly respected — in these tests, the Reynolds number seldom exceeds 20000 —, such visualizations provide informations, assuredly of qualitative nature, but often well appreciated by theoreticians looking for a basic physical scheme for their calculations, or by experimenters wishing to have precise data on flow patterns. That has been understood and put into practice by aerodynamicsts of many countries apart France, such at the US (Northrop, Boeing), Canada [11], the USSR [12], Germany (MBB), as well as Japan [13].

REFERENCES

1. WERLÉ H. *Le tunnel hydrodynamique au service de la recherche aérospatiale.* ONERA Publ. n° 156 (1974).

2. WERLÉ H. *Applications aérospatiales, industrielles et maritimes de la visualisation des écoulements.* ATMA (1975), ONERA Film 812 (1975).

3. WERLÉ H. *Méthodes de visualisation hydrodynamique des écoulements.* Annual Review of Fluid Mechanics (1973), ONERA Film 757 (1973).

4. WERLÉ H. *Sur l'éclatement des tourbillons.* ONERA N.T. n° 175 (1971).

5. SALMON M. *Concorde et la recherche aéronautique.* L'Aéron. et l'Astro., n° 11 (4-1969).

6. WERLÉ H. & GALLON M. *Contrôle d'écoulements par jet transversal.* L'Aéron. et l'Astro., n° 34 (1972-2), ONERA Film 649 (1972).

7. WERLÉ H. *Visualisation hydrodynamique d'écoulements instationnaires.* ONERA N.T. n° 180 (1971), ONERA Films 666 (1971) & 485 (1964).

8. WERLÉ H. & GALLON M. *Sur l'écoulement autour d'une prise d'air.* La Rech. Aérosp. n° 1975-2.

9. WERLÉ H. *Visualisation de l'effet de sol à basse vitesse autour d'une maquette d'avion.* La Rech. Aérosp. n° 1970-2, ONERA Film 412 (1971).

10. ONERA Film 888 (1977) - *Ecoulement autour d'un avion delta.*

11. PEAKE D.J. *Controlled and uncontrolled flow separation in three dimensions.* NRC, NAE LR 591 (July 1976).

12. GOLOVATIUK G.I. & TETERIUKOV Ia.I. *Flow patterns of fuselage wing models at supercritical angles of attack.* TSAGI Uchenye Zapiski, vol. 4, n° 1 (1973).

13. NAITO Yasuo. *Flow study by visualization using floating oil - 2nd Symposium on Flow Visualization ISAS* Univ. of Tokyo. July 1974 (A-8).

STEREOSCOPIC VISUAL STUDIES OF COMPLEX TURBULENCE SHEAR FLOWS

ROBERT S. BRODKEY*

The existence of coherent structures in turbulent shear flows and in flows with complex geometries places new demands on visual methods. The need for three-dimensional (stereoscopic) viewing in addition to the more conventional two-dimensional or pseudo three-dimensional viewing of the flow field is apparent. Additional considerations involve marking part or all of the fluid motions both on a fine and a gross scale, the need to follow particle paths which do not follow fluid motions, the use of convected versus non-convected views, and simultaneous visual and anemometry measurements.

1. INTRODUCTION

Flow visualization studies have been historically two-dimensional views of what is in many cases a three-dimensional flow. Although such viewing is appropriate for two-dimensional flows, for a three-dimensional flow it could be misleading. Two-dimensional flow visualization techniques that have been used to generate a streak sheet or line have involved hydrogen bubbles from a fine wire in a liquid, smoke particles from a heated wire, particles in the flow that follow the fluid motion but are illuminated by a narrow light beam, dye injection along the surface of solid materials, and others. There are many permutations and variations of these techniques that have been used by researchers in order to better understand the flow mechanism. Realizing the limited nature of such two-dimensional information, a number of workers have made use of either dual cameras or mirror arrangements to provide two views of the flow at right angles. However, the fluid markers used were often a streak sheet or line so that only limited three-dimensional information was obtained. Nevertheless, with such information and a careful and prolonged study of the flow pictures one can often put together a reasonable three-dimensional picture of what the flow must be.

As implied above, both the means of viewing and what is viewed are of importance. The camera arrangement, the means of lighting, and the method used to mark the flow must all be considered. Two additional important questions that can be asked are should the camera be stationary or moved and would there be value in having simultaneous measurement of the velocity by an anemometer.

2. VIEWING THE FLOW

There are several means of recording what is viewed. This is usually accom-

*Professor of Chemical Engineering, The Ohio State University, Columbus, Ohio, U.S.A. 43210

plished by photographic means either by still or by movie cameras. Because the camera will record everything it sees (with due consideration for depth of field of the lens), an easily interpretable limited two-dimensional view is obtained by only lighting two dimensions of the flow. A very limited three-dimensional view of a complex flow can be obtained by lighting the entire flow, but only marking the flow in two dimensions. The movement of marked fluid particles in and out of the plane provides the limited three-dimensional view. Such two-dimensional markings have been hydrogen bubbles from a fine wire in a slightly conducting liquid flow, helium-soap bubbles in air, smoke particles from a heated wire, and dye injection. In all cases, the experimenter has a choice of a continuous streak sheet or a series of streak lines. In order to gain more information on the limited three-dimensional view one can use two cameras or a mirror arrangement. Usually two simultaneous views at right angles are obtained: one view providing the main flow direction information and the other providing the limited information in the third direction. Although certainly not easy, a considerable amount of information about the three dimensionality of the flow can be obtained. Since such viewing is not a direct three-dimensional view, we call it pseudo-three-dimensional viewing. Limitations associated with the marking methods that can be used will be reserved until later.

Contrasted with pseudo-three-dimensional viewing would be full simultaneous three-dimensional visualization of the flow field as would be done with one's eyes. Three-dimensional techniques available involve stereoscopic viewing and holographic methods and have been discussed by Praturi et al. (4). Due to severe limitations on system stability and the availability of high levels of light required for high speed photography, holographic techniques do not seem feasible at the present time as has been discussed by these authors. Stereoscopic photography holds the most hope for a full three-dimensional visualization of the flow.

Stereoscopic visualization involves taking two simultaneous pictures and preferably registering them side by side on the film frame. The two pictures are taken at an angle and separation that would correspond to that observed by the human eye. This means that the two views are separated by approximately six centimeters and the angle of view depends on the distance from the object. The technique is described fully by Praturi et al. (4) and will not be repeated here, except to indicate that the most successful pictures are obtained by specially designed twin lens systems which are fully independent, matched lenses that put the two views side by side upon one frame of the film. Far less successful are adaptors that are placed before the lens on a single camera lens to accomplish the same end. The difficulty here is that the best part of the lens is lost. Stereoscopic viewing of the entire flow means the entire flow must be illuminated and marked. For this the use of small particles that move with the flow are most useful.

As with any experimental technique there are complicating factors associated with stereoscopic viewing. Let us first consider a flow field in which there are points of reference for the viewer, as an example a mixing tank with baffles. In these cases the viewer, using a stereoscopic view, will easily have a complete three-dimensional view of the flow field. Actually, even with one eye, one can judge in part the three-dimensional nature of the flow because of the existence of perspective. It is by means of perspective that three-dimensional views are represented on a two-dimensional flat surface. Thus even a single view of such a flow field will give the viewer a sense of three dimensionality because of perspective. However, the viewer will have trouble ascertaining whether flow elements are moving toward or away from him unless the flow particles are large enough so that their relative size changes becoming larger as they move toward the viewer. Unfortunately, particle motions that are observed by the use of scattered light at right angles do not change their size appreciably. Thus the viewer using only two-dimensional viewing has difficulty ascertaining actual flow directions in a complex three-dimensional flow field in spite of a sense of three-dimensionality due to perspective. In contrast, stereoscopic visualization allows the viewer to readily ascertain the direction of the flow of any particles regardless of their relative size due to the angle of view changing as the particle moves toward or away from the viewer. The combined physical picture of perspective and the stereoscopic viewing enhances the three-dimensionality and one can easily obtain a dramatic three-dimensional viewing of the flow field.

There is one situation in which stereoscopic viewing is difficult. If perspective is not available, such as may occur in photographing particle motions along a flat plate in the boundary layer, then it is difficult to immediately obtain a complete stereoscopic visualization of the flow. Julesz (2) has dramatically demonstrated this for computer generated three-dimensional geometrical objects. A random noise coverage was used so that the viewer could not tell the nature of the object without stereoscopic viewing. With stereoscopic viewing the object was apparent but sometimes it would take several minutes of concentrated viewing before the object began to stand out in three dimensions. It is important to realize that in this case one does not have perspective to aid in one's viewing. If one closes one eye, the three-dimensional figure disappears and on opening the eye the figure will return. In situations where perspective is not available, one can only obtain three-dimensionality by use of stereoscopic viewing. It is this fact that makes it difficult to study a field of particles alone in three dimensions. Our technique, that we have found successful, is to make short film loops of specific events and to view the events for extended periods of time over and over again. After several minutes one becomes familiar with many of the individual particles and the three-dimensional nature of the flow field can be observed.

In stereoscopic viewing there are precautions necessarily associated with the use of polarized light. The two views are polarized and are projected on a screen. The screen itself must be of a metalized type otherwise the polarization and separation of the two images will be destroyed. Glass beaded screens, painted walls, and paper screens are usually inadequate.

3. LIGHTING

In all flow visualization work careful attention must be given to lighting requirements. Enough light must be available to expose the film. This can become quite difficult especially for high speed photography. The intensity available from scattered light from particles is quite small compared to that available from direct lighting; however, direct lighting for flow visualization work seems more the exception than the rule. A great deal of light can also be lost in using slits. Our experience has been that every time a major change in the type of picture is desired, the lighting system must be redeveloped.

4. MARKING THE FLOW

Should dye or smoke injection be used, generation of hydrogen or helium-soap bubbles, or particles carried throughout the flow? The question amounts to whether part or all of the flow is to be marked. For a full view of the flow, the use of small suspended tracer particles has been extensively discussed by Brodkey et al. (1) and by Nychas et al. (3). The reasons for the preference of small particles suspended in the flow to injection techniques was discussed in the first article in some detail and will not be repeated here. The points involved are mainly: 1) possible flow disturbances and 2) injected flow markers form a streak line or sheet in an unsteady flow field and then spread to assume various configurations (i.e., all structures are equally marked at the injection line and can be followed for only a short distance down stream before being mixed with the fluid). Questions

119

that arise in using an injection technique involve developing a marker field in an unsteady velocity field, an uneven marked field in time where high concentrations remain in low velocity areas and high velocity regions are no longer marked, the existence of regions that were never marked, and the difficulty in interpretation.

One must also establish the scale of motion to be marked. Very fine particles will mark all scales whereas larger particles will only follow fluid motions of a larger scale. In some situations there is a desire to follow particle motions rather than fluid motions where the particles are known not to follow the fluid because of larger density differences. The motion of large, heavy particles is often the case for mixers used in heterogeneous chemical reactions involving catalysts and in crystallization. The stereoscopic technique with proper lighting and particles in the entire flow can be used to delineate these motions. In fact, very fine particles can be used to show the fluid motions at the same time the larger and heavier particles are viewed. Thus, interactions can be observed. It is also possible to use more restricted lighting in conjunction with stereoscopic viewing to illuminate a subarea and still view in full three dimensions.

Combined techniques are of interest. For example, there is the possibility of using combined dye injection and particle marking together with stereoscopic visualization. In a boundary layer, particles can be used to mark all of the fluid motions both inside and outside of the boundary layer; but, it is extremely difficult to tell where the boundary layer edge lies. Simultaneous injection of a dye in front of the trip for the boundary layer can be used to fill the layer with dye and thus provide a mark for the edge of boundary layer. Caution must be taken because too high a dye concentration will obscure the particles inside the boundary layer and too little concentration will not allow adequate definition of the boundary layer edge. As a further example, in a mixing vessel it is possible to study the fluid motion by fine particles, solid particle motions involved in processing by the larger particles characteristic of their size, and the interaction between the fluid and the larger particles by means of dye injection.

Finally, one should not forget the possibility of using color to mark particles that are essentially the same. This will usually be restricted to the larger particles, but can be of use in flows where the solids concentration is high.

5. CONVECTED OR NON-CONVECTED VIEW

There is another major consideration that must be established before satisfactory pictures can be taken; should a stationary (non-convected) or convected view of the flow be taken. In situations where the flow is developing in a local region, a non-convected view will be adequate. If, however, the structure to be viewed passes a point, the details of the structure can be seen more clearly in a view convected with the flow. Of course, moving with the flow is more difficult, but can be necessary and is the essential ingredient for the study of developing structure in a boundary layer. To stop a convected flow structure in a stationary view will require higher filming speeds and correspondingly more critical lighting in order to obtain adequate exposure. In the convected view, the filming speed can be reduced considerably and the lighting becomes far less critical. As another example, one can use the fluid motions within a mixing vessel. A stationary view of the flow as a whole is often quite adequate. If one is interested in the flow field in the vicinity and within the mixing blades, the camera can be rotated at the same speed as the impeller and thus maintain a given area in view within the impeller for long periods of time.

6. SIMULTANEOUS VELOCITY AND VISUALIZATION MEASUREMENTS

In the study of the coherent structure of turbulent shear flows, there is a need to know exactly the nature of the velocity signal that occurs during passage of coherent events. This can be done by simultaneously photographing the structure and using anemometry measurements. Of course, it is far simpler to do this in a non-convected view because if the measuring probe is convected too fast, it could interfere with the flow. Convection at less than the flow velocity can be used to advantage and still not interfere with the flow. There is the possibility of combining laser doppler anemometry with visual studies and thus not interfere with the flow. The importance of simultaneous visual and anemometry measurements should be emphasized. The signal signatures of an event need to be recognized so that more conventional measurements using anemometry methods can be used to establish meaningful averages by pattern recognition procedures or conditional sampling.

7. CONCLUSIONS

There can be advantages in using stereoscopic visualization over the more classic two-dimensional visualization or limited three-dimensional visualization. It is important to note that in using particles to mark the fluid, very fine particles can be found to follow the fluid, but there are many cases in which the motions of larger and heavier particles are important. Stereoscopic flow visualization can be used to delineate both of these simultaneously. Often dye injection can help along with particle motions. It is important to consider convected versus non-convected views and the fact that certain types of information are more readily available from a convected view. Obtaining pictures and interpretation of the flow can be far easier in a convected view. Finally, it is becoming necessary to obtain simultaneous visualization and anemometry measurements.

8. REFERENCES

1. Brodkey, R.S., H.C. Hershey, and E.R. Corino (1971) "Turbulence in Liquids", J.L. Zakin and G.K. Patterson eds., p. 129, Dept. of Chem. Eng. Cont. Ed. Series, Univ. of Missouri, Rolla.

2. Julesz, B. (1974) Amer. Scientist, 62, 32.

3. Nychas, S.G., H.C. Hershey, and R.S. Brodkey (1973) J. Fluid Mech., 61, 513.

4. Praturi, A.K., H.C. Hershey, and R.S. Brodkey (1977) "Turbulence in Liquids", J. L. Zakin and G.K. Patterson eds., Science Press, Princeton, N. J. (Proc. of the 4th Biennial Symp. on Turbulence in Liquids, 1975).

VISUALIZATION OF ARTIFICIAL TRANSIENT WATER WAVE

T. HIRAYAMA,* M. NAGAI,** and I. UENO***

Spatial characteristics of Artificial Transient Water Waves were considered through the prediction of the wave forms. Experiments were made in the towing tank and the small glazed chunnel. Comparison between the wave forms near the concentration and the predicted ones calculated from the time history before concentration indicates good agreement by use of "phase correction".

Introduction

Since the effectiveness of Transient Water Waves(T.W.W.) was pointed out in the study of the ship response test or wave impact pressure test, T.W.W. have been used for many experiments -Davis,M.C. and Zarnic,E.C.(Ref.1), Takezawa,S.(Ref.2-5 for example). On the other hand the study about T.W.W. has been continued, and the characteristics of T.W.W. in the linear field have been made clear almost all (Ref.6,7). But even in the linear field there are few reports about the spatial characteristics (Ref.6,8) of T.W.W. on account of the troublesomeness of confirming the unsteady spatial characteristics. To study wave impact pressure the spatial characteristics are very important point, and the ship motion depends on them , too. Thus the lack of the information about the spatial characteristics must be avoided. It will be, therefore, significant in its own way to adopt the method of predicting the wave form or the locus of water particle to understand the spatial characteristics of T.W.W.

Here we make a little description about T.W.W. Utilizing the difference in phase velocity of each component wave , we can easily make all component waves in phase as if they were one shot pulse. This phenomenon is called a concentration and illustrated in Fig.1. (The T.W.W. , used in this report, concentrate at 25m distant from the wave generator.) In spite of the transition of the wave form, T.W.W. contain the same component waves or amplitude of the Fourier spectrum does not change.Thus making experiments at the point of concentration makes it easy to obtain the frequency response of a ship because obtained time histories are very short. (Actual experiments are made just before the concentration to avoid the non-linearity which appears near the concentration.) Fig.2 presents the block diagram of the ship response test using T.W.W. Fig3 and Fig4 show typical examples of T.W.W. in Fourier spectrum and in time histories. T.W.W. in Fig3 is called "constant amplitude T.W.W." and Fig4 is called

*Associate Professor of the Department of Naval Architecture,Faculty of Engineering ,Yokohama National University,Tokiwadai-156,Hodogaya-ku,Yokohama,Japan
**Hull Designing Section, Ship Designing Department, Mitsubishi Heavy Industries, Midori-cho 1-1, Nishi-ku, Yokohama-shi, Japan
***Mitsui Ocean Developement and Engineering Co.,Ltd, 2-5,Kasumigaseki 3-chome, Chiyoda-ku,Tokyo,Japan,

"constant steepness T.W.W" which means that each component wave has the same steepness. In spatial use they must be arranged by use of the transfer function of the water waves in the towing tank and the wave generator system.

Experiment

Experiments were made both in the towing tank (LxBxD=50mx3.6mx3m) and the small glazed channel (5mx0.36mx0.3m) in the perfect darkness because of the convenience of photographing. Fig.5 presents the block diagram of experiments. Wave probes were set before and after concentration to grasp the transition of Fourier spectra. Photo. 1 shows the arrangement of the wave probes and the grid in the towing tank. The distance between·two probes is 2-m, and the grid space is 10-cm. To catch the changing of wave form, paper tape was used as a medium(Photo.2).

Wave heights and the plunger stroke of the wave generator were measured and the pulses from the strobo oscillator were recorded in the magnetic tape at the same time (Fig.6).

The technique of visualization of wave surface in the towing tank without side glass is primitive but very effective in this case. In spite of the camera angle, wave heights can be measured from the photographs directly since the parallax is compensated by the grid. The grid,as a co-ordinate , never disturbed the waves and media because the multiplex exposure technique was used. Flash interval of stroboscope is 0.2 sec in photo-2. Fig.6 presents the time histories of No-1 and No-2 wave probes.

The same analysis but the different method was used in the small glazed channel. As the grid was marked on the side glass of the channel, the parallax is negligible. The grid space is 1.0 cm. Fig.7 presents the arrangement of stroboscope and camera. Photographs were taken utilizing the total reflection pure optically and no medium was used. Thus this technique makes it possible to catch not only the simple wave form but the breaking wave shown in Photo.3. Photo.4 shows the concentration of T.W.W in the small glazed channel. So called "Angle 120°" can be seen in the Photo.3.

About the locus of the particle on water surface, experiments were made using the floating stylrol sponge as a medium(as shown in Fig.14 it also indicate local wave slope). Photo.5 , enlarged from Photo.1, shows the motion of the styrol sponge.

Analysis and Results

Recorded data were analyzed using Fast Ship Motion Analyzer, a kind of spectrum analyzer combined with mini-computer, and Fourier spectra were obtained. Fig.8 presents the block diagram of analyzation. Fig.9 presents Fourier amplitudes and Fig.10 presents the amplitude ratio and the phase difference deviation between two records shown in Fig.6. If linear, each must be unity and zero.

The amplitude ratio can be regarded as unity but it leaves some room for consideration about the phase difference because of large deviation. Space forms are predicted from the Fourier spectrum by No-1 wave probe by use of inverse Fourier transformation (see Appendix). Nonlinearity appeared in Fig.10 was approximated by a cubic curves, and taken into consideration. Fig.11 presents the comparison between the actual wave form measured from Photo and the predicted one. Case.1 means the prediction was made linearly. Nonlinear effect was considered in Case.2. Case- 2 prediction indicates good agreement. (The difference between the vertical scale and the horizontal one should be taken note of) Fig.12 presents the results of the experiments in the small glazed channel. The inclination coincides with the actual wave form.

Fig.13 shows the comparison about the locus of the water particles. Styrol sponge we used in the experiments is shown in Fig.14. The form of the particle is considered to indicate the position and the steepness at the same time. The steepness is presented in Fig.14. There is a little difference about the locus but the velocity of the particle is seemed to be in good agreement. In Fig.13,"Photo" is corresponded to the Photo.5. In the Fig.14, the parallax in the Photo.5 is corrected.

Conclusion

By working out the way of fluid surface visualization which is primitive and easy in method but quantitatively analyzable, the possibility of theoretical prediction, using time history, about the spatial characteristics of Transient Water Waves were confirmed.

Acknowledgement

The authers would like to express their thanks to Prof.S.Takezawa for his significant suggestions and to K.Miyakawa and T.Takayama who supported our experiments.

References

(1) M.C.Davis,E.C.Zarnick, Testing Ship Models in Transient Water Waves,proc.5th Symp.on Naval Hydrodynamics. (1964)
(2) S.Takezawa,M.Takekawa,T.Hirayama, Advanced Experimental Techniques for Testing Ship Models in Transient Water Waves, Proc. 11th Symp. on Naval Hydrodynamics. (1976)
(3) S.Takezawa, A Practical Method for Testing Ship Models in Transient Water Waves, Selected Papers J.S.N.A. Japan (1972-9) (in English)
(4) S.Takezawa,M.Takekawa, Testing Ship Models on Longitudinal Ship Motions by Transient Response Method, Bulletin of Faculty of Eng. Yokohama National University (1974)No 23. (in English)
(5) S.Takezawa,S.Hasegawa, On the Characteristics of Water Impact Pressures Acting on a Hull Surface Among Waves, J.S.N.A.Japan(1974) No.135.
(6) S.Takezawa,T.Hirayama, On the Generation of Arbitrary Transient Water Waves, J.S.N.A. Japan(1971) No.129.
(7) T.Hirayama, On Non-linear Characteristics of Transient Water Waves Used for Ship Response Tests(Part 1 and 2 in Japanese),J.S.N.A. Japan(1974,1976) No.136,137.
(8) M.Nagai,I.Ueno, Spatial Characteristics of Transient Water Waves, Graduation Thesis of Yokohama National University(1975). (in Japanese)

Appendix

Wave form prediction from a time history is given as

$$\eta(x)\Big|_{t=t_0} = \frac{1}{2\pi} \int_{-\infty}^{+\infty} F(\omega,x) e^{j\omega t_0} d\omega$$

Where

$$F(\omega,x) = A_1(\omega) \cos\left\{\omega t_0 + \Psi_1(\omega) - \frac{\omega^2}{g}(x-x_1) + \Delta\Phi'\right\} d\omega$$

$$\Delta\Phi' = \frac{x-x_1}{x_2-x_1} \Delta\Phi_{21}(\omega)$$

$$\Delta\Phi_{21}(\omega) = \Psi_2(\omega) - \Psi_1(\omega) + \frac{\omega|\omega|}{g}(x_2-x_1)$$

$$A_1(\omega) e^{j\Psi_1(\omega)} : \text{Fourier Spectrum of time history obtained by No-1 wave probe}$$

Velocity at the point of(x,y) is shown as

$$u = \frac{1}{\pi} \int_0^\infty A_1(\omega) \omega e^{\frac{\omega^2}{g}y} \cos\left\{\omega t + \Psi_1(\omega) - \frac{\omega^2}{g}(x-x_1) + \Delta\Phi'(\omega)\right\} d\omega$$

$$v = \frac{-1}{\pi} \int_0^\infty A_1(\omega) \omega e^{\frac{\omega^2}{g}y} \sin\left\{\omega t + \Psi_1(\omega) - \frac{\omega^2}{g}(x-x_1) + \Delta\Phi'(\omega)\right\} d\omega$$

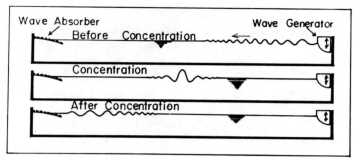

Fig.1
Concentration of a Transient Water Wave.

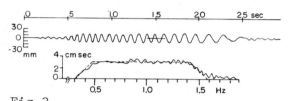

Fig.3
An example of Constant Amplitude Transient
Water Wave.Time history and spectrum.

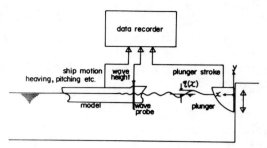

Fig.2
Block diagram of the ship response
test using Transient Water Waves,
and the co-ordinate.

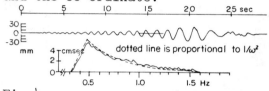

Fig.4
An example of Constant Steepness
Transient Water Wave.

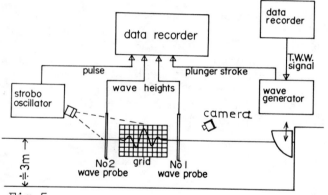

Fig.5
Block diagram of experiments in the large
towing tank witout glass walls.

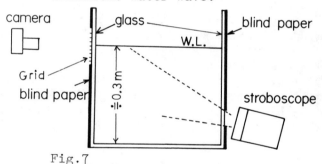

Fig.7
Arrangement of the experiments in
the glazed small channel.

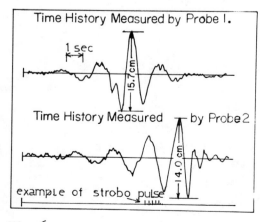

Fig.6
Measured time history and pulses
from wave probes and strobo osc-
illator.Corresponds to Fig.11

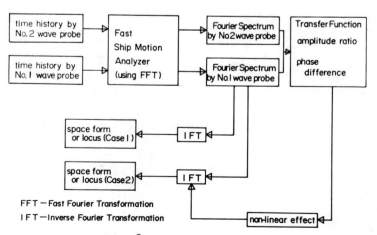

Fig.8
Block diagram of analyzation

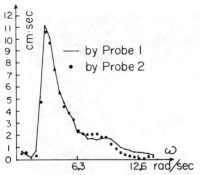

Fig.9
Amplitude parts of Fourier spectrum.
Corresponds to Fig.6.

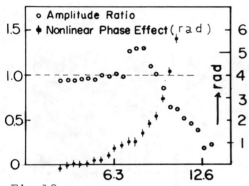

Fig.10
Amplitude ratio and phse difference deviation between the
two spectrum corresponding to
Fig.6

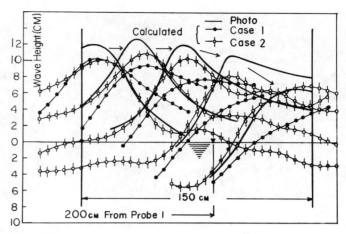

Fig.11
Comparison between the actual wave form
measured from a photo like Photo.2 and the
predicted one. The wave moves left to right.
Time interval is 0.2 sec.

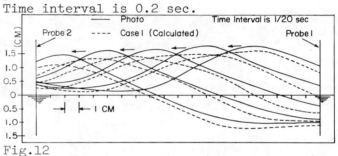

Fig.12
Comparison of the photo.4 and the calculation. The wave moves right to left. Time
interval is 0.05. Nonlinear correction
is not made.

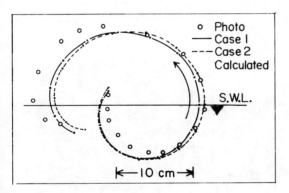

Fig.13
Comparison about the locus of the
float on the water surface.
Corresponds to Photo.5. Time interval is 0.1 sec.

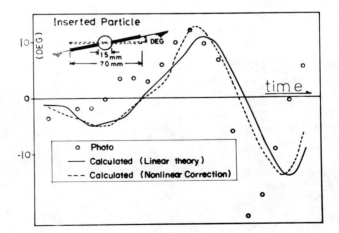

Fig.14
Comparison of the steepness. Corresponds
to Photo.5. The parallax of the photo
is corrected.

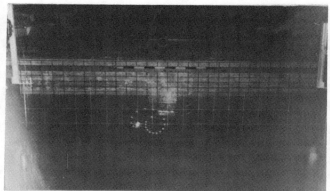

Photo.1
Arrangement of the experiments.Wave probes
are seen in both side.Distance of the grid
,setted on the center line of the tank, is
10cm. In the back part, side wall of the
towing tank is seen.

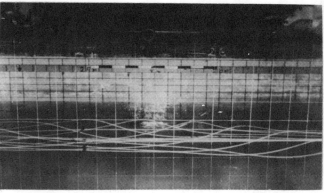

Photo.2
Example of visualization of T.W.W. transition. Flashing interval is 0.2 sec.
Wave profile of the middle is concentrated one. Wave generator is on the right
hand of the picture.

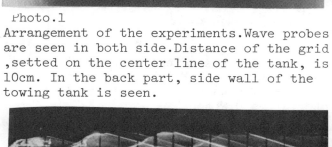

Photo.3
Example of wave breaking in the deep water of glazed channel. The distance of the
grid is 1.0 cm. The bottom of the channel
is not seen in this picture. Flashing interval is 0.05 sec.

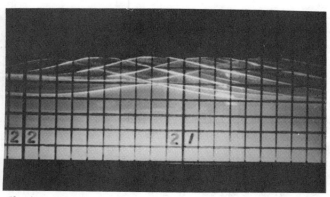

Photo.4
Concentration of Constant Steepness Transient Water Wave in the glazed channel.
Flashing interval is 0.05 sec. The wave
moves right to left.

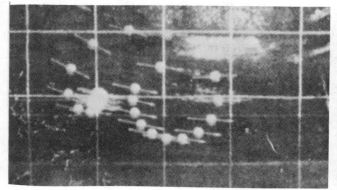

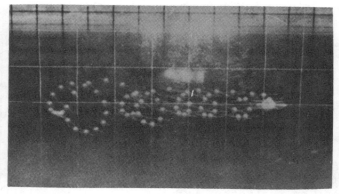

Photo.5 (Enlarged from photo.1)
Locus of a float with arms on the surface
of Constant Steepness T.W.W. Corresponds
to Fig.13. The grid distance is 10 cm and
the flashing interval is 0.1sec. Wave
slope can be observed at the same time.

Photo.6
Example of a locus of a float on the surface of Constant Amplitude T.W.W. Mass
transportation can be conjectured.

VISUALIZATION OF GASDYNAMIC PHENOMENA IN TWO-PHASE FLOW WITH CONDENSATION

R. CONRAD,* B. KRAUSE,* and G. WORTBERG**

Scattering of light by submicroscopic particles makes feasable the visualization of density fields in two-phase flow and the visualization of typical heterogeneous reactions as condensation, particle growth and evaporation. Making use of a light section procedure any desired plane of the flow field may be examined, and a large number of gasdynamic phenomena as shock waves, expansion waves and wakes may be studied. Even quantitative measurements of particle size number density and density of the carrier gas may be performed.

Introduction

A feature common to all the well known methods of interferometry, schlieren and shadow photography used in gasdynamic research is that these techniques yield results of the density or its spatial derivatives respectively which are values integrated along the path of the measuring light beam. Point by point examinations of the flow field however may be performed by a light section procedure which utilizes the scattering of light by submicroscopic particles contained in the fluid and moving with the flow. Moreover this method is very suitable for visualization of condensation, particle growth and evaporation in two-phase flow. Even quantitative measurements of particle size, number density and density of the carrier gas may be performed.

Light Scattering

The scattering of light by particles much smaller than the wave length λ of the incident light beam may be described by the well known formulas of Rayleigh, which show that the relative intensity I of the scattered light is proportional to the number N of particles in the scattering volume and to the sixth power of the particle radius r. Therefore an independent simultaneous determination of particle size and number density is not possible. For particles larger than about $1/20\ \lambda$ the rather complicated theory of Mie is applicable. However, if the refractive index n of the scattering particles is close to one and if these particles are non-absorbing, the following simple formulas after Rayleigh-Debye (Ref. 1) may be used to obtain information on number density and on size as well

*Graduate Student
**Professor, Lehrgebiet fuer Allgemeine Mechanik, Technische Hochschule, 5100 Aachen, Germany

$$I_1 = N \frac{64 \pi^4 r^6 (n-1)^2}{9 \lambda^4 R_o^2} P(\theta)$$

$$I_2 = N \frac{64 \pi^4 r^6 (n-1)^2}{9 \lambda^4 R_o^2} \cos^2 \theta \, P(\theta)$$

$$P(\theta) = 9 \frac{(\sin u - u \cos u)^2}{u^6} \; ; \quad u = 4\pi \frac{r}{\lambda} \sin(\theta/2)$$

The equations describe the relative intensity I observed at a distance R_o from the scattering volume when the incident light is polarized perpendicular to the plane of observation (I_1) or parallel to that plane (I_2). Fig. 1 shows experimental data and theoretical curves of the scattering polar diagram. The direction of the inci= dent light beam is indicated by arrows. Fig. 1 refers to spherical ice particles (n=1.33) and r/λ=0.07. For comparison one of the polar curves according to Rayleigh's original theory is shown by a dashed line (circle).

The radius of the scattering particles may be obtained from the "shape factor" $P(\theta)$ and the number of scattering particles from the relative intensity at a speci= fied angle θ. Note that in Rayleigh's original theory $P(\theta)$=1. Therefore according to that theory an independent determination of particle size is not possible.

Fig. 2 shows a probing along the axis of a Laval nozzle made of lucite by a light beam. By a suitable arrangement of mirrors a laser beam was directed along the nozzle axis. The flow is from bottom to top and the direction of the incident light beam is inverse. The onset of condensation a little bit downstream of the nozzle throat and the particle growth (indicated by the increasing brightness of the scattered light) may be seen.

Fig. 3 shows a sketch of the small wind tunnel used for this experiment. Moist air is drawn from the atmosphere and water vapor condenses rapidly due to the high cooling rate which is about 10^6 K sec^{-1} and due to the small relaxation time of water vapor condensation (Ref. 2).

Light Section Procedure

A beam of laser light may be expanded by two parallel cylindrical lenses. A thin plane sheet of parallel light is formed which may be used for probing any desired plane section of the flow. In contrast to the usual arrangement (Ref. 3) we placed this light section parallel to the direction of the flow of a free jet coming out of a supersonic nozzle. Whereas condensation is a rapid phenomenon with very short relaxation time the relaxation time of reevaporation is rather long. This means that the number of condensed particles per unit mass of the carrier gas (air) and that particle size remain almost constant if condensation has been completed inside the nozzle. Therefore the intensity of the light scattered from a fixed volume is proportional to the density. This fact may be used to visualize shock waves and expansion waves around an obstacle placed in the free jet. Fig. 4 shows as an example a cylindrical body (axis perpendicular to the flow direction). The detached shock wave, expansion waves and recompression waves may be seen. As the flow in the wake of the body is rather slow evaporation has time to procede and the wake is indicated by a lack of scattering particles and therefore by a dark area. Fig. 5 shows a sketch of the experimental set up. The flow is again from bottom to top. As the light beam is directed in the inverse direction below the obstacle a shadow is seen. Fig. 6 shows a similar experiment with a wedge (indicated by white ink) placed in the jet. An attached shock wave, expansion waves and recompression waves are seen. Fig. 7 shows an overexpanded jet (pressure in the bell jar(see Fig.3) much lower than the pressure at the nozzle exit). Expansion waves are reflected as compression waves at the boundaries of the jet. The same wave structure causes the periodic varying brightness of the probing beam downstream of the nozzle exit in Fig. 2.

References

(1) M. Kerker, The Scattering of Light, (1969), Academic Press, New York
(2) R. Conrad, Bildung und Wachstum von Kondensationskeimen in einer Düsenströmung, (1977)., Dissertation, Technische Hochschule Aachen, Germany.
(3) I. McGregor, The Vapor-Screen Method of Flow Visualization, J. Fluid Mech. 11 (1961) 481-511.

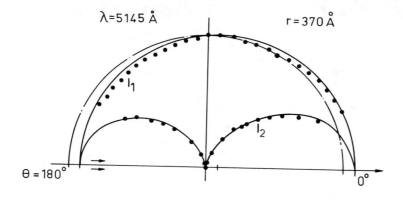

Fig. 1: Relative intensity vs. scattering angle, polar diagram

Fig. 2: Probing along nozzle axis

Fig. 4: Free jet with cylindrical
obstacle

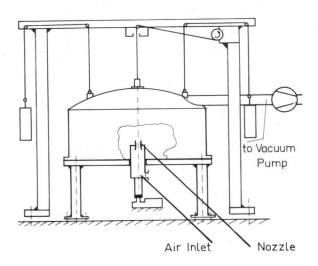

Fig. 3: Wind tunnel

Fig. 6: Free jet with wedge ─────────▶

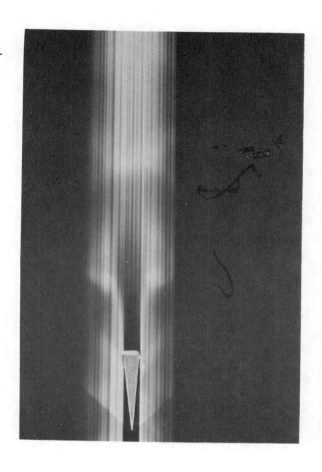

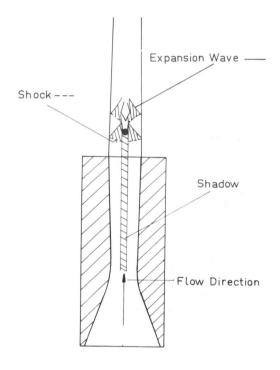

Expansion Wave ───

Shock ───

Shadow

Flow Direction

Fig. 5: Sketch of experimental set up

Fig. 7: Overexpanded jet ─────────▶

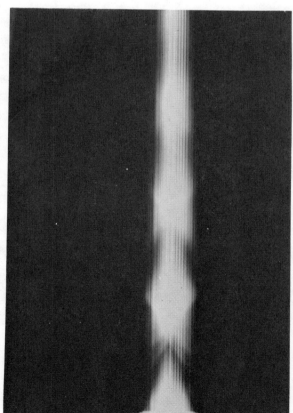

TRANSPORT PHENOMENA OF VENTILATING FLOWS IN A RECTANGULAR ROOM

M. AKIYAMA,* M. SUZUKI,* and I. NISHIWAKI*

Present study deals with an experimental investigation on the transport phenomena of the flow which ventilates a rectangular room. The iso-stream lines and velocity distributions are obtained by using an aluminum particle method and a milk solution method for water, and a combined zinc-stearate particle and tuft grid method for air in addition to a hot wire anemometry system.

The results encompass from laminar to turbulent flow, and it is found that the flow patterns can be divided into four distinct regimes considering the influence of the growth and decay of large scale vortices and the turbulent-eddy structure on the overall flow field.

The effects of important characteristic parameters are determined and the consequence of the results are well applied to explain the air ventilation and circulation system.

INTRODUCTION

Predictions of transport phenomena of the forced ventilation system has received increasing attention in recent years. Experimental measurements of velocity field of models and numerical solution of Navier-Stokes equations are of important tools to attack these problems at present. For the prediction of real velocity field, however, the use of local velocity profiles obtained from experimental model can not always satisfies after applying the dynamic simirality at hand. It is known that the upwind approximation for advectiv terms in finite difference methods for solving Navier-Stokes equations with appropriate boundary conditions gives us a stable solution /1,2/, but the result does not necessary give accurate solutions. Cooling in small electronic device, for example, in which the flow may become laminar can be predicted rather easily. Whereas the ventilation for buildings and factories which is characterized as the turbulent flow may become difficult for prediction, and extensive works establishing the turbulent model are being carried out recently/3,4/.

The purpose of the present investigation is to report on visualized flow results in a ventilated square room. The categolization of flow patterns for this model experiment is established and results intended to indicate a general distinction of the each flow regime. Scale effects of the model are also studied. The results can be used for direct comparison with the numerical solutions. The knowledge of the air movement can aid designers in establishing criteria necessary for ventilation system.

EXPERIMENTAL APPARATUS AND PROCEDURE

<u>Flow Visualization System for Air</u>   The general arrangement of the system is shown in Fig.1. Air from outside is sucked into the system by a centrifugal blower located at the end of the system. Fine zinc-stearate particles are suspended in front

*Faculty of Engineering,Utsunomita University, Ishii-cho, Utsunomiya 321-31, JAPAN

of the settling chamber. Air with the particles passes through the settling chamber which consists of 7 stages of grid and a set of guide vanes. Then,it is introduced to the test section where the flow patterns are photographed and the velocity distributions are determined. After passing through the test section,air moves into an outlet strainer where any possible irregularity of the flow due to the blower can be eliminated. At a filter,the zincstearate particles are removed from air and the flow rate of air is measured by providing a metering orifice and pressure difference is measured by U-tube manometer and confirmed by particle velocities at the intake air.

   Test Section   The ventilation system is such that the square chamber has an intake opening and is placed at the center of the upper surface with the dimension of one-nineth of the total width of the upper surface, and the outlet of the same size is located at the bottom corner of the right side wall. In order to test the two dimensional flow,it is necessary to eliminate the effects of front and back walls. Thus the depth of the room is five times that of one side wall length. Dimensions of the test section are shown in Fig.1. Test apparatus are placed in such a way that the cross section of the test domain is located horizontally. By this way any possible sedimentation of particles on the bottom of measuring section is avoided. Halation by this sedimentation and the effect of falling speed due to the gravitational force acting on the particles are eliminated.

   Test Procedure   Depending on the value of Reynolds number,a tracer method with zincstearate particles and a combined method by using tuft grid and hot wire anemometry are used.

   Tracer Method   An intermittent light-beam emitted by a 16mm movie projector was foucused on a test region through a slit of 4mm. Photographs of the trajectories of of particles passing through the high-intensity light-beam were taken on Tri-X film with shutter speed of 1 to 1/8 second. Other consideration regarding the accuracy of the photographed data used were the same as those developed by Aihara and others /5,6/. The flow field in terms of velocity and stream line are obtained by using the photographs.
   When one takes a picture with the intermittent light-beam,a tracer at high velocity shows a streak of broken line. With decrease of the particle velocity,the intervals of the broken line decrease,and finally for low velocity,the line becomes solid curve presenting a pathline. One can determine the values of higher velocity by measuring the pitch of stroboscope images and the values of lower velocity with the length of solid line and the shutter speed of the camera. Therefore,with only one appropriate photograph,quite a wide range of velocities ranging from 5 mm / sec to 1 m / sec can be determined.
   For the high velocity region,one has a velocity V=m×l×p (m/sec) where m is magnification of image, l is a pitch of image (m),and p is number of images per second (1/sec),and Reynolds number $Re=b×V_c÷\nu=b×m× l×p÷\nu$ where b is the width of the inlet (m), V_c is the inlet velocity (m/sec),and ν is the kinematic viscosity (m²/sec).

   Combined Method by Using Tuft Grid and Hot Wire Anemometry   Grid wires of 0.29mm diameter are stretched between the both sides of the frame. On the nodal points of the grid wires,shiny silk threads of 25mm length are sticked to form a kind of depth tuft grid. Minimum velocity at which one can depict the flow direction is found to be about 10cm/sec.
   The hot wire anemometry system used is Kanomax Ser. No.21-1000. Regular measuring stations numbered and located at each nodal point being with intervals of 20mm.

   Flow Visualization System For Water   In order to test the fluid property effect,water is used as a working fluid. The size of the test section is reduced to the half of that for air so as to test the scale effect. The circulation system used is a closed one with a bypass loop. Aluminum dust particles are used to visualize the flow.

<u>Residence Time Study</u>   A simple residence time study is carried out. To record the residence time, the test chamber is filled with a solution of powder milk. Then, the system is quickly started and fresh water is pumped into the test section. A series of photographs is taken during the period as the white milk-solution is being cleared out. The photographs taken on 35mm film are analized by a micro-reading-photometer No.3 of Shimazu Mg.Co..

RESULTS AND DISCUSSION

<u>Flow Patterns</u>   The flow patterns can be categorized into four distinct regimes. Explanation of the characteristics of each regime is in order.
1)   Laminar Flow Regime (Re=10~100)
The main flow issued at the intake behaves as a two-dimensional laminar free-jet as shown in Fig.3. It comes down almost straightly from the intake to the center of opposite floor. Because of the friction developed on its periphery, it integrates some of the surrounding fluid while spreading itself outwards and decreasing its velocity. After making two stagnation points, one on the floor and the other at the right-side wall, the main flow converges rapidly towards the outlet. Outside of the main flow, two circulations are formed. It is noted that the width of the main flow is wider in smaller Reynolds number region and is getting narrower in larger value of Reynolds number.
2)   First Transition-Flow Regime (Re=100~200, Stable Secondary Circulation)
Decrease of the main-flow width with increase of Reynolds number is seen to be minimal in this regime. This fact indicates that the two circulation regions which occupy the volume similar to that for the laminar flow regime must contain a larger amount of kinetic energy. The higher energy in the circulation regions has a potential of splitting a vortex into two or more parts to make a more uniform energy distribution over these regions. This situation allowes that, besides the two main circulations, the regions can have additional vortices. This is indeed the case. A new circulation tend to appear at the left upper corner first. In higher Reynolds number, the main circulation appeared in the right-hand side tends to constrict to form a new vortex at the upper part of the corner. Fig.4 shows a map of the iso-streamlines with Re=160 where one can find two stable circulations at the right corner.
3)   Second Transition-Flow Regime I (Re=200~300, Periodical Growth and Decay of Circulation)   This flow regime is characterized by periodical growth and decay of additional vortices at the both circulation regions of the flow system. Fig.5 shows one cycle of such a motion starting from the initiation stage to the decay of a secondary vortex located at the upper right corner of the room. The oscillation is seen to be stable in time and scale. As shown by the inset of the figure, the secondary vortex behaves as a three dimensional swelling but the apparent velocity in the transverse direction is unpredictable.
It is true in this region that the centrifugal force will act upon the vortex, but at this stage it is difficult to correlate the three dimensional nature of the vortex to the standing wave due to the Goertler type instability.
When the value of Reynolds number is about 300, the fluctuation of main flow appears close to the outlet.
4)   Second Transition Flow Regime II (Re=300~1000, Main Flow Fluctuation)
By increasing the values of Reynolds number further, the disturbed region of main flow initiated at the outlet develops towards the entrance region. On the other hand, different from the wide spreading of the laminar jet with a weak vortex-shear for a low Reynolds number region, the intense vortex shear appeared at the peripheral part of the jet core makes a visible vortex sheet near the inlet and goes up to the half way of the jet. Fig.6 shows a result of this type. The two different modes of disturbance described above are related each other. The two-dimensional shear-vortices with a selective frequancy at the peripheral part of the jet near the inlet increase their amplitude and the static pressure of this portion increases compared with the

one at the core region. After attaining a certain magnitude of the vortex energy with higher static pressure,there sets in a period where the energy will be transferred to the high frequency of the three-dimensional wave and the wave will be reached to the central region of the jet due to the static pressure difference and the diffusion. This final portion corresponds to the fluctuation observed near the outlet.

5) Turbulent Flow Regime (Re=1000∼8000)   Fig.7 shows typical pathlines of this flow regime. It is seen that the pathlines indicate frequent appearance, cracking and decay of the secondary circulations. While the detail structure of the turbulent flow is of importance,another fact that the mean flow pattern does not change appreciably with increase of Reynolds number is an important consideration for this regime. The mean flow pattern can be determined by using a tuft method as shown in Fig.8. Fig.9 illustrates a result of the velocity vector distribution measured with the aid of a hot wire anemometry and the tuft method for Re=5060. It is well to note that the velocity profiles for a full range of this Reynolds number regime were examined to ensure rather a weak dependence of the velocity profiles on Reynolds number.

Residence Time   The next aim of the flow visualization is a determination of residence time of the particles. Fig.10 shows two photographs of water flowing in the chamber where the white portion represents milk solution in circulation zones and the black part is of fresh water. For Reynolds number Re=103,the flow is laminar and no disturbance of the main flow is seen,whereas for Re=478,the flow is of transition and the fluctuation of the flow can be seen clearly. It is true that,with the lapse of time,the milk solution in circulation region will be replaced by fresh water. The photographs,however,indicate that the rate of entrainment of the circulating fluid into the main flow is not so large. Thus,for a long period of time it makes the circulation region and the main flow region separate to each other regardless of the flow regimes. When there is some contaminants in the flow,the residence-time distribution of the particles is an important variable. Though,it is not shown here,the local distributions of the relative concentration of the milk solution were measured. Another important factor is a critical residence time τ_c which may be defined as the time required to clean up the whole region of the uniformly contaminated room. The critical residence time can be served as an indicator of a ventilation or removal efficiency of the contaminant. Fig.11 shows the critical residence time of milk solution against the Reynolds number. As one can see from this figure,a strong relationship exists between the critical residece time τ_c and the flow patterns.

It is interesting to note that by rearranging the critical resident time in terms of a critical nondimensional time T_c,the experimental facts can adequately be explained for the laminar and turbulent flow regimes including the transition regime. The critical nondimensional time T_c used here is defined as follows: $T_c=\tau_c \times V \div l$,where τ_c is the critical residence time,V is the mean inlet velocity,and l is the width of the chamber. As one can see from the figure,the critical nondimensional time is increasing with increase of the inlet velocity in the laminar flow regime. It is obvious that the critical nondimential time for a plug flow is a constant and unity. When the flow is steady and laminar,any pathline should not merge with any other pathline. In this case,the diffusion of particles can be identified as the main cause of mixing. Therefor,if the flow patterns were identical in all the laminar flow regimes, the critical nondimintional time would be almost a constant. In reality,the occupied area by the main flow is large for low Reynolds number and small for high Reynolds number due to the difference in magnitude of inertia. It means that,except the initial stage of the evacuation process,an amount of milk solution which has to be removed mainly by diffusion process is small in low speed and large in high inlet-velocity. Consequently,in the laminar flow regime the critical nondimensional time increases with increase of the Reynolds number.

Further increase of Reynolds number leads to the first transition stage where the stable secondary circulation flow creates a new mixing process and this tends to lower the rate of increase of the nondiminsional evacuation time.

At the initial stage of the second transition zone (Re=200∼300),a clear difference in the critical nondimensional time from the first transition regime is observed exhibiting a decrease of the value with increase of the Reynolds number.

A significant amount of the decreasing rate of the critical nondimensional time is observed in the final stage of the second transition regime (Re=300∼1000). This is due to the main-flow fluctuation which strengthens the entrainment velocity of the circulatory fluid carried into the main flow. It may be desirable to confirm the result at Re=1000 in the light of additional experimental evidence on the full turbulent regime of Re≥1000.

Besides the direct observations, some implications of the results on the transport phenomena of the ventilation in a square room are listed in the conclusions.

CONCLUSIONS

By using air and water as working fluids, a flow visualization study was carried out to investigate the transport phenomena of ventilating flow in a square room and the following conclusions are obtained.

1) The flow regimes are divided into four distinct regions. These are: (1) Laminar Flow Regime (Re=10∼100),(2)First Transition Flow Regime (Re=100∼200,Stable Secondary Circulation),(3)Second Transition-Flow Regime (Phase I for Re=200∼300, Periodical Growth and Decay of Circulation,and Phase II for Re=300∼1000,Main Flow Fluctuation),and (4)Turbulent Flow Regime (Re=1000∼8000).

2) The residence time of the particles in the circulation region has close relation with the flow patterns observed. The graphical results of the critical residence time as a function of Reynolds number shown in Fig.11 can serve to evaluate the ventilation efficiency of the given flow pattern.

3) No scale and property effect is found in the range of experiments.

4) Only the Reynolds number effect is of importance in the range extended from the laminar to the second transition.

5) Since the flow pattern in the turbulent flow regime does not change appreciably,a so-called number of ventilation,the net flow-rate divided by the volume of a room ($m^3/h/m^3$),which is commonly used to indicate an effectiveness of ventilation system of a room may be regarded as a rough but good measure.

6) Suggestions can be made on numerical predictions of the flow field in ventilated rooms of this type:(1)The laminar flow regime is rather limited to a low Reynolds number (Re≤100) where accurate solutions are required. No laminar solution is needed for higher Reynolds number regime (Say Re≤200 is enough). (2)A single turbulent model may well be developed to explain the full turbulence regime (Re≥1000).

7) Studies on fluctuation quantities are further required for a better understanding of this type of ventilation system.

We are grateful to Messrs.Y.Odaka,H.Sugiyama,I.Urai,H.Kotado,and H.Koyanagi for their assistance in obtaining the data. This work was supported by the Education Ministry of Japan through a Special Research Grant in the fiscal year of 1976.

REFERENCES

1. A.Yoshikawa and K.Yamaguchi,Numerical Solution of Flows in Rooms,Trans.of Society of Heating,Air Conditioning and Sanitary Engineers of Japan,48,1,(1974),5-17
2. M.Kaizuka,Flow Pattern in Rooms and Air Distribution System,Dr.Eng.Thesis,(1971)
3. A.Yoshikawa and K.Yamaguchi,Numerical Solution of Flows in Rooms (Turbulent Case), Trans.,Soc.of Heat.,Air Condi. and Sani. Eng.of Japan,48,10,(1974),899-913
4. B.H.Hjertager and B.F.Magnussen,Numerical Prediction of 3-dim. Turbulent Buoyant Flow.,Summer Semi.,Turbulent Buoyant Conv.,(1976),Hemisphere Pub. Co.,429-441
5. T.Aihara and E.Saito,Measurment of Free Convection Velocity Field around the Periphery of a Horizontal Torus,Trans.of the ASME,Series C,94,1,(1972),95-98
6. T.Asanuma ed., Handbook of Flow Visualization,(1977),202-206

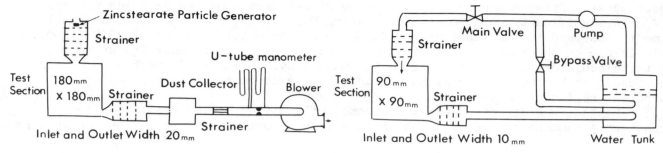

Fig.1 Experimental Set Up for Air

Fig.2 Experimental Set Up for Water

Fig.3 Laminar Flow Regime (Re= 91)

Fig.4 First Transition (Re= 160)

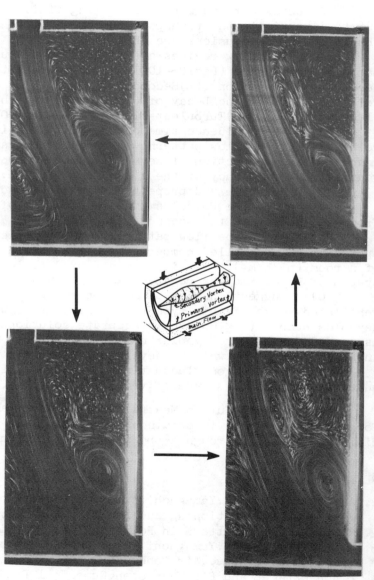

Fig.5 Periodical Growth and Decay
of Circulation (Re= 241)

Fig.6 Second Transition Ⅱ (Re= 350)

Fig.7 Turbulent Flow Regime (Re= 1280)

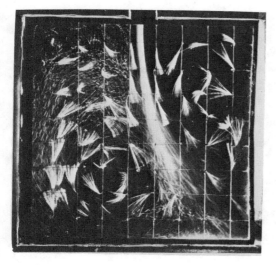

Fig.8 Turbulent Flow Regime (Re= 1280)

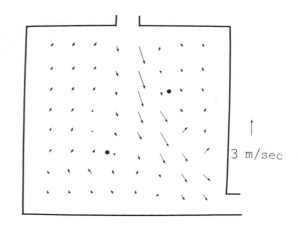

3 m/sec

Fig.9 Velocity Vector (Re= 5060)

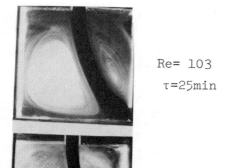

Re= 103
τ=25min

Re= 478
τ=2min

Fig.10 Residence Time

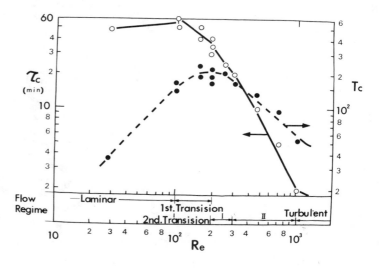

τ_c
(min)

T_c

Flow
Regime

Laminar | 1st.Transision | | II | Turbulent
2nd.Transision

R_e

Fig.11 Critical Residence Time

FLOWS AROUND RECTANGULAR PRISMS: NUMERICAL CALCULATIONS AND EXPERIMENTS

A. OKAJIMA,* K. SUGITANI,** and T. MIZOTA***

To furnish some fundamental information about the vortex-excited oscillation of bluff cylinders, flow patterns around rectangular prisms which are stationary and oscillating transversally in a uniform flow were visualized by the aluminium dust methods and compared with ones obtained by numerical calculations. There was consequently a good agreement between them.

Introduction

Information about unsteady flow patterns about bluff cylinders is of considerable practical interest in aeroelastical instability as well as in the basic understanding of fluid mechanics.

As is generally known, the rectangular cylinder, if supported elastically in a uniform stream, shows the so-called vortex-excited oscillation in the vicinity of the so-called resonance speed which is the stream speed where the natural frequency of the cylinder coincides with the frequency of the shedding vortex system. Also, apart from this normal vortex-excited oscillation at around the resonance speed, aeroelastical instabilities occurs at a half resonance speed for a rectangular section with the side ratio 2. However, the mechanism of the aeroelastical instabilities of the cylinders at around resonance speed and a half one seems to have been left obscure as yet.

The objective of the present paper is to examine the flow characteristics of the rectangular cylinder subjected to transversal oscillation in uniform flow, in detail. So we visualize the flow-patterns around the rectangular cylinders and also compare these results with ones of numerical solutions of the Navier-Stokes equations by the finite difference analogue. The cylinders of a square cross-section and a rectangular cross-section with the side ratio 2 are used in our experiments and calculations.

Experiments

Our experiments were conducted at the towing water channel. The visualization technique of the aluminum dust methods were employed to investigate flow patterns

---

  * Associate Professor, Research Institute for Applied Mechanics, Kyushu University, Fukuoka.
 ** Technician
*** Research Associate

around the stationary models and the sinusoidally oscillating ones in a heaving mode.

The experiments were designed to yield path line photographs of aluminum dust particles suspended into a water channel through which the model was towed. Since the velocity vector defined by the aluminum particles must lie tangent to a stream-line, a direct comparison with the streamline plots of numerical calculations is possible. The sizes of the model with a square section used in our experiments were H=5mm in width and C=5mm in chord-length, and the rectangular cylinder with a 8mm wide by 16mm long cross-section was used. The span-length 280mm of each model was submerged in water. The range of the towing speed U of the models was 0.6-1.25cm/s. The Reynolds number based on the height of a rectangular section $Re=UH/\nu$ for the present studies was 50 - 100.

During experiments of the forced oscillation, we chose the amplitude a and the forced frequency f_c of the oscillation as follows:

$a=0.05H$, $f_c=U/7.3H$, $U/6.0H$.

The incidence angle of the cylinders was kept to be zero relative to the uniform stream.

Numerical Calculations

Subject to the assumptions of incompressibility and two-dimensionality, the Navier-Stokes equations may be expressed as Poisson's equation and the vorticity transport equation. Before a finite-difference formulation of these two equations can be made, the finite difference mesh should be constructed to fit the problem that the stationary and the oscillating rectangular cylinders are moved in still liquid. A rectangular coordinate system is applied and the grid used during computation consists of 90 grid spaces in the stream-wise direction and 60 grid spaces in the cross stream direction. The grid point density is increased in the near flow-field and gradually decreased in the far flowfield. Such non-uniform nature of the mesh structure requires a special formulation of the finite-difference approximations to the particle derivations. It is necessary to obtain finite difference expressions for the first and second derivations, in order to formulate the governing equations in finite difference form. In this specific case, the expressions for the derivations of $\partial f/\partial x$, $\partial^2 f/\partial x^2$ are given as, respectively,

$$\frac{\partial f_{i,j}}{\partial x} = \frac{\Delta x_{i-1} f_{i+1,j}}{\Delta x_i(\Delta x_{i-1}+\Delta x_i)} - \frac{\Delta x_i f_{i-1,j}}{\Delta x_{i-1}(\Delta x_{i-1}+\Delta x_i)} - \frac{(\Delta x_{i-1}-\Delta x_i)f_{i,j}}{\Delta x_i \Delta x_{i-1}}$$

$$\frac{\partial^2 f_{i,j}}{\partial x^2} = \frac{2f_{i+1,j}}{\Delta x_i(\Delta x_{i-1}+\Delta x_i)} + \frac{2f_{i-1,j}}{\Delta x_{i-1}(\Delta x_{i-1}+\Delta x_i)} - \frac{2f_{i,j}}{\Delta x_i \Delta x_{i-1}}$$

The derivations of $\partial f/\partial y$, $\partial^2 f/\partial y^2$ are completely analogous to the above derivations with Δx replaced by Δy and index j varying.

The boundary conditions to be satisfied on the surface of the cylinder are that there is no cross flow through the surface and also no slip flow along it. The boundary condition imposed upon the flow field infinitely far from the cylinder is that the velocity of flow asymptotically tends to the velocity $U-2\pi f_c a \sin 2\pi f_c t$ as the distance from the cylinder becomes infinite. So we specify the boundary conditions of the flowfield for our calculations. At the upstream boundary, the flows are uniform; the vorticity is zero and the deviation of the stream function from the uniform flow is zero. The upper and lower boundaries are located three times as far as the cylinder height from the center line. The outside of these boundaries is assumed to be inviscid flows. The relations between the x- and y- components of velocity u_b, v_b on these boundaries are thus,

$$u_b = -\frac{1}{\pi}\int_{-\infty}^{\infty}\frac{v_b}{\xi-x_b}\,d\xi \qquad , \quad v_b = \frac{1}{\pi}\int_{-\infty}^{\infty}\frac{u_b}{\xi-x_b}\,d\xi$$

144

where ξ and x_b is on the upper or lower boundary. And the downstream boundary is sufficiently far located from the region of interest, i.e. *50H-60H*. The vorticity on the surfaces of the cylinder is computed to satisfy the no slip conditions. While computing the vorticity of four corner points, however, the upstream corner points are treated as if they lay on the front surface of the cylinder and the downstream corner points are assigned two values of vorticities on the upper or lower surface and the rear one.

The familiar and simple upwind-differencing technique is applied to the vorticity transport equation, and an implicit differencing scheme is used to determine the vorticity value at each point at new time.

Compatible values of the stream function are obtained by solving Poisson's equation by an iterative technique called successive over-relaxation method.

Results

(a) Flows around a stationary cylinder of a square section

Fig. 1 shows the visualized and the calculated flow patterns around a stationary cylinder of a square cross-section in uniform flow at Reynolds number *Re=50*. Flows are observed to separate at the downstream corners of the square section. A pair of standing vortices is formed behind the cylinder. With an increase of Reynolds number, the points of flow separation are removed upstream. For *Re=80* as shown in *Fig. 2*, flows separate in the neighbourhood of the upstream corners. In experiment, however, for *Re=80*, the wake is artificially stabilized to prevent the Karman vortex street from shedding by placing a thin splitter plate along the centerline behind the cylinder.

The visualized and the calculated flow patterns are clearly seen to coincide very well with each other, as shown in *Fig.1* and *2*.

(b) Flows around an oscillating cylinder of a rectangular section

Flows around a cylinder of a rectangular section with the side ratio 2 which is forced to oscillate in a direction perpendicular to the uniform stream at *Re=100* are shown in *Fig. 3*. The forced frequency of the cylinder is nearly equal to the frequency of the natural vortex shedding for a stationary cylinder. The flow patterns are found to change with the same frequency as the imposed one of cylinder and thus the oscillations of cylinder and wake are synchronized. There is a very good agreement in the unsteady flow patterns obtained by the visualized technique and the calculations. This good coincidence verifies the validity of some assumptions used in our calculations. *Fig. 4* is the case that the cylinder is forced to oscillate with greater frequency than that of Karman vortex street f_k, i.e. *1.3f_k*. These figures show the flow patterns over about three periods of the cylinder oscillation and these patterns change very complicatedly, since the forced frequency is outside the frequency range of synchronization. The visualized patterns, however, agree well with the plotted ones. By comparison of the two patterns at the time *t=36.50* and *t=54.75* in *Fig. 3*, the interval between which is about two and a half periods of the cylinder oscillation, these flow patterns seem to be almost up-side down in shape, each other. It is found that the flow patterns vary with five times periods of the cylinder oscillation. The periodic motion of flows with the forced frequency is restricted in the close vicinity of the oscillating cylinder. Such phenomena that the motion of flows around the rectangular cylinder can be strongly modulated by cylinder oscillation also occurs for the case of a circular cylinder (Ref. 1). This finding from visualization technique and calculation agrees with that from direct velocity measurements by Toebes(Ref. 2) or direct lift force measurements by Bishop and Hassan (Ref. 3).

References

(1) Okajima, A., 3rd Sympo. on Flow Visualization ISAS Univ. Tokyo (1975) pp.121-124 (in Japanese).
(2) Toebes, G.H., Trans. ASME, J. Basic Eng., Vol.91 (1969) pp.493-505.

(3) Bishop, R.E.D. and Hassan, A.Y., Proc. Roy. Soc. A. Vol.277 (1964) pp.51-75.

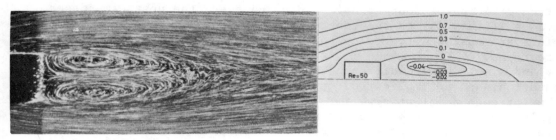

Re=45   *Re=50*

Fig. 1   Flow patterns around a square section

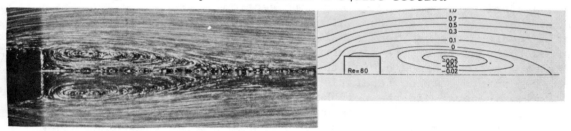

Re=80   *Re=80*

Fig. 2   Flow patterns around a square section

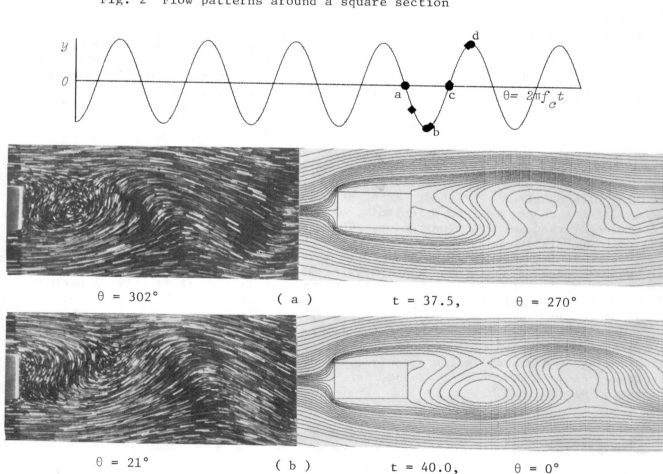

$\theta = 302°$ (a) $t = 37.5,$ $\theta = 270°$

$\theta = 21°$ (b) $t = 40.0,$ $\theta = 0°$

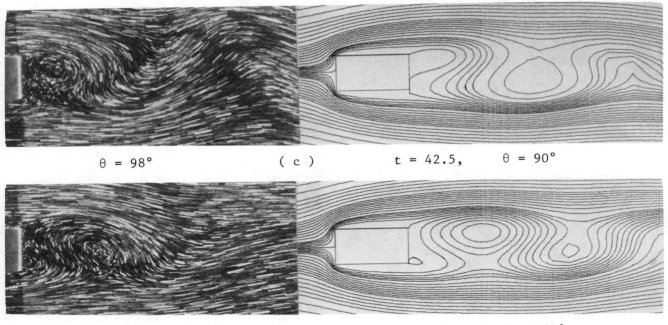

θ = 98°   (c)   t = 42.5,   θ = 90°

θ = 183°   (d)   t = 45.0,   θ = 180°

Fig. 3 Flow patterns around a rectangular cylinder, $f_c/f_k = 1.0$.

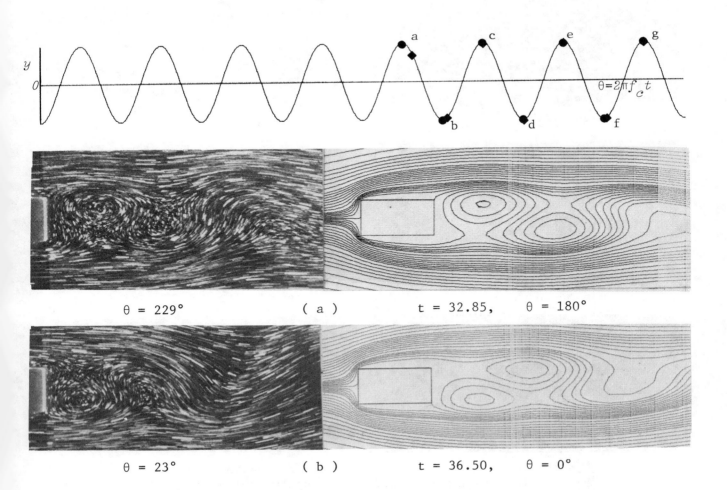

θ = 229°   (a)   t = 32.85,   θ = 180°

θ = 23°   (b)   t = 36.50,   θ = 0°

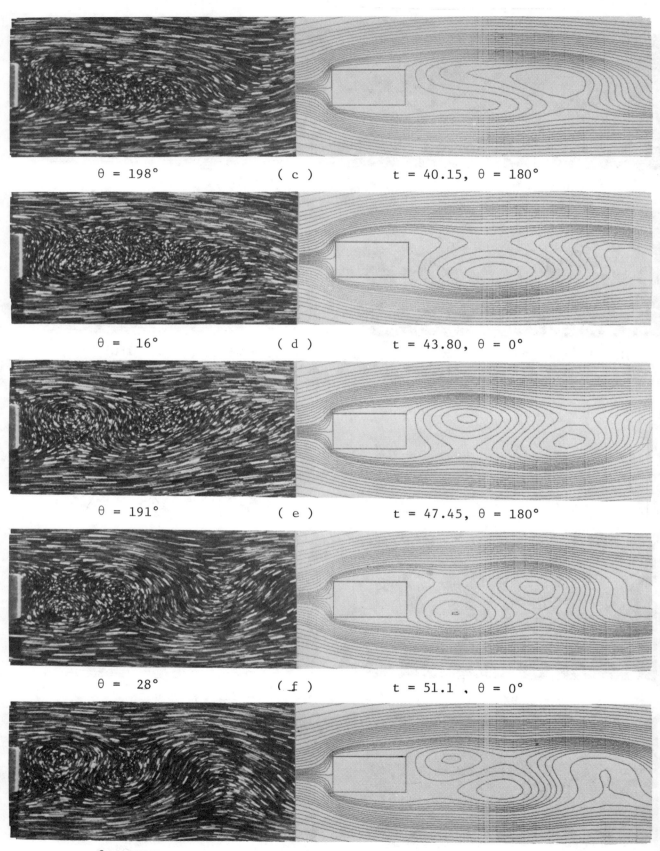

θ = 198°   (c)   t = 40.15, θ = 180°

θ = 16°   (d)   t = 43.80, θ = 0°

θ = 191°   (e)   t = 47.45, θ = 180°

θ = 28°   (f)   t = 51.1 , θ = 0°

θ = 191°   (g)   t = 54.75, θ = 180°

Fig. 4 Flow patterns around a rectangular cylinder, $f_c/f_k = 1.3$

FLOW VISUALIZATION AROUND TURBULENCE PROMOTER IN PARALLEL, CONVERGENT AND DIVERGENT CHANNELS

Y. SHIINA,* T. TAKIZUKA,* and Y. OKAMOTO*

Flow patterns around turbulence promoters in parallel, convergent and divergent channels are visualized by surface streak line method with aluminum-oil mixture as a tracer to study an effect of promoters on heat transfer.

The vortex flows were visualized by streak line pattern. Shedding vortices behind the promoter became weaker along the flow in case of accelerated flow. However, vortices were spreading downstream over the channel in case of decelerated flow.

1. Introduction

Heat transfer augmentation techniques are often used in the gas cooling systems, such as high temperature gas cooled reacters and gas-gas heat exchangers, because of lower heat-removal capacity of gas coolant than in case of liquid. It is already reported that turbulent heat transfer at high flux heating and accelerated flow decreases in a value of laminar flow, caused by laminarization (Ref.1,2). Transition and laminarization phenomena of gas flow cause serious problems which lead to fuel melt-down and failures in gas cooled reactors. Several augmentation techniques have been used to improve these deterioration phenomena. Among them, turbulence promoters are considered as useful devices. The promoters attached on heating surface markedly increase heat transfer rate.

However, Hishida et al. reported that heat transfer coefficient of the wall with turbulence promoters decreases at Reynolds number Re_w below 1000, where a hydraulic diameter was defined by twice the channel height, in case of the ratio of channel width to promoter diameter w/d=3 (Ref.3).

In order to study these results, flow mechanisms around the promoters are visualized by using a surface flow techniques. Vortices around the promoters are observed by Al-oil trancer system.

The transition and break-down behaviours of vortex are studied systematically and quantitatively in case of parallel, accelerated and decelerated flows.

*Japan Atomic Energy Research Institute, Tokai-mura, Naka-gun, Ibaraki-ken, Japan

2. Experimental apparatus and method

The schematic diagram of the apparatus used in this experiment is shown in Figure 1. Open channel flow circuit, 2130 mm in side length, 55 mm in depth and 100 mm in channel width, was used to visualize a vortex flow around the promoters. Test channels, each of which is parallel, convergent and divergent, consist of acrylic plate 1500 mm in length. We used Spinox oil as a fluid. A rotating screw, joined to variable-speed induced moter, circulated the oil in the channel circuit. Mixture of oil and aluminum powder is used as a surface tracer.

Figure 2 shows supplying grid of the aluminum-oil mixture. The oil-powder mixture pourred along the suplier suspends on the oil surface and forms streak line pattern. The upper stay of the comb-type supplier is made of bakelite plate and lower part of the supplier is steel. Steel combs, the number of which is from three to nine, are bound to the bakelite stay in accordance with channel width. In case when the amount of the Al powder in the mixture is too much, the Al-oil trancer does not float on the oil surface and submerges in the oil. On the other hand, in case when the powder is not so much, the streak line pattern does not become clear.

Velocity of flow was calculated from displacement time of the suspended aluminum powder to pass through marked points, one meter in length. In case when flowing speed is too low, oil surface became wavy caused by moter oscillation. On the other hand, in case when velocity becomes higher than about 10 cm/sec, fluid surface around the promoter becomes uneven. In both limiting cases described above, we do not succeed in visualizing the exact flow pattern.

In our experiment the visualization studies around the promoters had been carried out over the range of Reynolds number Re_w from 300 to 2000.

Figure 3 shows configuration of the promoters used in the experiment, cylinder, square and trapezoid.

A few promoters contact along one-side wall are inserted in the flow. The experiment is conducted with Reynolds number, channel width, shape of promoters, number of promoters, the ratio of pitch to promoter diameter p/d and inclined angle of the channel as parameters.

3. Experimental results

3-1. Vortex flow in parallel channel with promoters

Figure 4 shows the experimental result of vortex flow in parallel channel around a promoter in case of w/d=3. In case of Reynolds number $Re_w > 1000$, induced vortices are shedding repeatedly from the promoter and flowing downstream. But a steady vortex is attached itself at the back side of the promoter in case of Reynolds number $Re_w < 1000$. Fluid near the wall with the promoter is continuously mixed with the mainstream fluid by shedding vortices at Re_w over 1000. But in case of Re_w below 1000, fluid near the steady vortex is almost stagnant and little mixed with main stream. This result indicates that the deterioration of heat transfer at low Reynolds number Re_w 1000, reported by Hishida (Ref.3), is mainly caused by the stagnant flow at $Re_w < 1000$. The vortex changes its behaviour from steady to shedding at $Re_w \sim 1000$ in case of w/d=3. We call it a transition Reynolds number. The transition Reynolds number Re_{wt}, expressed in the channel height as hydraulic diameter, increases with increase in the channel width, but transition Reynolds number Re_{dt}, expressed in the promoter diameter as the hydraulic diameter, is almost constant and independence of the width. This result indicates that the flow pattern around the promoter mainly depends on the velocity of main flow and the promoter diameter, and it does not depend on the total flow rate. The transition Reynolds number does not depend upon the configuration of the promoter, but mainly depends upon size and intensity of the vortices. The vortices around the trapezoidal promoter grow more rapidly and stronger than those around the cylindrical and square-promoters. On the other hand, the vortices around the square promoter are weak and growing slowly compared with the case of cylindrical and trapezoid promoters.

Figure 5 shows the flow pattern around two pieces of promoters in the parallel channel at Re=1009 and w/d=2. In case of p/d 7, the vortex pattern between each promoter is not hardly affected by the pitch ratio of the promoter.

The number of flowing vortices around the promoter for ten seconds is shown in table 1 with p/d as a parameter. In case of p/d=5, the vortex flow behind the square promoters become almost stagnant, but vortex flow, behind trapezoid promoters does not become stagnant and vortex frequency is constant over the range of Reynolds number in this experiments. As already described, the fact indicates that the vortex flow around trapezoid promoter is formed just behind the promoter edge and growing rapidly.

3-2. Vortex flow in convergent and divergent channel with promoters

Figure 6 and 7 shows flow patterns in convergent and divergent channels around the promoter. The value of acceleration and deceleration parameters $K=(\nu/u^2)du/dx$, where x denotes the distance from entrance of test section and ν denotes kinetic viscosity, are shown in the capture of the figures. A steady vortex is attached behind the promoter at low Reynolds number $Re_w<900$ in case of w/d=3 in the same manner as in parallel channel. When Reynolds number becomes larger than the value of Re_{wt}, vortices behind the promoter are mixing and shuffling the fluid near the wall. Heat transfer rate is improved by these shedding vortices at $Re_w>Re_{wt}$.

The transition Reynolds number in convergent channel Re_{wt} becomes larger than in case of parallel channel. On the contrary, Re_{wt} in divergent channel becomes smaller.

Table 2 shows the data of transition Reynolds number Re_{wt} in parallel, convergent and divergent channels. The difference between transition Reynolds number Re_{wt} between each channel is seemed to be caused by velocity distribution of main flow, influenced by the vortex flow around the promoter.

Disturbances of induced vortices in the convergent flow becomes smaller due to flow acceleration as the flow goes down the stream as shown in Fig.6. On the other hand, disturbances in the divergent channel is spread over the channel and vortices are growing largely through the channel by deceleration of flow as shown in Fig.7. Comparison between the Fig.6 (c) and Fig.6 (d) indicates that flow disturbances behind the promoter are increasing with decrease in the value of w/d, such as in case of the divergent channel as shown in Fig.7.

Table 1 indicates that number of vortices in the divergent channel flowing down behind the promoters for ten seconds becomes smaller than in case of the convergent and parallel channels. As there are some cases when the flow has already separated before the promoter, it should be neccesary to account the induced rate of vortices more accurately, considering the relation between flow separation and deceleration.

Figure 8 shows the flow pattern around two promoters in case of p/d=5 at Re_w = 1350 and 737. We cannot observe a clear vortex flow between square promoters, even in case of the convergent and divergent channels. However, we can clearly observe the vortex flow shedding down the stream around the trapezoid promoters.

In these experiments, we can analyse the flow mechanism around the promoters and its behaviour caused by flow acceleration and deceleration using Al-oil mixture as a tracer. The relations between heat transfer at laminarization and flow pattern around the promoters were qualitatively studied considering the vortex behaviour, but, further studies are needed to make these mechanism clear.

4. Conclusion

Several conclusions are derived in the followings

1. The behaviour of the vortex flow pattern in convergent and divergent channels

is same as in the case of parallel channel; a steady vortex attached behind the promoter at $Re_w < Re_{wt}$, shedding vortices flowing down the stream at $Re_w > Re_{wt}$.

2. There is little mixing between a steady vortex and main-stream at $Re_w < Re_{wt}$. On the other hand, there is large mixing between shedding vortices and mainstream at $Re_w > Re_{wt}$.

3. The transition Reynolds number Re_{wt} of acelerated flow becomes larger and the Re_{wt} of decelerated flow becomes smaller than in case of the parallel channel.

4. Induced vortices in convergent channel with promoters are supressed or vanished and becomes smaller due to decrease in turbulence of accelerated flow. On the other hand, the vortices in the divergent channel become and grow larger due to flow deceleration.

5. The flow pattern around the promoter is possible to classify by using Re_d, rather than Re_w.

[Reference]

(1) W.M. Kays, et al, Heat Transfer to the Highly Accelerated Turbulent Boundary Layer With and Without Mass Addition, Trans. ASME. Ser. C 92[70]499

(2) C.A. Bankston, The Transition from Turbulent to Laminar Gas Flow in a Heated Pipe, Trans. ASME. Ser. C 92[70]569

(3) M. Hishida, et al, Enhanced Heat Transfer of Fuel at Low Reynolds Number, European Nuclear Conference 1975.

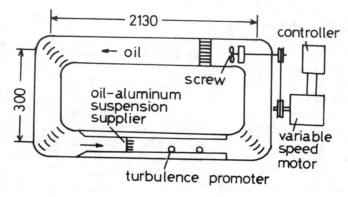

Fig.1 Schematic diagram of apparatus

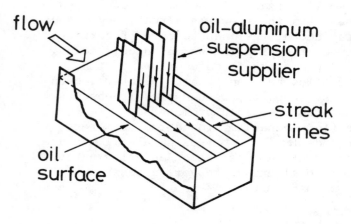

Fig.2 Outline of oil-aluminum suspension supplier

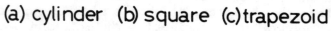

Fig.3 Configuration of promoter shape

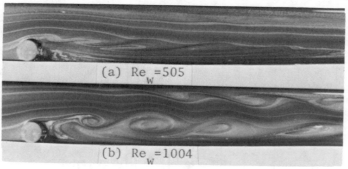

(a) Re_w=505

(b) Re_w=1004

Fig.4 Flow pattern around a promoter
in parallel channel in case of w/d=3

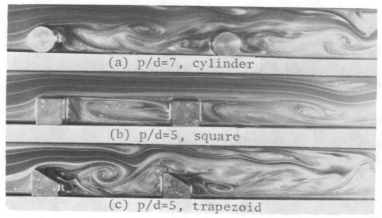

(a) p/d=7, cylinder

(b) p/d=5, square

(c) p/d=5, trapezoid

Fig.5 Flow pattern around two promoters
in parallel channel in case of w/d=2 at
Re_w=1009

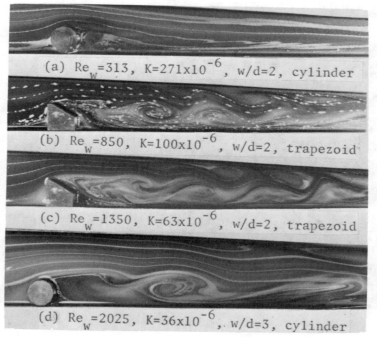

(a) Re_w=313, K=271x10^{-6}, w/d=2, cylinder

(b) Re_w=850, K=100x10^{-6}, w/d=2, trapezoid

(c) Re_w=1350, K=63x10^{-6}, w/d=2, trapezoid

(d) Re_w=2025, K=36x10^{-6}, w/d=3, cylinder

Fig.6 Flow pattern around a promoter
in convergent channel

type	configuration	number of flow vortices			
		$p/d=\infty$	7	5	3
parallel	square	10	7~8	0	0
	trapezoid	10~12	13	13	0
convergent	square	8~10	7	0	0
	trapezoid	13~14	11~12	11~12	0
divergent	square	7	7	0	0
	trapezoid	7~8	8~9	7~8	0

Table.1 Number of Vortices flowing down between promoters for ten seconds at $Re_w \sim 1050$ and w/d=2

type	transition Reynolds number Re_{wt}	
	w/d=2	w/d=3
parallel	690	850
convergent	850	930
divergent	460	<840

Table 2 Comparison of the transition Reynolds number Re_{wt}

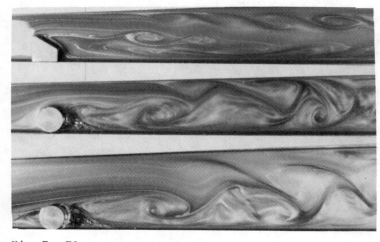

Fig.7 Flow pattern around a promoter in divergent channel

(a) Re_w=460, K=-175×10^{-6}, w/d=2, trapezoid

(b) Re_w=1390, K=-52×10^{-6}, w/d=2, cylinder

(c) Re_w=1956, K=-37×10^{-6}, w/d=3, cylinder

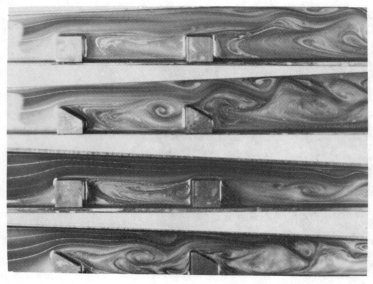

Fig.8 Flow pattern around two promoters in convergent and divergent channels in case of w/d=2 and p/d=5

(a) Re_w=737, K=-98×10^{-6}, square

(b) Re_w=737, K=-98×10^{-6}, trapezoid

(c) Re_w=1350, K=63×10^{-6}, trapezoid

(d) Re_w=1350, K=63×10^{-6}, trapezoid

A METHOD OF FLOW VISUALIZATION USING OPTICALLY ACTIVATED TRACERS

V. DELITZSCH* and D. W. SCHMIDT**

A method of flow visualization by the use of tracers has been developed in which both, the production or activation of the tracers and their observation, are done by optical means. This method has the special advantage that for its application no disturbing inserts or measuring probes are needed within the flow under test. It is, therefore, useful in many cases of flows where such submerged rigid bodies would disturb the flow to an inadmissible degree, or where the insertion of probes is difficult or even impossible. As a representative case for such a flow system, the instationary flow was taken which develops in a cylinder completely filled with liquid, after a sudden stop of rotation around its axis.

Introduction

Many special techniques have been developed for tracing the fluids in flow research. Of greatest value among these techniques are those which give a direct insight into the flow phenomena by optical observation. In many cases, such insight is needed for adequate understanding and interpretation of the results of measurements.

An essential demand for any method of measurement which might be useful in fluid dynamics is that it must not disturb the flow under test. If rigid probes must be used for the measurements, this demand is usually tried to be fulfilled by making the probes very small. But, as is wellknown, the smallest sizes which can be realized technically, prove often to be still too large. The tracer methods dealing with in this lecture do not need any probe in the flow. The tracers used for the observation of the flow are activated remotely by noncontact optical means.

Method of measurement

The method of measurement which has been newly developed may be demonstrated by applying it in a characteristic manner to a special instationary flow. Fig. 1 shows the main parts of the test setup used: The flow which is to be observed is produced within a cylindrical vessel completely filled with liquid. This transparent lucite vessel is at first put into stationary rotation about its axis by means of a stepping motor and is then (partly or even fully, in general) stopped suddenly. It may now be asked to observe the further development of the flow in the horizontal middle plane of the vessel. This is clearly a problem for which the application of methods of measurement which need probes located within the rotating system is

* Diplom-Physiker, ** Dr. rer. nat., both at Max-Planck-Institut für Strömungsforschung, D 34 Göttingen, Germany (BRD)

not attractive. Instead of this, it has been considered if one of the wellknown tracer methods (see Ref. 1 for literature on the subject) might come into question, but all of them proved to have more or less severe restrictions concerning their application to problems of our kind.

In our tracer method several of the possible reasons of trouble or inconvenience have been avoided. It makes use of single floating phosphorescent particles which can be activated by irradiation of high intensity light, originating for instance, as seen in the picture, from an UV-laser. The paths of the tracer particles in the flow are observed and recorded by an electronic camera, combined with an electronic image intensifier if necessary for higher sensitivity. As is shown in a block diagram in fig. 2 two delays are used for flexible timing of each procedure of measurement: Following a stop pulse (which initiates each procedure of measurement) by a delay I, the UV-laser is caused to transmit a single light pulse illuminating particles which are just within the laser beam. Thus being activated, the phosphorescent particles remain visible for several seconds during which - after a second adjustable delay, if wanted - records of the paths of the tracers can be taken. Performing also the recording by electronic means instead of the usual photographic techniques has the advantage that one gets the data directly as electric values which is most convenient for further data handling.

Production of the phosphorescent tracer particles

The main condition for the applicability of solid particles as tracers in a fluid is that their density must not differ from that of the surrounding medium. If this condition is fulfilled, one has, surely, no problems with disturbances by buoyant forces. Unfortunately, the only materials which come into question as tracers in our measuring system with regard to their phosphorescent properties, have densities near $4 g/cm^3$. This value is much higher than that of even the heaviest transparent test fluids available. In most practical cases it would be desirable to use water as the test fluid, which raises the demand for phosphorescent tracer particles of exactly the density $1 g/cm^3$.

This problem has been solved by combining the high density phosphorescent crystals of copper doped zinc sulfide, which are about 10 micron in diameter, with plastic foam having a density of $0.03 g/cm^3$. This can easily be done by glueing the crystals to small foam particles, or, as has been realized, by embedding them homogeneously in the foam. This latter method requires the crystals to be added to the plastic material before or at least during the foaming process. A procedure which gave good results was as follows: 90 g of Hostyren N 7000, a commercial solution of polystyrene containing an aliphatic hydrocarbon for foaming, was mixed homogeneously with 10 g of the zinc sulfide powder. By the action of water vapour, this mixture was foamed to small spheres having medium densities near the desired value of $1 g/cm^3$. These spheres have then been grinded under liquid nitrogen, and from the fragments particles of diameters between 0.5 and 1.0 mm have been selected by meshes. Finally, this selection in size was followed by a selection in density. For this purpose the particles have been suspended in water, and those which remained floating for a time of several hours have been sucked off carefully through a small pipe. The tracer particles thus selected with respect to their size and density proved to be useful for practically unlimited times.

Test of the method

Using the experimental arrangement I already shown in the first slide, the applicability of the phosphorescent tracers to problems of flow research could be demonstrated successfully for time of observation up to about 10 seconds. During such long times the intensity of light emitted by the tracer particles which have been activated by a single laser shot decreases by a factor in the order of 10^4. To obtain traces of constant brightness and minimum line width on the display, an automatic electronic brightness control which compensates for these large intensity variations should be further added to the electronic circuit such that one gets the same standard brightness of a tracer particle at any position and time.

In order to avoid the restrictions concerning the usable time of observation given by the limited time during which the decaying phosphorescence of the tracer particle remains detectable, an improved version of the experimental apparatus as shown in fig. 3 was further used for the test research. In this arrangement, the plane under observation is illuminated through a slit, followed by a shutter, during the whole time of observation. Thus, the tracer particles used must no longer be phosphorescent, that is afterglowing over rather long periods, but they have only to be fluorescent in order that, during activation by the irradiated light, their monochromatically emitted light may be observed through a special highly selective filter. The application of the filter makes the measurements nearly independent of disturbances by light of any other origin than that of the desired fluorescent emission.

The light barrier indicated in the slide is used for switching off the step motor such that the cylindrical vessel under observation comes to rest always at the same angular position shown in the topview. In this position the "optical correction device" enables all illuminating light rays to enter the interior of the vessel without refraction, thus providing a homogeneous illumination of the whole flow field under test.

The chain of events which now accompanies a period of measurement may be described with the aid of the electronic block diagram shown in fig. 4: Starting from a state of constant velocity of rotation of the step motor (its value being measured by an RPS-meter as indicated), a measurement is initiated by closing the "arming switch". This enables the next following output pulse from the light barrier to trigger a pulse which stops the motor and, simultaneously, to start the run of a time counter and open the shutter which controls the illuminating light. The appearance of a fluorescent tracer particle within the illuminated light sheet causes the electronic camera, a 50 by 50 diodes matrix system, to deliver pulses which, as voltage V_1 in the picture, are fed to the brightness control input Z of the cathode ray oscilloscope display. This results in a continuous reproduction of the path of the tracer particle on the screen. Via a "particle detector" the appearance of an illuminated particle is also communicated to the "delayed switch actuator" which, at an instant lying within the time during which the tracer is recorded visibly on the screen, connects a DC voltage V_2 to the Z-input of the scope and, simultaneously, stops the time counter. Starting from this wellknown instant after the initiation of a test cycle by stopping the motor, the trace of the observed particle is brightened entirely by a recognizable degree, thus providing one exact time mark on the trace. Further, periodic pulses of suitable frequency and duration, labelled as voltage V_3 in the figure, are used to produce additional time marks along the whole length of the tracer path recorded. A typical result of this combined electronic timing is to be seen in fig. 5 which shows an example of a recording obtained by the procedure described. Several pictures of this kind, all showing paths of tracer particles at times around 20 s after stopping the initial rotation of the cylinder at 9. 8 Hz, but at different radii, have been evaluated to obtain the result shown in fig. 6: The angular velocity of the flow is - fairly good - seen to vary linearly with the radius with the exception of a certain region near the wall of the vessel.

Summary

Finally, the advantages of the tracer method outlined in this lecture may be shortly summarized as follows:

1) Density of the tracer particles exactly equal to the density of the flow medium: No disturbing influence of buoyant forces;

2) No probes in the flow under test;

3) Observation of single particles: No disturbances by scattering of the illuminating light;

4) Optical selection of the monochromatic useful signal: Measurements highly independent of disturbing ambient light;

5) Direct electronic recording (no photographic interstage): Immediate and easy availability of the data as electrical values;

6) Electronic time marking;

7) Applicability to the investigation of instationary flow problems.

Reference

(1) V. Delitzsch, Methoden zur Sichtbarmachung von Strömungen mittels optisch aktivierter Tracer, Max-Planck-Institut für Strömungsforschung, Bericht 24/1976, Göttingen, Oktober 1976.

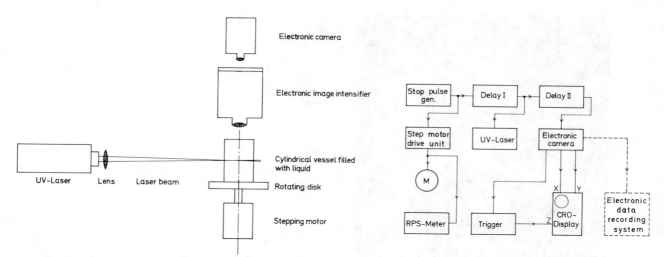

FIG.1: EXPERIMENTAL ARRANGEMENT I FOR THE DEMONSTRATION OF A METHOD OF FLOW
VISUALIZATION BY OPTICALLY ACTIVATED PHOSPHORESCENT TRACER PARTICLES

FIG.2: ELECTRONIC BLOCK DIAGRAM OF EXPERIMENTAL ARRANGEMENT I

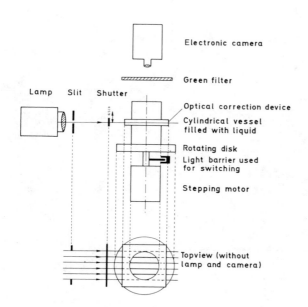

FIG.3: EXPERIMENTAL ARRANGEMENT II FOR FLOW MEASUREMENTS IN A ROTATING
SYSTEM USING OPTICALLY ACTIVATED FLUORESCENT TRACER PARTICLES

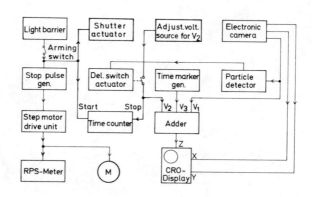

FIG.4: ELECTRONIC BLOCK DIAGRAM OF EXPERIMENTAL ARRANGEMENT II, INCLUDING
A CIRCUITRY FOR TIME MARKING THE DISPLAYED RECORDS OF TRACERS

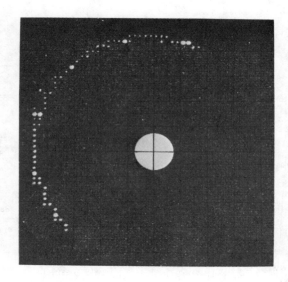

FIG.5: TYPICAL RESULT OF AN ELECTRONIC RECORDING (BY A DIODE MATRIX CAMERA)
OF THE PATH OF A FLUORESCENT TRACER PARTICLE IN AN INSTATIONARY
WATER FLOW, WITH ELECTRONICALLY PRODUCED TIME MARKERS

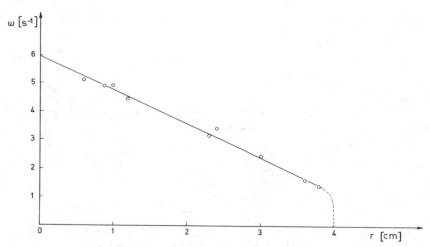

FIG.6: DEPENDENCE OF THE ANGULAR FREQUENCY OF THE TRACER PARTICLES ON THE RADIUS
OF THEIR PATHS IN THE CYLINDRICAL TEST VESSEL 20 SECONDS AFTER STOPPING
ITS 9,8 HZ-ROTATION (WALL OF THE CYLINDER AT R=4CM)

THREE-DIMENSIONAL MEASUREMENT OF FLOWS IN FLUIDIC ELEMENTS BY THE PULSE-LUMINESCENCE METHOD USING HIGH POWER NITROGEN PULSE-LASERS

N. NAKATANI,* Y. OHMI,** and T. YAMADA***

To visualize and measure quantitatively the flow velocity distributions, the velocity directions and the turbulence intensities in fluidic elements in three dimensions, the pulse-luminescence method is investigated. The apparatus is made by using two cross nitrogen pulse-lasers for high power excitation and VTR system for easy observation. By measuring flow in a turbulence amplifier and transient flow in a pipe, it is demonstrated that this technique is a powerful tool for three-dimensional measurement of stationary and non-stationary flows in fluidic elements.

1. Introduction

To visualize and measure quantitatively flows in fluidic elements, a hydrogen bubble method and a flow birefringence method have been used. The former method requires the generators of a tracer to be inserted into the flow, and the latter method necessitates time consuming integrations of measured velocity gradients to obtain velocities. The pulse-luminescence method developed by the authors has proved promising for instantaneous visualization and quantitative measurement of the velocity distribution of liquid flow. In the previous papers (Ref. 1, 2) we reported about the methods of visualizing and measuring the velocity distribution and the diffusion of turbulent flow in two dimensions. In this study the pulse-luminescence method is investigated for measuring flows in fluidic elements in three dimensions. The excitation is made by one nitrogen pulse-laser beam for measuring the flow velocity distribution and it is made by two cross nitrogen pulse-laser beams for measuring the velocity direction and the turbulence intensity. VTR system is used for the easy observation of the emission pattern. Using this method, flow in a turbulence amplifier and transient flow in a pipe for step input are measured.

2. Mehtod

Luminescent particles of tracer are well dispersed in the liquid fluid. In the measurement of the flow velocity distribution the luminescent particles are excited by one nitrogen pulse-laser and the movement of the emission part is observed from the direction perpendicular to that of the inci-

---

*Assistant Professor, **Graduate Student, ***Professor
Department of Precision Engineering, Faculty of Engineering,
Osaka University, Yamada-Kami, Suita, Osaka 565, Japan

dent laser beam as shown in Fig. 1 and two other directions. In the measurement of the velocity direction and turbulence, the luminescent particles are excited by two cross nitrogen pulse-laser beams and the movement and diffusion of the emission part in the intersection are observed from three directions. At various times elapsed after the excitation each emission part is recorded by a video tape recorder. For the elapsed times shown in Fig. 2 (a) the emission parts indicate flow velocity distribution patterns shown in Fig. 2 (b) and diffusion patterns shown in Fig. 2 (c). The luminescent particles are excited by the laser beam with a finite width. As a result when the emission part in the intersection is observed from the perpendicular to the excitation, the pattern is superimposed by the relative diffusion patterns on some layers in the depth direction of the observation. By superimposing several patterns at the same elapsed time, Taylor diffusion pattern is obtained. The relationship between the diffusion width $\overline{Y^2}^{\frac{1}{2}}$ obtained experimentally and the turbulence intensity $\overline{v_1^2}^{\frac{1}{2}}$ (where v_1 is fluctuating velocity in the direction of the diffusion width Y) is given by equations (1) - (4).

When $T \approx 0$,

$$K = \overline{v_1^2}T \qquad (1)$$

$$\overline{Y^2}^{\frac{1}{2}} = \overline{v_1^2}T \qquad (2)$$

and when $T \gg T_*$,

$$K = \overline{v_1^2}T_* \qquad (3)$$

$$\overline{Y^2}^{\frac{1}{2}} = (2\overline{v_1^2}T_*)^{\frac{1}{2}}T^{\frac{1}{2}} \qquad (4)$$

where K is the turbulent diffusion coefficient, T the diffusion time and T_* the life time of the vortex.

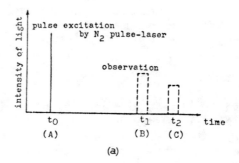

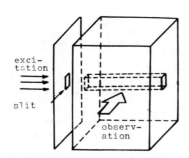

Fig. 1 Schematic diagram of the measurement.

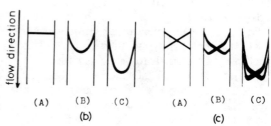

3. Apparatus

The schematic diagram of the apparatus for the measurement of velocity direction and turbulence intensity in three dimensions by the pulse-luminescence method is shown in Fig. 3. Two nitrogen pulse-lasers NL1 and NL2 are made for making intersection in flow field and for instantaneous and high power excitation. Either of those is used for

Fig. 2 (a) Schematic relation in pulse excitation by nitrogen pulse-laser and observation, (b) schematic patterns of velocity distribution at each time elapsed after the excitation (with one beam excitation), (c) schematic patterns of emission part, diffusing from the excitation to the observation (with two beams excitation).

the measurement of flow velocity distribution. The power of the lasers is about 1 MW at maximum output and about 100 - 200 KW in ordinary use. Their pulse time (a few nanoseconds) is so short that the distance which a lumi-

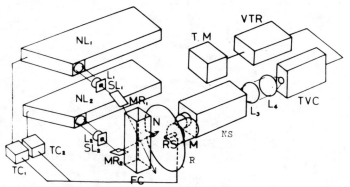

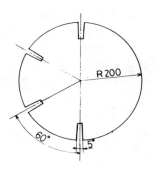

Fig. 3 Schematic diagram of the apparatus for measuring flow velocity direction and turbulence intensity in three dimensions by the pulse-luminescence method. NL_1, NL_2: nitrogen pulse-lasers, SL_1, SL_2: slits, L_1, L_2: condenser lenses, TC_1, TC_2: trigger switches, RS: a rotating switch, R: a rotating disk, N: a notch, MR_1, MR_2: mirrors, M: a motor, FC: a flow cell, NS: a night vision scope, L_3, L_4: relay lenses, TVC: a television camera, VTR: a video tape recorder, TM: TV monitor.

Fig. 4 Schematic diagram of the rotating disk R with the four notches.

nescent particle moves during the excitation can be neglected. The luminnescent particles are excited by the two cross laser beams through two slits SL_1 and SL_2. A rotating switch RS is used as a switch of the laser triggers TC_1 and TC_2. Observation at time t elapsed after the excitation of the luminescent particles is made through the notch in the rotating disk R. t is adjusted by setting the angular width between the rotating switch RS and the notch in the rotating disk, and by the rotation speed of the disk. By the use of a rotating disk shown in Fig. 4, one exposure can be made with a notch and some exposures can be made continuously with some notches. The latter is useful for observation of transient flow. In order to record a luminescence image of low light intensity by one exposure, a night vision scope NS is used, which has a light amplification of 12000 ×. The scope output image is recorded in a video tape recorder VTR through relay lenses L_3 and L_4 and a television camera TVC, and is reconstructed on a television monitor TM.

The flow liquid used is water or water-glycerine solution, and the luminescent particles LC-G1A (Ref. 1) of tracer in 0.05 - 0.10% wt concentration are well dispersed in the liquid. Because the density of the luminescent particles is higher than that of the water-glycerine solution, the particles sink to the bottom of the fluid tank by gravitational force, and the concentration of the particles in the fluid decreases in a short time. To keep the concentration of the particles constant, the bottom of the fluid tank is made funnel-shaped and the output pipe from the bottom is connected to the input pipe of the fluid supply pump.

4. Measurement of Flow in Turbulence Amplifier

The dimension of the turbulence amplifier used are shown in Fig. 5. The curve of output pressure P_O for supply pressure P_S without control flow is shown in Fig. 6. When the output pressure is measured, output flow is stopped by closing the output nozzle.

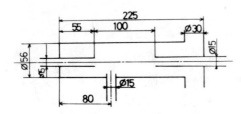

Fig. 5 Dimension of the turbulence amplifier.

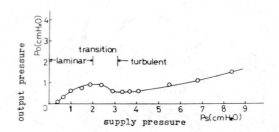

Fig. 6 P_O - P_S characteristics of the turbulence amplifier.

4-1 Measurement of flow velocity distribution

To clarify the relation between the pressure characteristics of the amplifier and the flow, the flow velocity distributions are measured. One laser beam enters to the central layer of the amplifier for the excitation and the emission part is observed from the direction perpendicular to the excitation, as shown in Fig. 1. The results are shown in Fig. 7 and 8. It is seen from Fig. 7 that the transition from laminar flow into turbulent flow occurs corresponding to the pressure characteristics of Fig. 6. At P_S = 1 cm H_2O the jet is laminar and impinges to the output port with a little diffusion. Consequently the output pressure becomes large. Over P_S = 2 cm H_2O the transition into turbulent flow occurs, and velocity distribution of turbulent flow is observed intermittently. At P_S = 3.5 cm H_2O and 7 cm H_2O the jet is turbulent completely, and its width becomes broad with diffusion near the output port, so that the output pressure decreases. The curve of the gain P_O/P_C (P_C: control pressure) for P_S is shown in Fig. 9. The gain becomes largest at P_S = 2 cm H_2O. Then the schematic patterns of the velocity distribution and the typical photograph shown in Figs. 10 and 11 are obtained. The jet is laminar without control flow as shown in Fig. 7, but it is turbulent in large scale with control flow as shown in Fig. 11 and diffuses largely. Consequently its width becomes broad near the output port in the same way as that of the turbulent jet shown in Fig. 7. As a result the output pressure decreases.

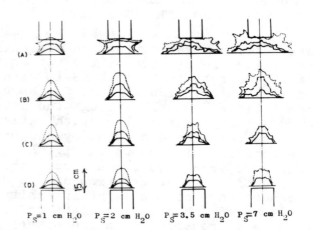

Fig. 7 Schematic patterns of the velocity distributions in the turbulence amplifier for various supply pressures P_S. ——: t=21 ms, –·—: t= 42 ms, – – –: t=83 ms, (t: time elapsed from the excitation to the observation).

P_S= 1 cm H_2O P_S= 7 cm H_2O

Fig. 8 Typical patterns of the velocity distributions in front of the output port, shown in Fig. 7.

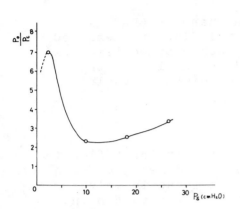

Fig. 9 The curve of gain P_O/P_C for supply pressure P_S in the turbulence amplifier.

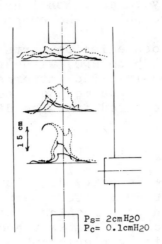

P_S= 2cmH2O
P_C= 0.1cmH2O

Fig. 10 Schematic patterns of the velocity distribution with control flow. ———: t= 21 ms, ———: t=42 ms, ———: t=83 ms (t: time from the excitation to the observation).

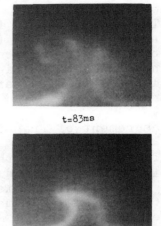

t=83ms

t= 42ms

Fig. 11 Typical patterns of the velocity distribtuions shown in Fig. 10.

4-2 Measurement of velocity direction and turbulence intensity

The luminescent particles are excited by two cross laser beams for the measurement of the velocity direction. To simplify the experiment, the laser beams are entered to the central layer near the control port. The cross emission part is observed from the perpendicular direction to the excitation. The mean velocity vectors obtained by superimposing the several patterns are shown in Fig. 13. The interaction between main jet and the control flow can

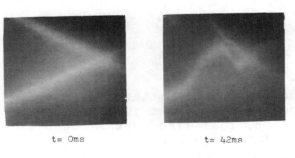

t= 0ms   t= 42ms

Fig. 12 Patterns of the cross emission part in each elapsed time.

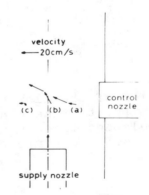

velocity
—20cm/s

(c) (b) (a)

control nozzle

supply nozzle

Fig. 13 Mean velocity vectors near the control port. P_S = 2 cm H2O, P_C = 0.6 cm H2O.

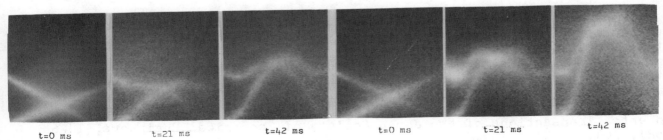

t=0 ms   t=21 ms   t=42 ms   t=0 ms   t=21 ms   t=42 ms

Fig. 14 Diffusion patterns of the cross emission part for each elapsed time. (a): laminar jet, (b): turbulent jet.

be observed clearly. Generally three dimensional velocity vector can be measured by observing the movement of the cross emission part from three directions.

The diffusions of the cross emission parts in laminar jet (P_S = 1 cm H₂O) and turbulent jet (P_S = 3.5 cm H₂O) without control flow are measured for obtaining turbulence intensity. The patterns for each elapsed time are obtained as shown in Fig. 14. By the method described in section 2, the turbulence intensity can be measured from the diffusion width of their patterns.

5. Measurement of Transient Flow in a Pipe

The variation of the flow velocity distribution for step input flow in the pipe of the inner diameter 20 mm is observed from the direction shown in Fig. 1. Step input is given by switching flow using a magnetic valve with three ports. The magnetic valve is switched by another contact point of the rotating switch RS having variable angle for that of the laser trigger. Consequently the actions of the laser trigger can be made at various times elapsed after switching of the magnetic valve and the flow velocity distributions can be observed. The flow patterns obtained are shown in Fig. 15. At the beginning the pattern of the flow velocity distribution shows uniform profile and after a while it gets to show parabolic profile. The experimental result is found to be very nearly equal to the theoretical result of Ref. 3.

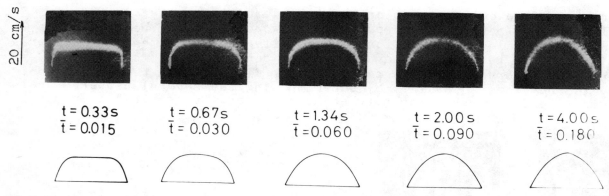

| t = 0.33 s | t = 0.67 s | t = 1.34 s | t = 2.00 s | t = 4.00 s |
| $\bar{t}$ = 0.015 | $\bar{t}$ = 0.030 | $\bar{t}$ = 0.060 | $\bar{t}$ = 0.090 | $\bar{t}$ = 0.180 |

Fig. 15 The variation of the flow velocity distribution pattern for step input flow in the pipe. t: time elapsed from the step input flow to the recording, $\bar{t}$: the elapsed time in non-dimension, full line curves are the theoretical flow velocity distributions calculated from Ref. 3.

References

(1) N. Nakatani, K. Fujiwara, M. Matsumoto and T. Yamada, Measurement of Flow Velocity Distributions by Pulse Luminescence Method, J. Phys. E: Sci. Instrum., 8-12 (1975), 1042.
(2) N. Nakatani, M. Matsumoto, Y. Ohmi and T. Yamada, Turbulence Measurement by the Pulse Luminescence Method Using a Nitrogen Pulse Laser, J. Phys. E: Sci. Instrum., 10-2 (1977), 172.
(3) H. Itō, Theory of Laminar Flow through a pipe with Non-steady Pressure Gradients, Transactions of the Japan Society of Mechanical Engineers, 18-6 (1952), 101.

SOME FUNDAMENTAL TECHNIQUES OF FLOW VISUALIZATION

J. SAKAGAMI*

In this paper, three techniques of visualization of air are described. I) Puppus wind vane: A wind direction indicator, made of a puppus and a single fibre, which works as wind vane even in very low wind speed and follows faithfully rapid fluctuations of wind direction. II) Single particle generator: It can produce 1 or 2 very light and flocky particles which are suitable for investigation of flow in Lagrangean way. III) Smoke layers at arbitrary distances from a surface can be produced which are able to investigate flow patterns at various heights.

I Puppus wind vane

In order to measure directions of air flow and their fluctuations, usually tufts or short wools are used. However, they do not show true direction of flow when wind speed is low, because of their stiffness and original figures, and also they do not follow rapid change of flow direction. To avoid these defects, we made a wind direction indicator with puppus of dandelion and a single fibre of Nylon, and we call it as 'puppus wind vane', though it has no vane in itself.

At the end of shortened stem (3~4 mm in length) of a puppus, an end of a single fibre about 5 mm in length is attached with diluted adhesive agent, then the another end of the fibre is fixed directly to a needle. If the wind direction changes widely, the end of the fibre is attached to a bottom of a small semi-closed tube, and this is put on the top of a needle, so the puppus moves freely and the fibre does not to be entangled to the needle. (Fig. 1). (Ref. 1). As the weight of the moving part of this vane is less than 0.07 mg, effect of gravity is very small compared to the drag of the wind, so when the wind speed is larger than 10 cm/s, it works as a wind direction indicator. When it is placed in a region of Kármán vortex street behind a cylinder, we can recognize its oscillation up to about 400 Hz. When the wind speed becomes larger, the puppus is folded up naturally, and decreases the drag, so in our wind tunnel, it could be used up to 14 m/s. Deflecting angles of the vane are determined by visual or photographical observation from the side or from above. If many vanes are set 3 dimensionally, we take stereograph or stereocinematograph of them, and analyse the photos by the procedure of multiplex used in photogrammetry. Fig. 2 shows a stereograph which was taken in the field in order to analyse the micro-turbulence in the atmosphere, and 720 vanes were set in a space of 70 mm x 30 cm x 21 cm. (Ref. 2). Fig. 3 (a) shows the flow pattern at a leeward position of a semi-elliptic cylinder, and in lower part, there appears a reverse flow in a dead water region. Fig. 3 (b) shows the similar pattern when another similar cylinder is placed closely to the

---

* Professor emeritus, Ochanomizu University, Bunkyo, Tokyo, Japan

former one. Fig. 3(c) shows that the reverse flow diminishes when the latter cylinder is placed windward at a distance of one long-axis of the elliptic cylinder. The range of angle of the vane during a certain time interval shows the order of the turbulence, and periods of change of wind direction can also be estimated. Rotational flow whose axis is along the mean flow direction, such as Görtler vortex near a concave surface, can also be noticeable, as it rotates when it is set in the flow.

II Single particle generator

To investigate flows in Lagrangean way, particles in small size and 1~2 in number at one time are necessary. One crystal of metaldehyde is put in a small nichrome spiral (2 mm in diameter, 4 mm in length) and then the spiral is heated electrically, so 1~2 small (2 mm in dimension) and very light and flocky particles like snow flakes, whose falling velocity is less than 1 cm/s, are generated. Fig. 4 shows a photograph (one of a pair of a stereophotograph) which was taken for the investigation of Lagrangean correlation of wind velocity in the atmosphere. Stroboscopic illumination (20 Hz and 4 ws/pulse) was added to continuous illumination. With such photographs, we could determine 3 dimensional positions of particles. The distances between two white spots corresponded to travelling distances in 0.05 s, so we could calculate the Lagrangean velocity, and then Lagrangean correlation directly. Furthermore, the standard deviations of position of the particles could also be determined, so the Taylor's diffusion formula was examined experimentally. (Ref. 3).

III Smoke layers at small distances from a solid surface

At present, the smoke wire method has been developed. However, it cannot be used to make horizontal smoke layers. In some cases, the horizontal smoke layers are also necessary, according to the circumstances of the experiment. A nichrome ribbon 0.1 mm in thickness and 0.8 mm in breadth is spanned with tension over two bridges, whose height is preassigned. (Fig. 5). Paraffine oil is smeared over the ribbon, then heating current is feeded. The smoke layer appears during several seconds. Fig. 6 shows the flow patterns in leeward region of a sphere on a plate. The smoke layers are produced at the heights of 2 mm, 5 mm and 8 mm above the plate. The patterns are very different each other (Ref. 4). Fig. 7 shows the behavior of Tollmien-Schlichting waves. Smoke layers are produced 0.7 mm for upper 2 photos, and 1.7 mm for lower 2 photos, above the plate. We can examine the difference of process of development of the waves.

References

(1) J. Sakagami, On the structure of the atmospheric turbulence near the ground, II, Natural Science Rep., Ochanomizu Univ., 2 (1951-11), 52

(2) J. Sakagami, On the structure of the atmospheric turbulence near the ground, IV, ditto, 6-1 (1955-7), 75

(3) J. Sakagami and M. Mochizuki, Lagrangean measurement of small scale atmospheric turbulence by floating fine particles, ditto, 8-1 (1957-11), 67

(4) M. Mochizuki, Smoke observation on boundary layer transition caused by a spherical roughness element, J. Phys. Soc., Japan, 16-5 (1961-5), 195

Fig 1

Puppus wind vane

Fig. 2

Stereograph for the investigation
or micro-turbulence in the atmosphere

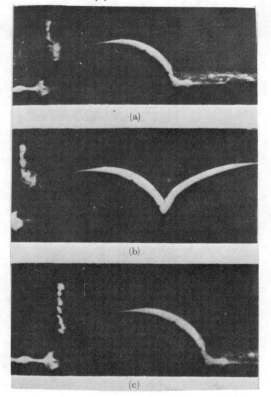

(a)

(b)

(c)

Fig. 3

Flow patterns behind a semi-
elliptic cylinder

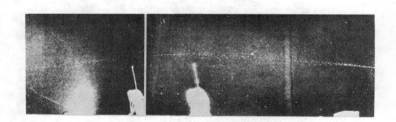

Fig. 4

Path line of metaldehyde particles

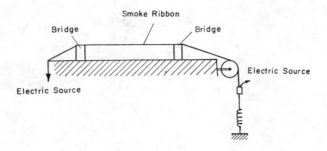

Fig. 5

Device of smoke layer generator

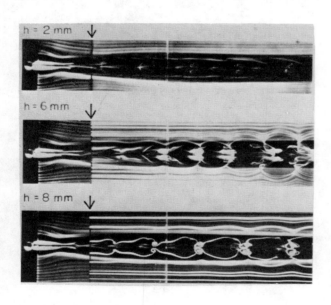

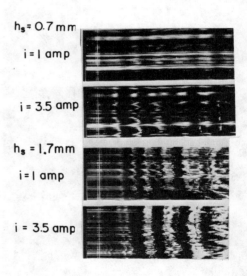

Fig. 6

Flow patterns in leeward region of a sphere. Right patterns from the arrow marks are elevated smoke patterns.

Fig. 7

Development of Tollmien-Schlichting wave. h_s is the height of smoke layer.

FLOW VISUALIZATION TECHNIQUES IN VEHICLE AERODYNAMICS

W.-H. HUCHO and **L. J. JANSSEN***

Flow visualization as well as flow simulation - both
closely related - are helpful aids during the aerodynamic design of vehicles. The different methods with which
the flow is made visible either on the surface of the
body or within the surrounding space are briefly
described. Emphasis is placed on the results to be
achieved with a specific method rather than upon
description of the related technical details.

1. Objectives of Flow Visualization in Vehicle Aerodynamics

The flow field around a road vehicle is extremely complex. It is characterized not only by a high degree of three-dimensionality but by separation,
reattachment, and vortex formation as well. In the course of the design of a
new vehicle this flow field has to be tailored such that various properties
of the vehicle, which often enough lead to requirements contradictory to the
flow itself, are realized: low drag, good handling, clean windows, sufficient
cooling air flow, etc. Normally this work has to be performed without
detailed knowledge about the flow field under consideration. Time limitations
frequently do not permit all the measurements - pressure distribution,
boundary layer flow, vortex formation - necessary to arrive at a quantitative
description of the flow around the vehicle.

Flow visualization, if properly applied, can help to overcome this situation.
With its various techniques the different flow regimes around a car can be
detected and localized. The influence of geometrical modifications on these
flow patterns and the interaction among them can be observed. Typical flow
patterns can be correlated with measured data of integral character, such as,
for example, drag. Thus at least a qualitative understanding of the flow around
the ear, its dependence on geometrical parameters and its effects on the car
can be generated. While in many other applications of fluid mechanics flow
visualization serves as a supplement when a flow field has to be described,
in vehicle aerodynamics it is the predominant tool for that purpose.

* Volkswagenwerk AG, Wolfsburg, GERMANY

2. Particularities of Flow Visualization in Vehicle Aerodynamics

In addition to the classical objective to make a flow visible which otherwise is nonvisible, flow visualization in vehicles aerodynamics must deal with flow phenomena which lead to visible effects during actual driving, such as water droplet flow on windows and dirt deposits on the body.

Particular problems arise when these flows are to be simulated either in the wind tunnel or on the test track because quantitative results are required. The mechanisms of droplet formation and whirl up of soil have to be investigated. Their correct reproduction is essential for the entire simulation.

In vehicle aerodynamics both classical flow v i s u a l i z a t i o n and flow s i m u l a t i o n are accompanied by specific problems because most actual design work is carried out with full scale models and cars. Large quantities of smoke, gas filled bubbles, powders, or colours are needed together with powerfull light sources. All techniques have to be refined to the point that they can be applied easily, quickly and reliably in a large wind tunnel without harming personnel, the tunnel or its equipment. On the other hand, visualization and simulation techniques applicable in actual highway testing are needed.

3. Flow Visualization Techniques

3.1. Visualization of Surface Flow

The flow on the surface of vehicles can be made visible by three different techniques. Attached flow as well as flow characterized by different types of separation can be made visible with wool tufts and with a surface oil film containing coloured or luminiscent pigments. Regions of separated flow, preferably those, where separation is quasi two dimensional, can be detected by talcum (or "dry" powders) introduced into the wake, when the surface of the model is dampened slightly with oil before commencement of the test. This latter method will be discussed in section 4.3. because it is preferably applied during soil deposit tests.

Figs. 1 and 2 are typical results of wool tuft tests. Regions of attached and separated flow can be clearly distinguished. Fig. 1 (fast back car) shows the vortex formation at the A-pillar at zero yaw.

Fig. 1: Wool tuft test on VW-Passat at zero yaw, wind from the right

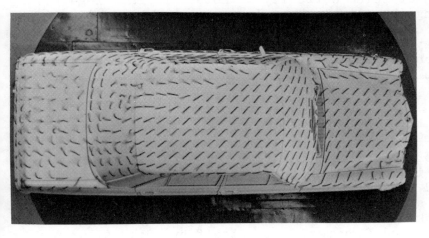

Fig. 2: Wool tuft test on
MERCEDES-BENZ 200, yaw
angle ß = 45°, wind from
the right

Fig. 2 (notch back car) indicates, that vortices start to roll up at the
windward front fender and the roof, both of which are "yawed leading edges"
in crosswinds. Furthermore the irregularities of the tufts on the rear window
and the rear bonnet demonstrate that separation takes place in a quasi two
dimensional manner.
Wool tuft techniques have several advantages: they are clean and can be
applied even during road tests. On the other hand a significant amount of
time is needed to place accurately the 600 to 800 tufts, required to generate
sufficient information. For this reason wool tufts have generally been
replaced by surface oil film methods in wind tunnel testing.

Figs. 3 and 4 give an impression of the amount of information contained in an
oil film picture. Both figures show the lee side of a VW-van (quarter-scale
model) at a yaw angle of ß = 20°. Fig. 3 depicts the sharp vertical leading
edge which causes separation, while the well rounded corner, shown in Fig. 4,
is passed without separation.

Fig. 3: Oil film picture using a white
pigment. Quarter-scale VW-van, simplified
geometry; sharp leading edges (R=0), yaw
angle ß = 20°

Fig. 4: Oil film picture, as Fig.3,
leading edges well rounded
(R=20 mm), yaw angle ß = 20°

Oil film patterns can only be interpreted correctly if notes are made regarding the direction of flow during the actual test. In regions of separation, oil film pictures do not always conform to wool tuft pictures. At locations where wool tufts flutter randomly the oil film still may show a regular structure indicating a well organized flow in the layer adjacent to the wall. This "secondary" flow seems to play on important role in the deposit of soil on body panels.

3.2. Visualization of Spatial Flow

Many flow phenomena can be better understood when flow patterns are known not only adjacent to the surface of the car but in the entire neighbouring space. On cars, regions of attached and separated flow are in significant interaction with one another. Figs. 5 and 6 show the different "smoke" techniques. The "smoke" may either be a real smoke - for instance from tobacco - or an oil vapour. The regular motion in a sound flow manifests itself by smoke filaments forming streamlines from which velocity vectors can be deduced. Difficulties arise in regions of high local velocity, see, for example, the connection of windscreen and roof, Fig. 5. There the streamlines are very close to each other. Neighbouring smoke filaments tend to merge due to their own turbulence, or they are dissipated in the turbulent boundary layer at the wall. Pictures of good contrast can be achieved, when the illumination is focused to the plane of interest. Three-dimensionally skewed streamlines are much more difficult to represent photographically.

Separation bubbles and wakes can be made visible, when smoke is introduced into the separated flow. Due to the turbulent motion, the smoke is distributed through out the entire space of separation, see Fig. 6. The reproduction of the wake's shape does not depend on the location of smoke introduction, provided that smoke is introduced into the specific wake under consideration. Separation at the leading edge of the hood, see Fig. 5, is clearly visible.

Fig. 5: Streamlines visualized by smoke, single smoke probe moved upwards in increments multi-exposure photo, illumination by narrow light band from top; sharp leading edge at the hood

Fig. 6: Separation bubble filled with smoke; identical car as in Fig. 5

Figs. 7 and 8, which originate from the development of a fastback, demonstrate that with smoke even large wake volumes can be filled. In Fig. 7 the line of separation is fixed artificially by means of a fillet at the rear end of the roof, thus leading to a squareback type flow regime with a large, almost prismatic wake. In Fig. 8 it becomes evident that the downwash induced by the two trailing vortices starting from C-pillar pushes the separation line rearward and forces the wake downwards.

Fig. 7: Rear wake filled with smoke, spot light from the rear

Fig. 8: Rear wake filled with smoke, as Fig. 7

Both smoke techniques - the one using streamlines, the other filling the wakes - can be applied simultanously, as in Fig. 9, which shows an unexpectedly small wake.

Fig. 9: Simultaneous visualization of streamlines and wake

If wakes and bubbles are illuminated by a narrow light beam the overall geometry of a separation zone can be mapped, when the light plane is moved incrementally and the subject photographed. This method however, yields little information about secondary motion within the separated flow.

This motion can be captured when soap-bubbles, filled with helium, are fed into the flow. Fig. 10 depicts the rear of a van and soap-bubbles which were fed into the flow ahead of the vehicle near the ground. Illumination was provided through a longitudinal window in the ground plane. The path of single bubbles can be easily followed. Within the wake two different zones can be distinguished: the "near wake", which is characterized by rotational motion and backward flow. This flow regime is responsible for dirt deposits on the rear of the van. Within the "far wake", flow is preferably parallel to the outer flow. Depending on the time of exposure, the history of single particles or the overall flow regime can be accentuated. The longer the exposure time the more the bubble picture resembles a smoke picture.

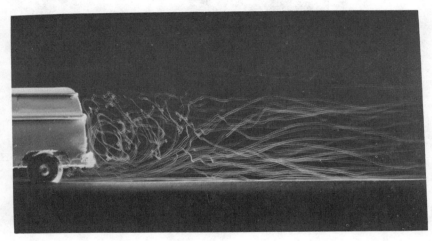

Fig. 10: Helium filled soap-bubbles fed into the airstream ahead of the car; model scale 1:20, illumination through a narrow window in the ground plane; exposure time 3/4 sec.

4. Flow Simulation

4.1. Necessity for Flow Simulation

During actual driving, cars are exposed to rain, mud or dust. The body and its components must be designed so that neither droplet flow nor the deposit of dirt or dust impairs function and safety of the car. This requirement calls for specific designs which must be studied very early during the course of development, when only models of the car under development exist. Related work has to be carried out in a wind tunnel. Techniques appropriate to simulate reality are needed for this purpose. In a later phase of development driveable cars exist but test time is limited. Reproducible "accelerated" procedures must be applied to discover areas requiring attention and to test alternative solutions.

4.2. Simulation of Water Flow

Fig. 11 reproduces a scene from rain simulation in the wind tunnel. Water, which is blended with a fluorescent additive, is sprayed into the undisturbed flow. Visibility is improved under ultra violett light. Photographs for the purposes of documentation must be taken while the test is actually in progress. The test engineer sitting inside the car can easily judge the quality of the wiper design under consideration.

Fig. 11 clearly indicates, the well known fact that heavy particles such as rain droplets do not always follow the direction of the air flow. The droplets running upwards along the windshield separate from the contour at the leading edge of the roof forming a curtain of moisture extending far downstream.

Fig. 11: Rain simulation in the wind tunnel; water blended with fluorescent additive ("leak detection fluid"); ultra violet diffuse light

Fig. 12: Path of rain streaks made visible by clear water and "Lithopone" (Trade Mark BASF), a white, water soluble pigment

Fig. 12 is an example of an alternative flow simulation technique, which is preferably applied when body details are to be optimized with the objective of keeping the windows free of droplets and moisture. Upstream of the detail in question water solvable paint is applied. Clean water is sprayed into the airstream ahead of the car and the streaks which manifest themselves, as in Fig. 12, can be studied after shut down of wind and water flow. The specific example originates from the design of the VW-Scirocco. The left half of the picture applies to an early design of the gap between hatch-door and roof. With slight modifications this gap was turned into a water trap which provides excellent rearward visibility under all weather conditions.

Both methods described above are compared in Fig. 13, which involves an investigation about the influence of outside-mirror location on moisture on the side window. Fig. 13a has been taken after a road test using the soluble paint method. Fig. 13b was taken from inside the car during a wind tunnel test applying water with the fluorescent additive. Both techniques ultimately lead to the same results.

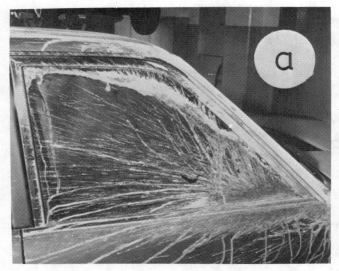 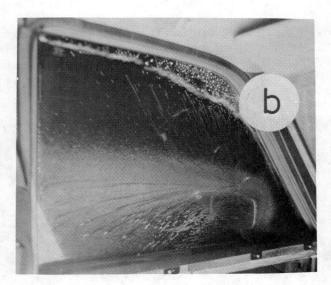

Fig. 13: Moisture pattern on side window

    a: "Lithopone"-method, photo taken after road test from outside, without mirror;

    b: with leak detection fluid, photo taken during wind tunnel test from inside, with mirror

4.3. Soil Deposits

Soil and mud deposits on cars can be annoying not only for aesthetic reasons. Safety can be affected when lights or windows are covered with dirt. Soil deposits must be dealt very early because the requirement to keep specific parts free of soil may lead to major changes in body geometry.

As long as only models are available two different techniques can be applied. The talcum method was prevously mentioned. Those regions of the body, where soil deposits are suspected, are sprayed slightly with oil. Then talcum is fed into the airstream in the same pattern as observed when the wheel is rolling on road, kicks up dust or mud. The talcum method is very sensitive. Those regions which remain clean during the test will not be soiled on road, but zones which are slightly covered with talcum may remain clean on the road.

The talcum method is also appropriate for quick testing on the road. Fig. 14 depicts the rear-end of a station wagon which was modified to conform to the shape of a prototype. Talcum was fed into the rear wake during actual driving. Without an airfoil this specific design is characterized by significant talcum deposit on the rear window (Fig. 14a). With an airfoil properly matched to the body the entire rear window can be kept clean (Fig. 14c).

Fig. 14: Soil deposits on rear window, talcum method, road test
  a: without airfoil;
  b: with airfoil, gap 20 mm;
  c: with airfoil, gap 40 mm

The other method of carrying out dirt tests on models is to simulate mud with clear water which is fed into the air behind the tyres. The parts under observation are equipped with small plates covered with a dry sponge, having equal weight at the beginning of the test. During the test a known quantity of water is injected into the air in the manner described above. After the test the weight of each plate is determined and provides a quantitative indicator of mud distribution.

Road tests can be carried out in a similar manner. A test track is first prepared with a specific soil. The test car and a reference car are driven over the track for several rounds. Afterwards the soiled plates are taken off the car and weighed. Fig. 15 contains typical results from such a test. The soil coefficient is the local soil weight per unit area, related to the soil weight per unit area in the middle of the rear window of the reference car. As can be seen from the diagrams in Fig. 15 the amount of soil as well as the soil distribution is significantly different among the cars compared. Running the test together with a reference car leads to results characterized by a high degree of reproduceabilty.

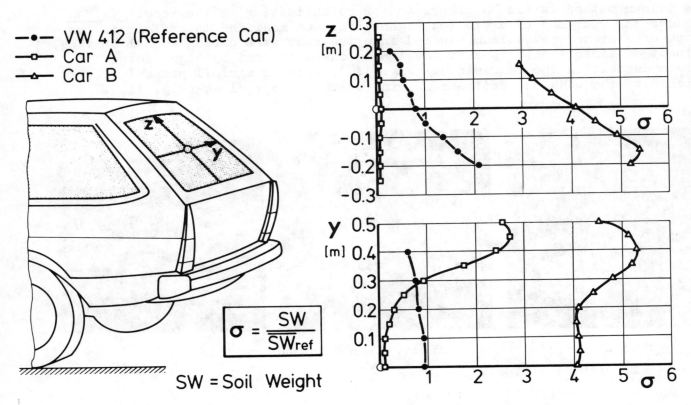

Fig. 15: Quantitative dirt measurement, road test on prepared track, using a reference car

5. Conclusions and Prospects

Flow visualization and flow simulation have proven to be important tools in vehicle aerodynamics. Many details of the flow field, which elsewhere in fluid mechanics are investigated by theory or detailed measurements, are made visible in vehicle aerodynamics to assist in generating a qualitative information about the ruling flow mechanisms. This information serves as a guide during the design of body shapes and details in the wind tunnel.

Techniques involving both visualization and simulation of flow require further improvement in order to provide more quantitative information. Special attention must be paid to ready applicability of the methods even in large wind tunnels and, if possible, on the road as well.

EFFECT OF FINITE AND INFINITE ASPECT RATIOS ON FLOW PATTERNS IN CURVED RECTANGULAR CHANNELS

K. C. CHENG,* J. NAKAYAMA,** and M. AKIYAMA***

Flow visualizations are applied to predict various types of flow patterns appeared in fully developed laminar flow in curved rectangular channels. Aspect ratios of the channel section examined are b/a=1,2,3,4,5,6,8,10 and 12, and curvature ratios are a/R=0.2 and 0.025.

The cigarette smoke for air, and milk solution for water are used to visualize the flow patterns.

For a wide range of finite aspect-ratios, the interactions between vortices due to Dean's hydrodynamic instability and boundary-value-type vortices are explained. The critical Dean number for the onset of the instability-type vortices under the influence of the boundary-value-type vortices are determined. Thus, it is the first time to answer directly to the questions of sidewall effects on the actual instability problem. The results are compared with the available theoretical results for the cases of the flow in curved square channel and of the instability flow in curved parallel-plate channel and the agreements are found to be good.

INTRODUCTION

It is our concern here to understand the phenomena of steady fully developed laminar flow of a vicous fluid in curved rectangular channels.

For the flow in curved rectangular channel with the aspect ratio approximately unity, a counter-rotating pair of vortices may appear as a results of retardation of the flow due to the friction of the two side-walls parallel to the acting direction of centrifugal force. The assumption that one pair of vortices should exist in a curved rectangular channels has been accepted in analytical solutions of boundary-layer approximation (Ref. 1,2). The flow results from numerical solution, however, show an additional counter-rotating pair of vortices near the central part of the outerwall (Ref. 3,4).

For the flow in a curved channel of infinite aspect ratio or between concentric cylinders, Dean's hydrodynamic instability aries (Ref. 5).

Unfortunately, no experimental confirmation for both cases seems to be available. Moreover, the full account of the interaction of the two phenomena has not been obtained in theory as well as in experiment so far. It seems that such a flow pattern can best be determined by direct flow-visualization techniques.

The objective of this work is to present;(1)the series of photographes which shows the interaction of Dean's hydrodynamic instability and boundary-value-type vortices for fully developed laminar flows in curved rectangular channels with a wide range of finite aspect-ratios,(2)the neutral stability curve for the onset of the Dean's instability under the influence of finite aspect-ratios of the channels.

*Professor,Dept.of Mech.Eng.,Univ.of Alberta,Edmonton,Alberta,Canada T6G 2G8
**Associ.Prof.,Dept.of Mech.Eng.,Fukushima Technical College,Iwaki 970,Japan
***Associ.Prof.,Dept.of Mech.Eng.,Utsunomiya Univ.,Utsunomiya 321-31,Japan

EXPERIMENTAL PROCEDURE

The general arrangement of the experimental apparatus for air is shown in Fig.1. The description of test sections is shown in Fig.2 and the dimensions of the sections are listed in Table 1. The apparatus for water is similar to that for air except the curvature ratio being a/R=0.025 (Ref. 6).

The essential quantities measured in this experiment are the volume flow rate and working fluid temperatures. The Reynolds numbers Re and the Dean numbers K are defined in terms of the mean velocity v and the corresponding length a or De as follows.

1st. case with the channel height a

$$R_{eA}=v \times a/\nu, \quad K_A=R_{eA}(a/R)^{1/2}$$

where ν is kinetic viscosity and R is radius of curvature. These definitions are suitable for instability problem.

2nd. case with the hydraulic diameter De=2ab/(a+b)

$$R_{eB}=v \times D_e/\nu, \quad K_B=R_{eB}(a/R)^{1/2}$$

where b is channel width. For the conduit with b/a to be about unity, these quantities are commonly used.

EXPERIMENTAL RESULTS AND DISCUSSION

Typical photographs are selected for the aspect ratios of b/a=1.0,2.0,5.0 and 12.0 to demonstrate the development of the two types of secondary flows in curved rectangular channels.

The Dean Number Effects   The effects of Dean number on secondary flow patterns are shown in Fig.4 for a curved square channel flow. Due to the existence of the two side-wall boundaries,a pair of counter-rotating vortices appears in low Dean number as shown in Fig.4(a). The existance of the boundary-value-type vortices is commonly understood. Fig.4(b) shows an appearance of additional counter-rotating pair of vortices near the central outerwall. It seems to be the first time to show this phenomena photographically and it confirms the theoretical prediction of the previous works (Ref. 3,4). The primary reason for the new type of vortices is due to the instability (Ref. 4). When the value of Dean number exceeds a certain value,an instability problem arises,and it leads to another type of laminar flow. The explanation is suited for the flow in a curved parallel plate channel.

In order to gain further insight into the actual phenomena,a new dimension of reasoning is essential by taking account the effect of the boundary-value-type vortices. Since it requires kinetic energy to make a room for the instability-type flow against the pair of vortices of the boundary-value-type which already occupy the whole domain,it is understood that the values of critical Dean number for the onset of the Dean's instability in a curved square duct is higher than that for in a curved parallel channel. However,the critical value for the curved square channes is rather low as compared with the case for b/a=2 and 3. When a high intensity of the boundary-value-type vortices is accomplished at a reasonable Dean number,the circulations tend to split to form more numbers of vortices due to maldistribution of circulation energy. A possible location for the additional eddies due to the energy from the boundary-value-type vortices is at the stagnation point and coincides with the one for the Dean's instability-type vortices. Moreover the direction of circulations are identical. Consequently,one has an additional pair of vortices at the central part of the outerwall with relatively low Dean number.

The siuation shown in Fig.4(c) is quite similar to that reported in Fig.2(b) of Ref. 4 where K_B=202 and a/R=0.01. With further increase of Dean number,the flow becomes complex as shown Fig.4(d). It is of particular interest to note that the complex flow pattern already appeares in a relatively low Dean number.

The Aspect Ratio Effects   The results of the secondary flow patterns for aspect ratios b/a=2,5 and 12 are shown in Fig.5,6 and 7,respectively. The secondary flow develops similar to that for square channel at b/a=2,but the critical Dean number is considerably high. Fig.5(d) indicates another significant fact that at higher Dean number the stable two pairs of the Dean's-instability-vortices appear in addition to the boundary-value-type vortices. It is noted that the boundary-type vortices still keep their high intensity near the both corners of the inner wall.

Fig. 6(a) shows the case with b/a=5.0,where the pair of the boundary-value-type vortices which reaches to the channel center with difficulty. The critical Dean number is lowered drastically. With nearly the same Dean number,some different flow patterns of instability type are possible as shown in Fig.6(b) and (c) for K_A=172.6.

As shown in Fig.7 for b/a=12 and a/R=0.025,the boundary-value-type vortices are confined to within 1 to 2 times of the channel height a. Beyond a critical Dean number, regular secondary flows of the Dean's instability type appeare.

ONSET OF DEAN'S VORTICES

The experimental data of the critical Dean number K_A and K_B from the flow visualization for a/R=0.2 are plotted against the aspect ratios b/a in Fig.8 and Fig.9,respectively. Below the value of the critical Dean number no detectable secondary flow of Dean's type is found,while beyond the value one can find the secondary flow of the Dean's-instability-type.

One notes that the theoretical prediction of the onset of hydrodynamic instability in curved parallel channel,b=∞,is known to have a critical Dean number $K_{A,crit}$=25.4 (Ref. 5) or $K_{B,crit}$=66.3. For the case of b/a=12 and a/R=0.025,the critical Dean number is found to be $K_{A,crit}$=27.1 and 6.7% off from the theoretical value for b/a=∞ and a/R≪0[1] (Ref. 5). Considering the fact that the results shown in Fig.8 and 9 are for a/R=0.2,the curvature ratio may effect on the critical value.

Nothing the diviation of critical Dean number at b/a=8 is less than 10% from the value at b/a=12,it is concluded that the sidewall effect is relatively small up to b/a=8.

Further decreasing the aspect ratio,the critical Dean number increases significantly and reachs to the maximum at b/a=2.0 and then decreases. The consequence of the boundary-value-type flow to the onset of instability-type vortices at the central region of the channel has explained in the previous section including the reasoning on the maximum critical Dean number.

CONCLUSIONS

1)   The fully developed laminar flows in curved rectangular channels for the cases of the aspect ratios b/a=1,2,3,4,5,6,8,10 and 12 with a/R=0.2 and the case of b/a=12 with a/R=0.025 are investigated by flow visualization techniques. The results are compared with the available theoretical results in good agreements:for the case of the flow in curved square channel (Ref. 4),and for the case of the Dean's instability flow in curved parallel-plate channel (Ref. 5).

2)   A neutral stability curve is obtained to determine the onset of Dean's instability under the influence of the boundary-value-type vortices (Fig.8 and 9).

3)   In the case for the aspect ratios 1,2 and 5,it is found that an additional counter-rotating pair of vortices appears near the outer stagnation point of the familiar boundary-value-type vortices at a certain higher Dean number depending on the aspect ratio. Further increase of the value of Dean number with the aspect ratio more than two develops another new pair of vortices making the two counter-rotating pairs of vortices near the outer region of the channel. The number of pairs of vortices may be different depending on the aspect ratios. The secondary flow patterns with the newly developed vortices have not been reported in the past and these phenomena lead to the appearance of Dean's instability problem.

4)   When the aspect ratio exceed more than eight,the secondary flow of the boundary-value-type dose not strengthen and is confined near to the both sidewalls, and the central region of the channel is left with no secondary flow. Above the critical Dean number,the secondary flow of the Dean's centrifugal stability type appears.

We are grateful to Mr.I.Urai of Utsunomiya Univ. who took the pictures (Fig.7).

REFERENCES

1. H.Ludwieg,Die ausgebildete kanalstömung in einem rotierenden system,Ingenieur-Archiv 19,(1951) 296-308.
2. Y.Mori and Y.Uchida,Study on Forced Convective Heat Transfer in Curved Square Channel(1st report. Theory of laminar region),Trans.Japan Soc.Mech.Engrs.,33,(1967) 1836-1846.
3. K.C.Cheng and M.Akiyama,Laminar Forced Convection Heat Transfer in Curved Rectangular Channels,Int.J.Heat Mass Transfer,13(1970) 471-490.
4. K.C.Cheng,R-C Lin and J-W Ou,Fully Developed Laminar Flow in Curved Rectangular Channels Trans.ASME.,J.Fluids Eng.,(1976-March)41-48.
5. W.R.Dean,Fluid Motion in a Curved Channel,Pro.Roy.Soc.London,Ser.A,121,(1928)402-420.
6. I.Urai,M.Akiyama,M.Suzuki and I.Nishiwaki,Visualization of Low Speed Water Flow, 4th Sympo.Flow Visualization ISAS Univ.of Tokyo,(1976,July) 93-96.

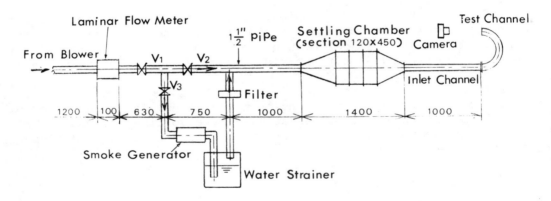

Fig. 1 General Arrengement of Test Apparatus

Fig. 2 A View of Test Apparatus

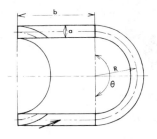

Fig. 3 Test Section

Table 1 Dimensions of Test Sections

R(mm)	θ(°)	a(mm)	b(mm)	a/R	b/a
125	180	25	25	0.2	1
125	180	25	50	0.2	2
125	180	25	75	0.2	3
125	180	25	100	0.2	4
125	180	25	125	0.2	5
125	180	25	150	0.2	6
125	180	25	200	0.2	8
125	180	25	250	0.2	10
125	180	25	300	0.2	12

(a) K_A=66.4,K_B=93.9   (b) K_A=124.5,K_B=176.1   (c) K_A=215.8,K_B=305.2   (d) K_A=319.5,K_B=451.9

Fig. 4 Secondary Flow Patterns for The Case b/a = 1

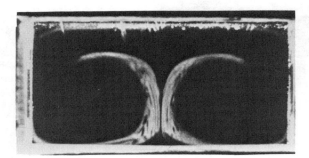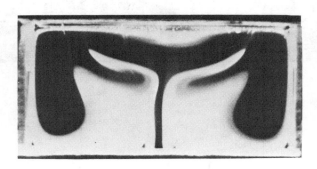

(a) K_A=41.5, K_B=78.2   (b) K_A=87.1, K_B=164.3

(c) K_A=136.9, K_B=258.3   (d) K_A=174.3, K_B=328.7

Fig.5 Secondary Flow Patterns for The Case b/a= 2

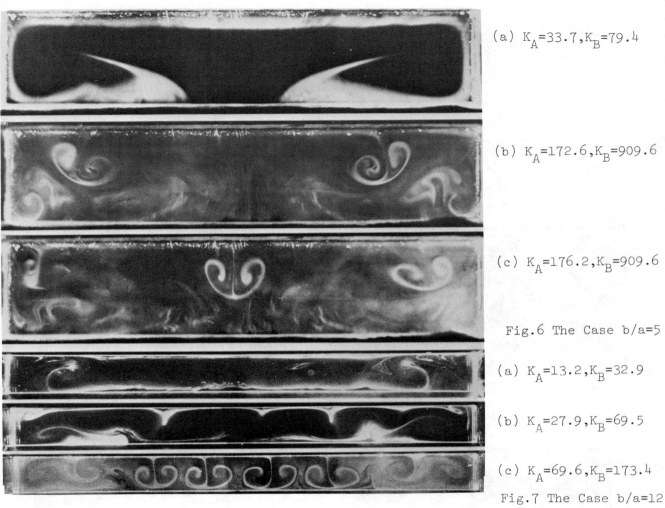

(a) $K_A=33.7, K_B=79.4$

(b) $K_A=172.6, K_B=909.6$

(c) $K_A=176.2, K_B=909.6$

Fig.6 The Case b/a=5

(a) $K_A=13.2, K_B=32.9$

(b) $K_A=27.9, K_B=69.5$

(c) $K_A=69.6, K_B=173.4$

Fig.7 The Case b/a=12

Fig.8 Critical Dean Number K_A

Fig.9 Critical Dean Number K_B

TUFT AND WALL
TRACING METHODS

VISUALIZATION OF GAS FLOW IN NATURAL CIRCULATION BOILERS

M. OBATA,* S. MIYAO,** K. KURATA,*** and K. KUSAKARI****

In order to confirm the gas flow characteristics of the basic design used for coal-fired boilers, air flow visualization tests were performed with a scale down model.

The tuft line method was applied to visualize the flow direction of the air flow pass through the furnace and convection heat transfer passes of the unit. The velocity field was measured by means of hot wire anemometer.

The overall gas distribution and velocity patterns obtained from these tests enable us to establish gas-starved areas, pressure drops and protection against ash erosion problems.

Introduction

In the field of coal-fired boilers, a better method to confirm gas flow characteristics of their basic design, such as the overall gas distribution, velocity pattern and draft losses associated with ash erosion problems, especially in heat recovery area, will make use of flow model tests applying flow visualization techniques because of their complexity in geometries.

Flow pattern tests for industrial use will be required to get the results easily and rapidly in allowable accuracy. For this purpose, the tuft method will be considered to be one of the more suitable methods; this method is fit for easy observation of flow patterns and because it does not require special photographic techniques and equipments, it is often applied to both air and water as a flow medium.

It is the object of this paper to describe the results of our latest model tests for a coal-fired boiler (Figure 1) as a possible example of the flow visualization by the air tuft method, and to show how gas flow patterns of the basic design can be confirmed.

Nomenclature

a = Sectional area of throat
G = Momentum flux
L = Reference length

K = Constant
M = Mass flow rate
ΔP = Pressure drop

---

* Reseach Engineer, Turbomachinery Research Dept., Research Institute, Ishikawajima-Harima Heavy Industries Co., Ltd., Koto-ku, Tokyo, Japan

** Design Engineer, Boilers Basic Design Dept., Energy Plant Division, Ishikawajima-Harima Heavy Industries Co., Ltd., Koto-ku, Tokyo, Japan

*** Deputy Manager, Turbomachinery Research & Development Depts.,Research Institute, Ishikawajima-Harima Heavy Industries Co., Ltd., Koto-ku, Tokyo, Japan

**** Manager, Turbomachinery Research Dept., Research Institute, Ishikawajima-Hariam Heavy Industries Co., Ltd., Koto-ku, Tokyo, Japan

Q = Volume flow rate
V = Reference velocity
X = Distance from imaginery top of jet

σ = Velocity deviation
μ = Viscosity coefficient
ρ = Fluid density

Suffixes

A = Actual boiler condition
M = Model or test condition
o = Flush
T = Total

mean = Mean
RH = Reheater side
SH = Superheater side

Consideration of Similarity

In the aerodynamic field, it is well known that the similarity of flow pattern between a model and an actual boiler is obtained by applying the equivalent Reynolds number (defined by eq.1) of the actual flow to the model under a geometrical similarity,

$$Re = \frac{\rho VL}{\mu} \qquad (1)$$

It is also empirically known that the flow pattern does not change any more by increasing Reynolds number larger than a certain approximate value, 1×10^4. On boiler model, however, it is usually impossible to simulate exactly by the isothermal fluid flow as cold air to the gaseous flow with extreme variations of temperatures including combustion process. Consequently, in order to obtain the similarity of the flow patterns in the boiler as far as possible, other special treatments should be made to bank elements and burner ports.

Flow patterns in bank elements are strongly affected by their pressure losses and it is almost impossible to make model elements with the geometrical similarity because of their complexity. However, they can be simulated aerodynamically by applying the coinciding Euler number (defined by eq.2).

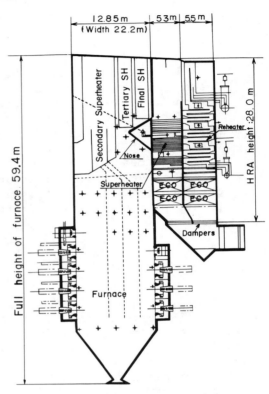

Fig. 1 660 MW Coal-fired Boiler

$$E = \frac{\Delta P}{\rho V^2} \qquad (2)$$

In the present model, all bank elements are replaced by plate fin-piled ones.

Flow patterns in the furnace become very complex by mixing process between induced swirling flows from burners and the inside hot gas flow. Therefore, the flow patterns of a furnace will be simulated by coinciding momentum fluxes for two flows. This is achieved by modifying the diameter of the burner throat of the model in accordance with the following formulae,

$$M = KX\sqrt{G\rho_x} \qquad M_o = \sqrt{G\rho_o a} \qquad (3)$$

Here will be summarized the main steps of the design procedure of the model.
(1) Decide a model size, flow medium, measuring method and fluid source in accordance with design data, requirements and purpose.
(2) Decide the arrangement, size and dimension of each heat transfer element and calculate its pressure losses by applying equation 2.
(3) Modify the burner throat ports by equation 3.
(4) Check Reynolds number of each part by equation 1.

This completes one approximation; if Reynolds number is lower than 1×10^4, or if there are other problems, these steps are repeated by new model size or fluid source to satisfaction.

Apparatus and Experimental Procedures

The arrangement of the test equipment used for the flow visualization experiments is presented schematically in Figure 2. The test set-up consisted of the air supplying and distributor system; the flow rate measuring system; the velocity measuring system, and the plastic test model.

The model was decided as 1/40 scale of the actual boiler and constructed on a frame work of visible acrilate plates or plastic resins for flow visualization tests. Figure 3 shows a sectional side elevation of the model and views through the furnace and convection sections with approximate dimension. The model represents one-half the full unit width because the actual boiler will be completely symmetrical. Among the equipments of the actual boiler, those which will give minor effects on flow pattens and gas velocity profiles, were omitted in designing the model. For nose section in the furnace, two different types, sharp and flat, were prepared.

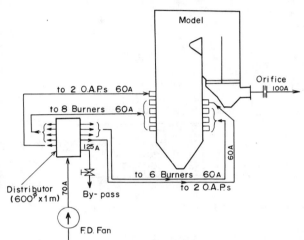

Fig. 2 Flow Diagram

One of the well-established methods of measuring flow patterns is the tuft method and the hot-wire anemometer method. Combination of both methods can be used to measure the magnitude and direction of flow velocities. Therefore, the tuft line method was applied to the flow visualization test. Tufts were made of silk fibers on a steel wire (0.8 mm dia.) lined with the pitch of 30 mm, and they were fixed in the test section during the flow visualization tests. Air velocities in the test section were measured with the miniature hot-wire probe.

The locations of tuft lines and velocity measuring points and those for static pressure measurements are shown in Figure 4 with the typical experimental conditions.

In the test, the firing conditions were achieved by changing the opening of dampers to make coincide the ratio of pressure drops for both passes of the model with that for the actual boiler, following the eqts.(4) and (5).

$$Q_T = Q_{SH} + Q_{RH} \qquad (4)$$

$$\frac{\Delta P_{RH}}{\Delta P_{SH}} = \frac{\Delta RH}{\Delta SH} \left(\frac{V_{RH}}{V_{SH}}\right)^2 = \left(\frac{L_{SH}}{L_{RH}}\right)^2 \left(\frac{Q_{RH}}{Q_{SH}}\right)^2 \qquad (5)$$

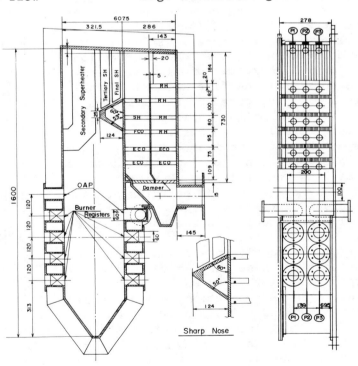

Fig. 3 Plastic Test Model

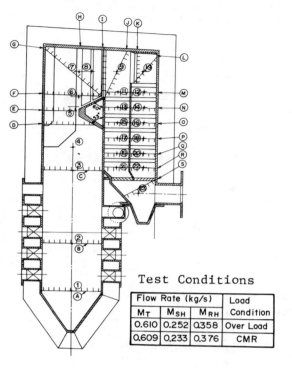

Test Conditions

Flow Rate (kg/s)			Load
M_T	M_{SH}	M_{RH}	Condition
0.610	0.252	0.358	Over Load
0.609	0.233	0.376	CMR

Fig. 4 Locations of Measurents

Test Results and Discussion

(1) Flow Patterns

Flow patterns observed from the tuft
lines were nearly the same for three planes
of the model – Ⓟ1 to Ⓟ3 in Figure 3.
Furthermore, no significant difference
appeared between the overload and cmr load
conditions. The result of flow patterns
in the model for the flat nose is shown in
Figure 5 with the observed tufts centre
line plane, Ⓟ2, as the typical flow pattern under overload and cmr load conditions of the actual boiler. Figure 6 shows
the typical patterns for the sharp nose, on
the cmr load condition and is similar to
Figure 5, except the flow patterns at the
upper slope of the nose section where a relatively large recirculating zone is recognized.

From the results presented it is apparent that the nose shape have little
effect on the flow patterns in the heat
recovery area (HRA) section, which reveals
the very good and uniform ones without
significant stagnation nor recirculation.

It is also clearly shown that the recirculating zone at the upper slop of the
nose becomes much smaller by means of the
minor change in the shape of the nose and

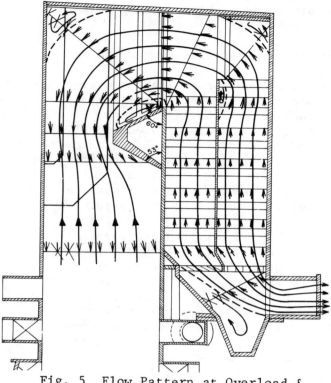

Fig. 5 Flow Pattern at Overload &
CMR Load (with Flat Nose in
Furnace)

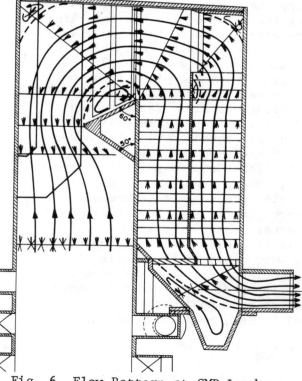

Fig. 6 Flow Pattern at CMR Load
(with Sharp Nose in Furnace)

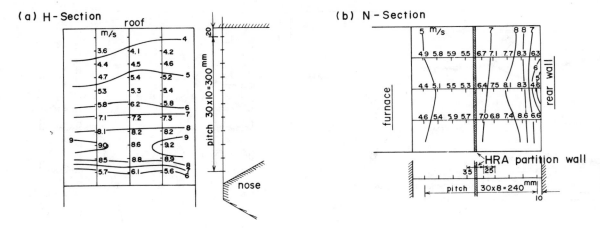

Fig. 7 Velocity Contour Map at CMR Load
(with Flat Nose in Furnace)

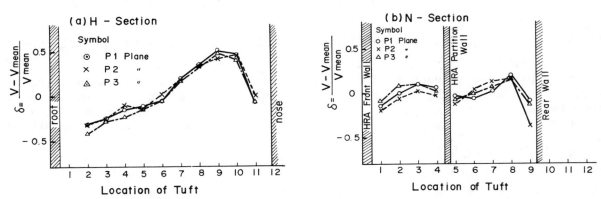

Fig. 8 Velocity Deviation of Figure 7

that flow patterns of the nose section are greatly improved.

A small recirculating zone, which was not stable but fluctuating in the test, is apparent at the inlet to the reheater behind the HRA partition wall screen. This recirculation may increase appreciably the velocity deviations and consequently ash erosion is likely to occur. Then, in actual boilers, in order to reduce the recirculating zone, the virtical reheater element of the HRA section was moved closer to the screen tubes of the partition wall.

(2) Velocity Distributions and Deviations

Typical velocity distributions data and their contour maps are shown in Figure 7 and their deviations in Figure 8. Where, the mean velocity at each cross section is taken as the arithmetic average value of the velocity distributions measured at the cross sectional plane. Inadequate zones, however, are excepted from the calculations; i.e. recirculating zone and boundary layers near walls.

From the results of measured velocity profiles and their deviations, it can be said that very good deviations were obtained from sections Ⓜ to Ⓡ in HRA passes, with maximum value of 25%. While, relatively higher values were appeared at the sections Ⓗ, Ⓘ and Ⓚ.

The maximum velocity associated with ash erosion problems at each section of the actual boiler will be estimated and confirmed by applying the present deviations to the average gas velocity, as follows the next relation:

(Average gas velocity in actual boiler) x (Max. value of deviation in model)
= (Estimated max. gas velocity in actual boiler)

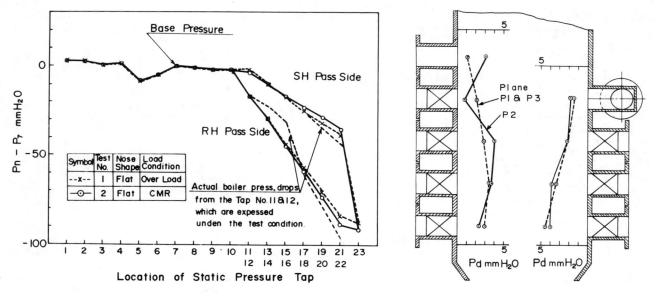

Fig. 9 Static Pressure Distribution   Fig. 10 Dynamic Pressure Distribution

(3) Static Pressure Distributions

The static pressure at each section of the model is confirmed as the difference of pressure drops based on the static pressure at the nose section shown in Figure 9, which approximately agree with those of the actual boiler. The conversion applied from the expected pressure drops of the actual boiler to those of test is as follows:

$$\Delta P_M = \Delta P_A \times \frac{K_M}{K_Z} \times \frac{\rho_M}{\rho_A} \left(\frac{V_M}{V_A}\right)^2 \qquad (6)$$

(4) Distribution of Flows Supplied from Burners and Over Air Ports

The flow distributions from the burners and the over air ports are presented by the dynamic pressure in Figure 10, which shows the confirmation of relatively uniform distribution of flow rates through them.

Conclusion

The flow similarity was confirmed satisfactorily through these tests and in applying the flat furnace nose to the actual boiler, we could obtain the satisfactory results in the flow patterns, velocity profiles and draft losses which would avoid the ash erosion.

The method and results obtained from these tests can be easily applied to the basic design of the same sort of coal-fired boilers.

Acknowledgements

This paper is given by kind permission of the Electricity Commission of New South Wales, Australia.

The authors express particular thanks to Mr. I. Kohi, the Associate Director of the Research Institute, IHI, for his help in preparing the paper and also acknowledge work of staff in the Turbomachinery Research Department.

Reference

(1) Thring, M. W. and Newby, M. P.: Combustion length of Enclosed Turbulent Jet Flames, 4th Symposium on Combustion, P. 789, Sept. 1952.

COMPUTER-AIDED FLOW VISUALIZATION

KOICHI OSHIMA*

Flow visualization used to be a technique in which some kinds of flow pattern at a certain moment is recorded and displayed on a paper, commonly a photographic paper. Now, we reconstruct this process into three steps; data acquisition, data reduction and data display, and each step is fully computerized using various kinds of hardwares and a new real time control system developed. Full discription of this system and some examples treated by this system are presented in this paper.

Introduction

In the system developed, mini- and micro-computer systems were fully utilized for flow visualization testing, which is considered to classify into three stages; the data acquisition, the data reduction and the data display stages. The whole system is supervised by a mini-computer, and in each stage of the flow visualization numerous micro-computers are used to control the various hardware systems. The software system was also developed, which is capable to operate several I/O devices in parallel and to accept system-interrupt in multi-level. This system is quite flexible and can be applied to any laboratory experiments, but here only a few examples of them will be shown.

The first one is the laser holographic interferometry of convective flow field, in which the data is recorded in a photographic plate as the image of phase contrast, and reproduced again as an optical image. During this reproducing process, the interference fringes are adjusted so that the fringe shift is directly related to the density deviation at a certain point, by a computer control. The second is the velocity field display of a vortex turbulent puff, which is a fluid blob bounded by a free boundary and has turbulent fluid motion in it. The data are acquired by common hot-wire apparatus, and the obtained raw data are computer-analysed to get various statistical characters of it, and displayed by the computer display system. The third is the flow field around an oscillating airfoil, the original data of which come from the numerical simulation of such flow field by a larger computer. The instantaneous stream lines and the equi-vorticity lines calculated by this mini-computer system are displayed on the display system. This saves the expensive large computer's CPU time, and also provides the flow patterns which are never obtained by any simple flow visualization methods.

---

* Professor; Institute of Space and Aeronautical Sciences, University of Tokyo. Komaba, Meguro-ku, Tokyo 153

System Description

In Fig.1, the computer system is shown, the main CPU is MELCOM 70 processor with 20 kwords core memory supported by two casett magnetic tape recorders, which serve as the mass memory as well as the system program storage. The system outputs are supplied to the four 12-bits DA converters, which are separately controlled and used for the visual diplays on the CRT tube and for the graphic presentation on the X-Y pen recorder. The CPU system accepts the 32-bits TTL level signals simultaneously, which come from the Kethley Digital Voltmeter, or from the two-channel high speed AD converter, which converts analog signals to 10-bits digital signals up to one conversion per 5 microsecond. The control signals for the digital voltmeter, that is, the strobe signal and the data-ready interrupt, are supplied to and from the CPU system, and those for the high speed AD converter, that is, the start of the data conversion, the start of read out to the CPU system, and all the related control signal of it, are also supported by the CPU system. Besides those, the CPU system controls the numerous switches, senses the many limit switches and accept the interrupt signals from the outside switches. Thus it supervises the total experimental sequence automatically. This drastically reduces the experimental timme and improves the reliability of the experiment.

In Fig.2, the ITV system is shown, which is fed TV signal from the ITV camera, stores it on the magnetic tape recorder. This TV signal is also stored in the image converter, which is controlled by the CPU system and reads out the image signal as the digital signal and modifies the image by write control. This TV system is wholly controlled by the micro-computer system, which is supervised by the CPU.

A new system program was developed, based on the real time monitoring program supplied by the manufacturer. This contains the numerous subroutines for the statistical treatments of the data and also the control programs of the every I/O devices. For the slow devices such as the pen-recorder display and the tele-type-writter, time sharing operation is provided. The real time control of the experimental event sequence is programmed by interrupt mode, thus it is capable to run simultaneously more than two experiments in parallel. Using the subroutine group for the I/O device control, the experimental program can be written down in Fortran language, but this point is not emphasized since it tends to limit the sys-tem flexibility. By this program, typically the correlation analysis, the fast Fourier transform or the conditional sampling of two time sequence data of 1000 are rather common job.

Fig.1 Main CPU and I/O devices   Fig.2 TV Contol System

Holographic Interferometry

It has been passed more than ten years since the laser holographic interferometry was introduced to the fluid-dynamicists, and the merit as well as the difficulty of it are well recognized. Therefore, only two points of them will be mentioned here. The one is; the laser light, if proper care is taken, has extremely long coherent optical path length. Then it can be interfere between two beams which pass through the different materials, such as air and water, or water and acryl plate. This gives the capability to measure the temperature fields in the transparent solid and in the surrounding atmosphere simultaneously. Fig.3 shows an interferogram of an acryl plate heated by a hot cylinder contacted on it. The natural convective flow around the cylinder is also visible on this picture. Also, combined with the three-dimensional observation capability of holography, the density fields in the solid plate and in the adjacent boundary layer are observable from a single photographic plate. Fig.4 shows such an example of the thermal boundary layer flows along an inclined acryl plate due to the buoyant force heated by the plate itself. The second point is; the interference fringe position is adjustable during the reconstruction process, that is, the zero fringe pattern can be oriented so that the fringe shift at a certain position is directly read out. Combined with the computer controlled devices, this process is fully automized, and is connected to the CPU system. Fig.3 is an example of such fringe shift adjusted to give a parallel, equally separated zero fringe, which gives the density profile of under-expanded jet.

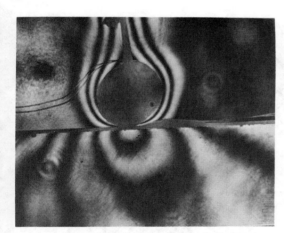

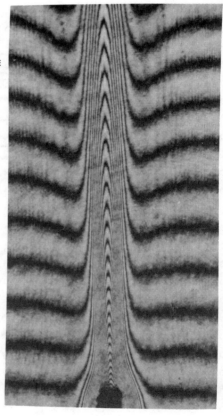

Fig.3 Interferogram of heated plate and buoyant flow

Fig.4 Thermal boundary layer along a heated plate

Fig.5 Interferogram of under-expanded jet.

Display of Turbulent Character

History of turbulent research is quite old, probably as old as the history of fluid mechanics itself. However, it is relatively recent that a break-through of it came around based on hot-wire measurement supported by sophisticated electronics and new data reduction technique, ensamble average and conditional sampling, supported by the powerful computer systems. Contrary to the conventional flow visualization, in which the stream lines or the streak lines are directly visualized as if it were seen, the values obtained in these data reductions are, for example, the correlation coefficients or the Fourier transformed coefficients, which may not be familier with our naked eyes. Since these are, of course, three- or four-dimensional, some kinds of visual display techniques are desirable to promote the understanding of the comp-lex turbulent characters. As an example of such problems, here the turbulent puff is taken, which is produced by instantaneously ejecting a finite amount of fluid mass into still atmosphere through an circular orifice covered by a turbulence-producing gauze. Fig.6 shows the conventional map presentation of the mean velocity field and the velocity fluctuation strength, which configured on the printed numbers. Note that these are ensamble averaged values which make it easy to see the overall characters of the field in spite of the fluctuating nature of the each event. Fig.7 is the various statistical parameters of the puff shown using the three-dimensional display, and it is seen, by this method, the characteristics of the puff are well revealed for easy understanding.

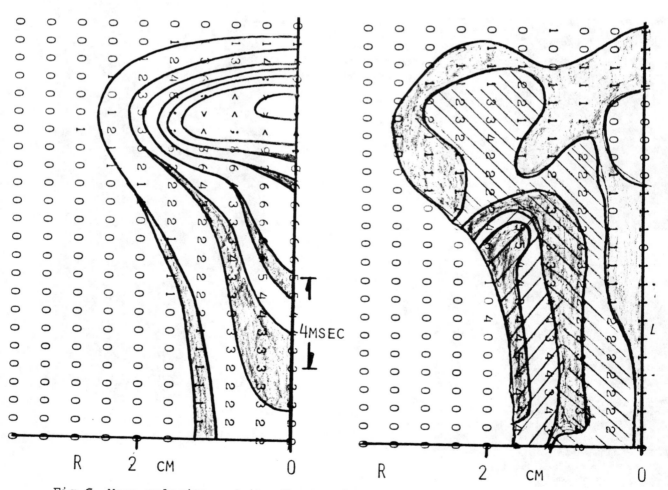

Fig.6 Mean velocity and its fluctuation distributions of the puff

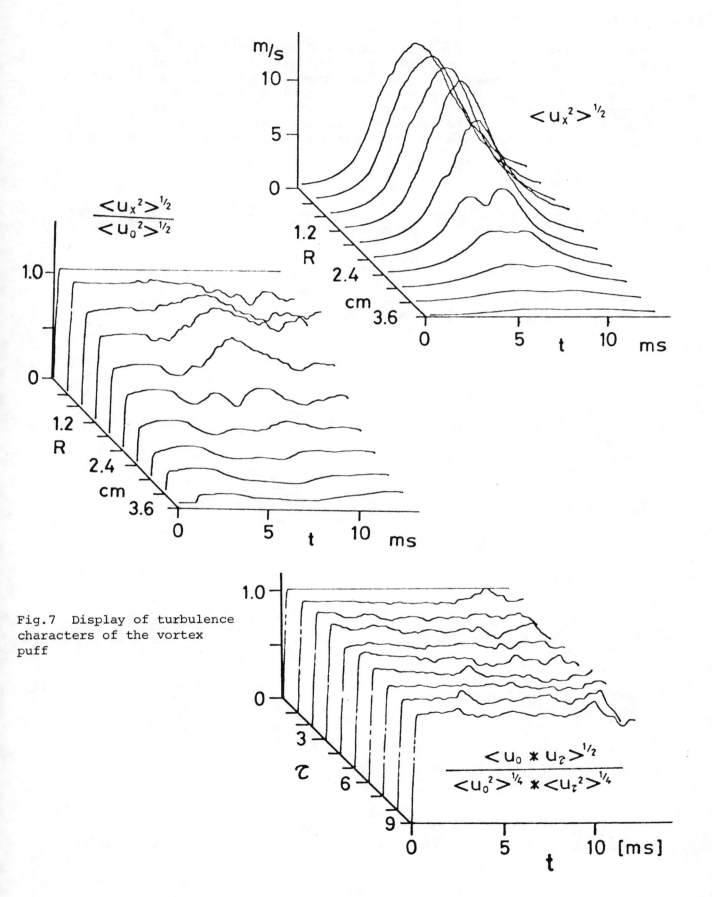

Fig.7 Display of turbulence characters of the vortex puff

Flow Field Around an Oscillating Airfoil

Fig.8 shows the instantaneous stream lines and equi-vorticity lines around an oscillating elliptic airfoil in time sequence. Through the computer aided data reduction, these lines are visualized. Otherwise, conventional visualization technique may not give such hypothetical lines as the instantaneous stream lines nor those directly connected to the velocity vector as the vorticity lines. Although the original data of those came from the computer simulation result, the conventional flow visualized pattern are also used as the original data and give the same useful pattern.

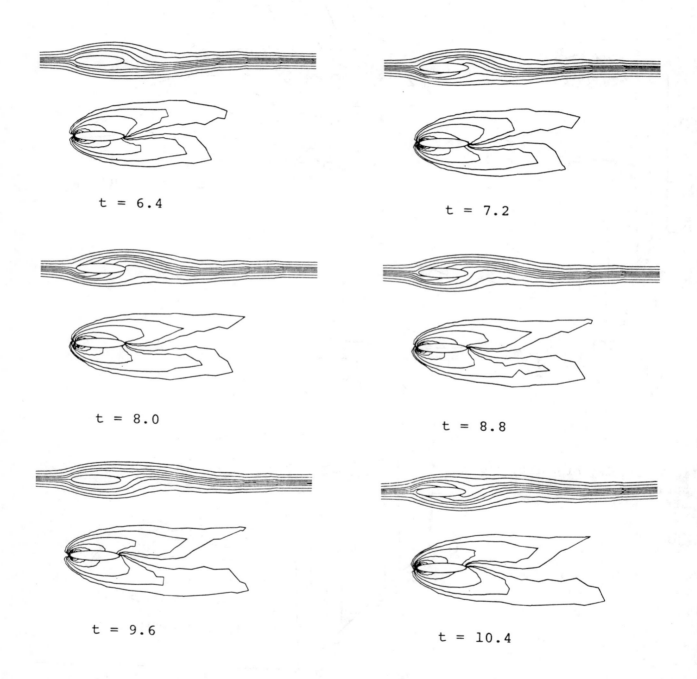

t = 6.4

t = 7.2

t = 8.0

t = 8.8

t = 9.6

t = 10.4

Fig.8 Stream lines and equi-vorticity lines around an oscillating airfoil

VISUALIZATION OF LONGITUDINAL VORTEX NEAR THE WALL BY VARIOUS METHODS

TETSUO TAGORI, KOMEI MASUNAGA, HISASHI OKAMOTO, and MASAMI SUZUKI

The visualization of longitudinal vortex near the wall was carried out, by the oil film method, soluble chemical film method, chemical film staining method, twin tufts method and tuft grid method, to compare each patterns and examine the accuracy of these methods. A flat plate was installed in the measuring part of circulating water channel as the wall, and its length was 1.8m. The longitudinal vortex was generated by the rotary guide vane (100mm diameter) or tigonal pyramid on the flat plate.

1. Introduction

Longitudinal vortices near the surface of body are frequently observed in various flows, around delta wings, blades of turbomachinery, building, hull of ship models, and in diffusers. In order to visualize these flows, many flow-visualization techniques had been adopted. These techniques were tuft gried, twin tufts, injecting tracer, oil film, soluble chemical film and chemical film staining method. Reports of the merits and demerits of them have apparently not been published to date. For this reason it may be necessary to reconsider the method which has been used.

A flat plate equipped with longitudinal vortex generator was installed in measuring part of a circulating water channel. Two kinds of the longitudinal vortices were generated by a rotary guide vane and a tigonal pyramid respectively. In this paper the visualization of them was carried out, in this channel, by the twin tufts method, tuft grid method, oil film method, soluble chemical film method and chemical film staining method,to examine the accuracy, responce, easiness and so on of these methods. The rotary guide vane generates the circular cross sectional longitudinal vortex without wake and the tigonal pyramid generates the elliptical cross section.

The results were shown by fig. 7, 8 and Table 1.

2. Apparatus

The circulating water channel used in this experiment is a horizontal type with open water measuring part (5.5m. in length, 1.5m.in width, and 1.1m. in water depth). A wooden flat plate was installed in the measuring part of this channel as a wall, and its length was 1.8m. The longitudinal

Department of Marine Engineering, Faculty of Engineering, University of Tokyo Hongo, Bunkyo-ku, Tokyo, 113, Japan.

vortex was generated by a rotary guide vane or a tigonal pyramid located near the flat plate. Fig. 1 shows the flat plate and the vortex generaters. The flow velocity in these experiments was 0.6, 0.9 and 1.2 m/s.

3 Methods and Results of Flow Visualization

3-1 Tuft Grid Method

Every tufts 50 mm long was attached on the grid of wires with 10 mm meshes. The meterial of tufts is nylon yarn. The grid was placed in normal to the measuring surface in the direction of the mean flow, and the tuft pattern was photographed from downstream. When the longitudinal vortex, generated by the rotary guide vane or the tigonal pyramid, was moving backward, we could express the facts as follows,

1' the center of vortex after installed tigonal pyramid separated from the body surface
2' the vortex by the tigonal pyramid became a circular far away from pyramid
3' the vortex by the rotary guide vane lead secondary vortex.

3-2 Twin Tufts Method

The twin tufts (Fig. 2) was installed like a matrix on the surface of the flat plate under investigation. The specific gravity of the tuft was prepared about 1. As the tufts were kept at some distance from the surface, the flow direction indicated by them may be different from the limiting stream line, by this method. The experimental results, showed in Table 1, is the average of ten photographs.

3-3 Oil Film Method

This is a quick and convenient technique as a detecting method of the separation and the reattachment lines if the flow field contains separated regions at the surface. When we used them in a flow from 0.5 m/s to 1.0 m/s, in the extreme case, this technique was not able to show correctly the pattern of the original flow. Because of that reason, we examined the kinds of pigments. Fig. 4 is an example of photograph. The result is given by Table 1.

3-4 Soluble Chemical Film Method

For this method, the benzoic acid film was used. A 100 per cent benzoic acid-acetone solution was sprayed on the surface of the flat plate by means of an ordinary paint spray-gun, with an air pressure from $6.5 kgf/cm^2$ to $7.5 kgf/cm^2$. This surface of the flat plate was covered with aceton-resisting paint. We adjusted the spraying pressure and nozzle distance in order to attain the fine grain surface. We could observe many finer streamline pattern on this surface. (Fig. 5)

3-5 Chemical Film Staining Method

The chemicals used in this experiment were the lead carbonate and the ammonium sulfide. The lead carbonate was powdered very fine, by grinding in a mortar. This lead carbonate was emulsified by adding thinner for oil paint. The surface of the flat plate was coated with the lead carbonate emulsion. An ammonium sulfide was injected into a stream near the surface of flat plate from a small ejecting tube (0.5mm inner diameter). After we tried to inject the ammonium sulfide from many direction to get good streaks. For example, this result is shown in Fig. 6 and Table 1. From the photograph (in Fig. 6), it could be observed, the streaks behind the tigonal pyramid were shorter than these of the rotary guide vane.

4 Comparison of Results

The experimental results which were obtained in the way described above are shown in Fig. 7, 8 and Table 1. Tuft grid was used as an supporting method. By this method, we could observe the cross section of longitudinal vortex near the wall. By twin tufts method, the correct direction of the limiting stream line could not be indicated, as the tuft is apart from the surface of the flat plate. The benzoic acid has low solubility and small roughness. The limiting streamlines were obtained by the soluble chemical film method, because this method was not affected by the gravity, viscosity and roughness of the surface.

The oil film method was affected by the density, viscosity, prescription, painting, flow velocity and water-temprature. The oleic acid is often used paint as an additive, dispersing agent, which controlls the size of the coagulating flocs. We obtained the following results from Fig. 3,
(1)   the response may be in inverse proportion to flow velocity, when the thickness and prescription is constant.
(2)   the larger the size of streak becomes, the thicker oil film on the wall is and the higher concentration of oreic acid in oil film paint is.
(3)   zonal painting method is given the influence by the buoyancy or gravity, when the oil drop is generated on the surface of the flat plate.
(4)   angle of streaks makes 30 - 50% errors sometimes.
(5)   the position of the separation line is downed by the weight of the pigment. The difference between streaks and limiting streamlines varies by position and is compricated. In this method the error of the flow pattern may very possibly be produced.

In the chemical film staining method, we cannot avoid the fact that the main flow is, to a certain degree, disturbed by the presence of the ejecting devices. The injected ammonium sulfiede diffused downstream ——— moved away from the surface, ———reattached to the surface, and made the change of color.

5 Conclusion

A flat plate was installed in measuring part of the circulating water channel as a wall. The longitudinal vortex was generated by the rotary guide vane or the tigonal pyramid located near the flat plate. The visualization of these flows was carried out by the tuft grid, twin tufts, oil film, soluble chemical film and chemical film staining method to examine the accuracy of these methods.

In effect, it was recognized that;
(1)   As to twin tufts method, the inclination angle of the tuft was smaller than the other method except the tigonal pyramid in 1.2m/s. Its responce was the quickest in all of the method.
(2)   At flow velocity from 0.5m/s to 1.0m/s, the oil film method depended strongly on the thickness and the prescription of its oil film paint. The deviations of the streak angle of oil film method from limiting streamline were about ±20 (tigonal pyramid), ±8(rotary guide vane) degrees.
For the chemical film staining method, the streaks behind the tigonal pyramid were shorter than these of guide vane. The in-

clination angle of streaks was smaller than limiting streamline, and it was thought that this deviation was due to the diffusion of injected ammonium sulfide.

6 References

(1) W. Merzkirch, Flow Visualization, (1966) Academic Press.
(2) T. Asanuma (ed), Hand-book of Flow Visualization (in Japanese), (1977) Asakura-Shoten.
(3) J. D. Main-Smith, Chemical Solids as Diffusible Coating Films for Visual Indications of Boundary-Layer Transition in Air and Water, Reports and Memoranda No.2755, February, (1950).
(4) T. Tagori, The 4th Symposium Flow Visualization (in Japanese), 7-12, July (1976).
(5) H. Murai, Journal of the J.S.M.E., (in Japanese), 104-110, (1971), Vol. 74, No. 634.

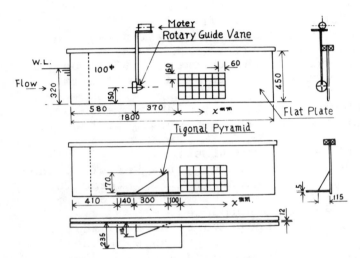

Fig 1, Arrangement of Flat Plate and Vortex Generators.

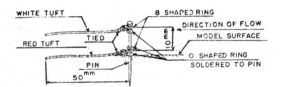

Fig 2 Mechanism of Twin Tuft

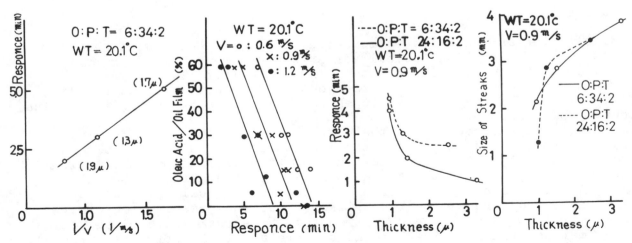

Fig 3 Results of relation among Thicknes of oil Film, Oil Film Responce, Size of Coagulating Flocs and Scale of Streak pattern

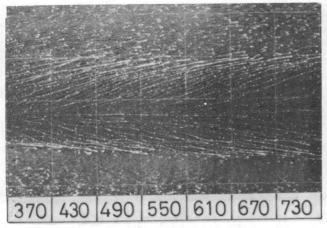

V=1.2 m/s Rotary Guide Vane (3.5 r p s)
O : P : T = 21:21:1   WT = 19.5°c

V=1.2 m/s Tigonal Pyramid
O:P:T = 24:16:1 WT 20°c

Fig 4   Pesult of oil Film Method

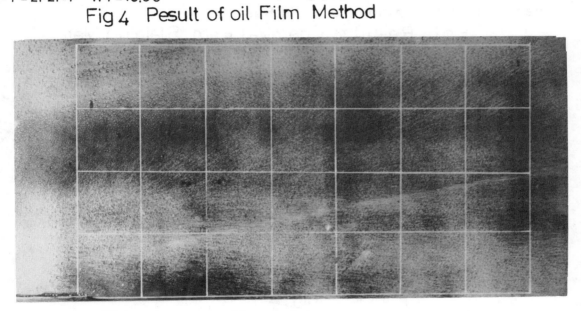

V = 0.6 m/s   Tigonal Pyramid
Fig 5 Soluble Chemical Film   method

Table 1

Flow Velocity (m/s)	Vortex Generator	Twin Tuft Method	Oil Film (O:Oleis Acid, P:Liquid Paraffin, R:Rouge, T:TiO₂, C:Carbon Black)					Soluble Chemical Film Aceton Benzoic Acid	Chemical Film Staining Lead Carbonate Ammonium Sulfide
			O:P:R 40:20:1	O:P:C 40:20:1	O:P:T 40:20:1	O:P:T 12:28:2	O:P:T 0:25:1		
0.6	Rotary Guide Vane (1.5 r.p.s.)	$\alpha=19°$ 10~30Hz	$\alpha=28.5°$ 20 min	$\alpha=20°$ 13 min	$\alpha=31°$ 10 min			$\alpha=24.5°$ 1 min	$\alpha=10°$ 12 sec
	Tigonal Pyramid	$\alpha=25°$ 10~30Hz		$\alpha=37°$ 7.5 min	$\alpha=32°$ 16 min	$\alpha=32°$ 23 min	$\alpha=38°$ 10 min	$\alpha=30°$ 50 sec	$\alpha=18°$ 11 Sec
0.9	Rotary Guide Vane (2.3 r.p.s.)	$\alpha=15°$ 10Hz	$\alpha=21°$ 16 min	$\alpha=22°$ 10 min	$\alpha=23°$ 5 min			$\alpha=23.5°$ 30 sec	$\alpha=15°$ 10 sec
	Tigonal Pyramid	$\alpha=24°$ 10Hz	$\alpha=32°$ 7.5 min	$\alpha=33°$ 8 min	$\alpha=35°$ 6.5 min	$\alpha=45°$ 15 min	$\alpha=23.5°$ 4.5 min	$\alpha=27.5°$ 30 sec	$\alpha=16°$ 10 sec
1.2	Rotary Guide Vane (3.5 r.p.s.)	$\alpha=18°$ 10Hz	$\alpha=27°$ 7. min	$\alpha=20°$ 4 min	$\alpha=22°$ 8.5 min			$\alpha=25°$ 30 sec	$\alpha=10°$ 10 sec
	Tigonal Pyramid	$\alpha=22°$ 10Hz	$\alpha=40°$ 8.5 min	$\alpha=28°$ 4 min	$\alpha=27°$ 3.5 min	$\alpha=30°$ 7 min	$\alpha=31°$ 2 min	$\alpha=20°$ 30 sec	$\alpha=32°$ 10 sec

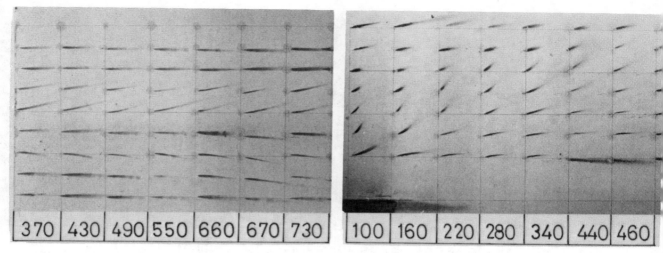

V=1.2 ᵐ/s Rotary Guide Vane (3.5rps)   V=1.2ᵐ/s   Tigonal Pyramid
Lead Carbonate   Ammonium Sulfide
Fig 6 Result of Chemical Film Staining Method

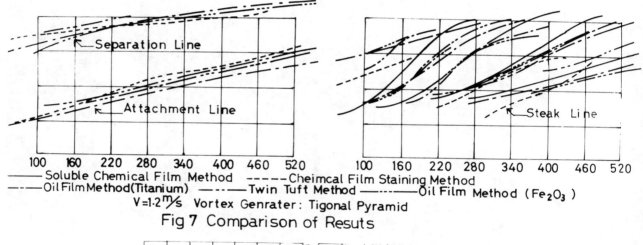

—— Soluble Chemical Film Method ----- Cheimcal Film Staining Method
---- Oil Film Method(Titanium) —··— Twin Tuft Method ——·——Oil Film Method (Fe₂O₃)
V=1.2ᵐ/s Vortex Genrater: Tigonal Pyramid
Fig 7 Comparison of Resuts

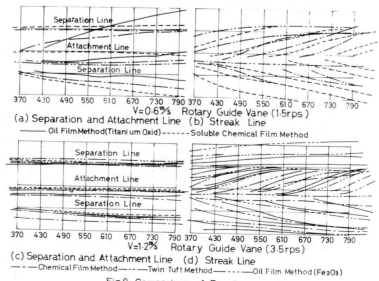

V=0.6ᵐ/s Rotary Guide Vane (1.5rps)
(a) Separation and Attachment Line (b) Streak Line
—— Oil Film Method(Titanium Oxid)------Soluble Chemical Film Method

V=1.2ᵐ/s Rotary Guide Vane (3.5rps)
(c) Separation and Attachment Line (d) Streak Line
———Chemical Film Method——·——Twin Tuft Method ——·—·—Oil Film Method (Fe₂O₃)
Fig 8. Comparison of Resuts

CHEMICAL REACTION AND ELECTRICAL CONTROL METHODS

THE ELECTROLYTIC PRECIPITATION METHOD OF FLOW VISUALIZATION

S. TANEDA,* H. HONJI,** and M. TATSUNO***

An electrochemical method of visualization of slow water flows is described. A new concept of integrated streaklines is proposed, which show the overall behavior of boundary layers. The method provides a proper means for visualizing the integrated streaklines in unsteady flows. Examples of flow patterns obtained by use of the method are presented.

Introduction

The purpose of this report is to describe an electrochemical method of flow visualization, and its relation to a concept of integrated streaklines in unsteady flows. First the concept of integrated streaklines is proposed. Next technical details of the visualization of the new streaklines are described in two separate sections. Final part of the report is concerned with case studies.

Integrated Streaklines

For simplicity, two-dimensional flows are considered. As is well known, streamline, streakline, and particle path do not coincide with each other in unsteady flows. Streamlines are lines to which the instantaneous direction of motion at any point is tangential, streaklines are the instantaneous loci of all fluid particles which have passed through certain points, and particle paths are the loci of certain fluid particles (Ref. 1). It should be noted here that in unsteady flows none of these three curves can show up the overall behavior of boundary layers. The only possible way to observe the behavior of unsteady boundary layers is to place tracer particles at the whole surface of the body.

Streaklines composed of all fluid particles which come out of the whole surface of a body differ from the usual streaklines, and may be called the 'integrated streaklines'. Here the 'whole surface' means a boundary contour of the body. The new streaklines show, in the vicinity of the body, the overall behavior of boundary layers, or equivalently the layers of vorticity formed on the body. Taneda, one of the authors, showed that such streaklines in a restricted sense play a fundamental role in studying unsteady flows (Ref. 2). Only these integrated streaklines do not miss any separation points which appear around the body. When the flow is steady, the integrated streaklines coincide with the usual streamlines, streaklines, and particle paths. The technique which will be described next is closely related to the integrated streaklines.

*Professor, **Assistant Professor, ***Research Associate, Research Institute for Applied Mechanics, Kyushu University, Fukuoka 812, Japan.

General Procedure of Visualization

It has been often noticed that faint colloidal smoke appears around a metallic body, when it is immersed in water. This phenomenon seems to be ascribed to a certain chemical process in the water. The process may be promoted by applying a voltage externally. The present method is based on the process just mentioned.

Electrolysis of water is made by applying a d-c voltage between electrodes, and the white colloidal smoke is generated on the whole surface of the anode. This smoke is used as a tracer material. The cathode may be located at any position where it does not disturb the flow. When the body is moved in the water, the smoke is shed downstream from the body. Streaks of the smoke are made visible by lighting up in a dark room. These smoke streaks show the integrated streaklines proposed previously.

Practically, the similar streaklines have been observed by placing dye smoothly on the surface of the body and allowing it to be shed. The present technique has the advantage of feasibility of controlling the tracer material electrically. It should be emphasized that the smoke visualizing the integrated streaklines must be discharged from the whole surface of the body.

Apparatus and Electrochemical System

A diagram of the typical set-up of apparatus is shown in Fig. 1, which is almost self-explanatory. A horizontal sheet of the flow behind a test cylinder is illuminated by a projection lamp for photography from top of the cylinder. A test body which is used as the anode may be of any shape. It can be made directly one of metals such as copper, iron, lead, tin, brass, and solder. In most cases, however, the body is made of brass, and the surface of its sectional portion is coated with solder smoothly. The other part of the surface is coated with an insulation. This is because solder is the most effective metal for smoke generation. Tin is also a good material, but relatively inferior are the other metals. When the body is made of non-conductors, its surface is wrapped up by a thin film of solder. In any case, the smoke comes out from the surface part which is exposed directly to the water.

For best photography, a quantity of the smoke to be produced must be controlled carefully. Electric current is a primary factor which affects the quantity of the smoke. This is a function of electrolyte concentration, the distance between electrodes, the kind of metal used for the electrodes, shape of the electrodes, and the velocity of flow.

In the present work, most of the experiments were made at velocities between about 0.1 and about 5 cm/s. Normal water at room temperatures was used as a working fluid. The distance between electrodes ranged from about 10 to 200 cm in towing tanks with the length of 2 or 6 m. For illumination of the flow field, a 1 kW slide projector provided the sufficient intensity of light focused with a lens. Usually, ASA 200 films and a 35 mm camera were used; the camera was mounted on a carriage of the test body or on a tripod at rest with respect to the tank. The exposure time was normally between 1/30 and 1 s. The voltage applied was less than about 30 V, and the current less than about 60 mA in most cases; 5 V and 10 mA were sufficient in some cases. In hard water no electrolyte additives were needed. In soft water sodium chloride has been found a satisfactory electrolyte.

The current must be increased to obtain the sufficient smoke density as the velocity is increased. When the velocity exceeds 5 cm/s or so, however, it is difficult to produce sufficient quantity of the smoke even if the ample electrolyte is added. Thus, the application of the technique is limited to the visualization of slow flows of water. Aging of the electrodes due to a chemical corrosion of the surface occurs in the course of repeated operations, and cleaning or renewal of the surface of the test body is often needed. In steady flow, however, this corrosion itself marks the surface of the body with certain lines. These lines sometimes provide useful information on the flow.

An experiment was made to test the current dependence of the rate of smoke production in a test tube. The relative transparency of a beam of light was measured by using a photo-transducer assembly. The decrease of the optical transparency means the increase of the quantity of the smoke produced. An example of the result is given in Fig. 2, which shows that the transparency decreases monotonically with the increase of the current.

The white smoke which is generated from the anode and precipitated in the water is composed of granulated metallic salt. The smoke particles were observed with a microscope, and it was found that the shape of each particle is approximately spherical. The average diameter of the particle was about 1 μm, when a solder rod was used with the current of 15 mA for instance. The smoke precipitated is completely opaque in the water. The composite particles reflect the projected light effectively and a striking contrast for photography is attainable. The present method may be called the electrolytic precipitation method in consideration of the procedure so far described.

NaCl is the most useful additive; NH_4NO_3 is useful as well if the anode is made of tin. Quantities of the additives should be determined by considering the current required. The velocity of sedimentation of each individual particle is less than about 0.001 cm/s. In still water, the particles generated from the anode tend to gather into lumps, each of which can have the larger velocity. In flowing water, however, no lumps are formed and the particle sedimentation has no appreciable influences on the flow visualization.

Examples of Flows

The following notation is used; U is the velocity of a test body, d, the diameter or the span of the body, T, the time elapsed from starting of motion, T_e, the exposure time for photography, and R (= Ud/ν), the Reynolds number with ν the kinematic viscosity. Representative pictures taken by using the technique are presented in Figs. 3 to 7.

Figure 3(a) shows the steady flow behind a circular cylinder. The twin vortices are clearly visualized by the white smoke streaks which show the integrated streaklines. The streamline pattern visualized with aluminum particles is superposed. The dead water regions visualized with these two different methods agree approximately with each other. Figure 3(b) shows the flow immediately after the acceleration of the cylinder. The streamline pattern which shows the instantaneous velocity field is completely changed to show a newly established potential flow, while a certain trace of the initial vortex layers is still preserved in the integrated streakline pattern. This indicates that the streamline pattern and the integrated streakline pattern are complements each of the other when studying unsteady flows.

Figure 4 shows a starting flow past a 90° flat plate. Rolling up of the integrated streakline behind the plate is clearly seen. As is usual with all the presented pictures, the smoke is shed from the whole surface of the plate. The present technique is used advantageously to visualize starting flows without disturbing a working fluid before the starting of motion of a test body.

Figure 5 shows a regular Kármán vortex street behind a circular cylinder moving at a constant velocity. A similar pattern can be observed by means of the so called condensed-milk method.

The technique need not necessarily be used in a towing tank. Figure 6 shows the vortex layer developed on the surface of a circular cylinder, which has started to make rotatory oscillations about its axis. The layer instability is due to the centrifugal force.

Figure 7 shows two successive patterns of the integrated streaklines shed from an elliptic cylinder, which is making back and forth oscillations in the direction

211

of the uniform stream. The integrated streaklines are seen clearly. It is not easy to pinpoint exactly their separation points. However, what is even more fundamental is the integrated streakline pattern itself, which shows up the overall behavior of the separated boundary layers.

Summary

The concept of the integrated streaklines is proposed, which are composed of all vorticity layers separated from the whole surface of a body. The integrated streakline patterns make it possible to observe the overall behavior of boundary layers and their separation points even when the flow is not steady. The electrolytic precipitation method provides a simple and proper means for visualizing the integrated streaklines.

Acknowledgements

We would like to express our sincere thanks to Professor Sir James Lighthill for helpful discussions. The assistance of Mr H. Amamoto and Mr K. Ishi-i is also gratefully acknowledged.

References

(1) P. Bradshaw, Experimental Fluid Mechanics, 2nd ed.,(1970),147, Pergamon Press.
(2) S. Taneda, Visual Study of Unsteady Separated Flows around Bodies, Progress in Aeronautical Sciences, 17-4 (1977), Pergamon Press; to be published.

Figures

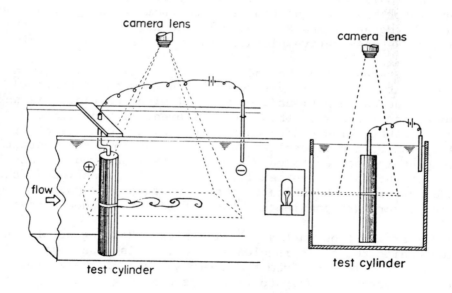

Fig. 1 Schematic of a typical experimental apparatus as installed on a water tank.

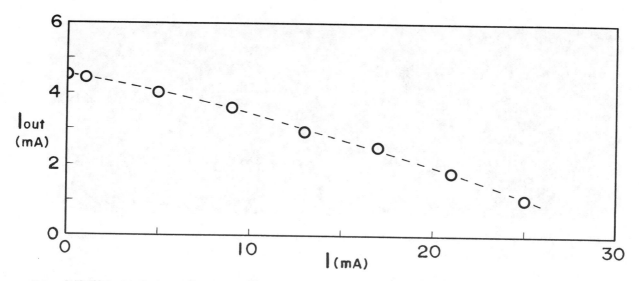

Fig.2 Decrease of optical transparency of a working fluid (Water 70 cc with additive of salt 0.35 g) with the increase of current applied. The anode was made of tin.

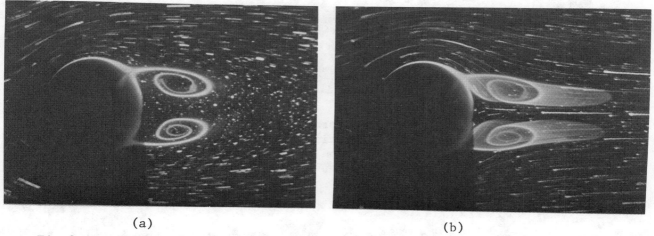

(a)   (b)

Fig.3 Variation of flow when a brass cylinder coated with solder is accelerated impulsively from U = 0.21 cm/s to 2.04 cm/s. R is changed from 17.7 to 174, T = 0.3 s.

Fig.4 Flow behind a flat brass plate started impulsively from rest. U = 0.23 cm/s, d = 5.0 cm, T = 18.7 s, T_e = 1/2 s, and R = 88.0.

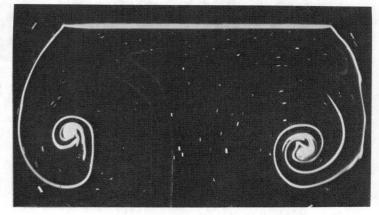

Fig.5 Vortex wake of a brass cylinder coated with solder. U = 2.14 cm/s, d = 0.50 cm, T_e = 1/15 s, Applied d-c voltage = 5 V, and current = 10 mA, R = 117.

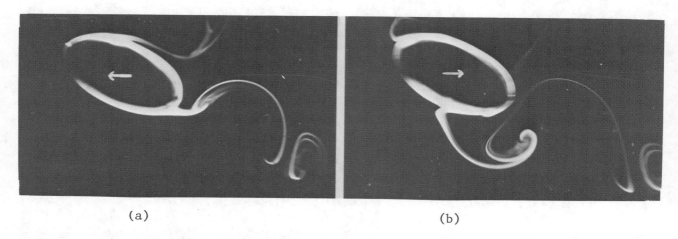

Fig.6 A series of columnar vortices formed on the surface of a brass cylinder making rotatory oscillations. Oscillation frequency = 0.25 Hz, Amplitude = 2 π , T = 7.8 s, and d = 1.0 cm.

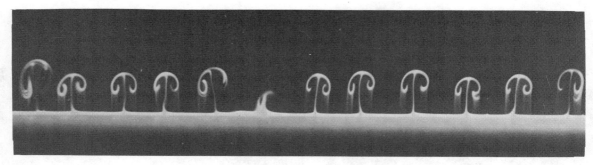

(a)   (b)

Fig.7 Flow around an elliptic cylinder making back and forth oscillations in uniform stream. The cylinder is made of plastics, and wrapped up by a thin film of solder. Length of major axis = 3.0 cm, Length of minor axis = 1.5 cm, Oscillation frequency = 0.1 Hz, Amplitude = 1.0 cm, Angle of attack = 30°, U = 0.48 cm/s. Distance displaced by the cylinder from its mean position for (a) = 0.56 cm upstream, and for (b) = 0.16 cm downstream. Arrows indicate the direction of displacement.

SOME REMARKS ON HYDROGEN BUBBLE TECHNIQUE FOR LOW SPEED WATER FLOWS

T. MATSUI,* H. NAGATA,** and H. YASUDA***

A fine tungsten wire of 5 to 10µ diameter was used to generate hydrogen bubbles, and a proper electric potential pulse
was applied between the wire as a cathode and an anode plate.
Thus, very fine bubbles can be generated to trace the water
flow at very low speed accurately, the maximum speed being
about 5 mm/s. A new cross-line type generating wire can be applied to several cases of low speed flows of water to observe
or to measure stream-lines, velocity and even vorticity by the
aid of photographs of crossing time-lines.

Some troubles due to the wake of a generating wire are
discussed.

1. Introduction

Hydrogen bubble technique is of great advantage to visualize water flows. In
this technique, small hydrogen bubbles generated by the electrolysis of water are
used as flow tracers. However, in low
speed water flows, the bubbles do not
trace the flows accurately because of
buoyancy. So, the technique has never
been applied to the water flows at a
speed lower than 10 mm/s.(Ref.1). When
boundary layers or wakes are to be investigated, however, it is required to
extend the application of this technique to the low speed water flows in
those regions. Studies of low speed water flows by the hydrogen bubble technique have scarcely been done except
some attempts by the present authors(
Ref.2,3).

In general, the ratio of the buoyancy to the drag working on bubbles is
considered to become smaller for smaller bubbles. Therefore, very small bubbles are expected to trace the flow accurately. In the present study, the

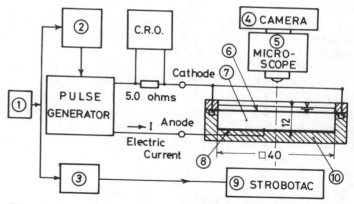

Fig.1 General layout of the experimental apparatus used for the observation of bubbles
generating from the surface of the wire.
1)Starting switch 2)Time control unit for
pulse generator 3)Delay time control unit
6)Generating wire 7)Working fluid (pH=4.0)
8)Anode plate 10)Small water tank

*Professor, Faculty of Engineering, Gifu University, Kakamigahara, Gifu 504,
  Japan
**Assistant, Faculty of Engineering, Gifu University
***Engineer, Pacific Industries Co., Ltd.

conditions under which as small bubbles as possible are generated were investigated by means of microscopic observation of processes of formation and evolution of bubbles on the surface of a generating wire. On the basis of the results, very small bubbles could be generated to trace the water flow at very low speed accurately, the maximum speed being about 5 mm/s. Several examples of observation or measurement of stream-lines, velocity, and even vorticity in low speed flows are shown.

Some troubles due to the wake of a generating wire are discussed when the wire is set in a low speed shear flow.

2. Behaviours of bubbles generated on the surface of a wire

2.1 Method of observation

The general arrangement of the apparatus is shown in Fig.1. A generating wire(6) is set beneath the free surface of the water in a small water tank(10). The appropriate electric potential pulse is applied between the wire as a cathode and an anode plate(8). The behaviours of bubbles evolved and detached from the surface of the wire are observed being enlarged by a microscope(5) and images of bubbles at 40 magnification were taken on films by a camera(4). A short electric pulse is generated at the moment when a starting switch(1) is closed as seen in Fig.2. After the delay time T_D, an electronic strobotac(9) flashes, and the conditions of bubbles at the moment are observed. The pulth width T, and the delay time T_D can be adjusted with an accuracy of ±0.1 ms. by the control circuits(2) and(3). As a working fluid, the buffer solution(7), potassium biphthalate($KHC_6H_4(COO)_2$) in pure distilled water, was used to keep the solution always at pH 4.0. In order to keep the surface of the electrodes clean, a small piece of stainless steel was used as an anode, and much care was taken to remove dirt oxide from the generating wire.

2.2 Evolution of bubbles

The conditions of bubble generation depend on many factors, i.e., the diameter d_w and material of the generating wire, the physical and chemical properties of working fluid, electric conditions such as feeding current, pulse width T, etc., and the type of flow. Among those factors, in practice, the electric conditions and the diameter and material of the generating wire can be selected rather at will. In the following sections, the relations between the conditions of evolution of bubbles and these two factors are discussed.

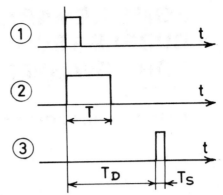

Fig.2 Time diagram.
1)Signal of the starting trigger
2)Pulse width T
3)Delay time T_D and flashing duration of strobotac T_S

(a) $T_D = 5$ ms.

(b) $T_D = 40$ ms.

(c) $T_D = 100$ ms.

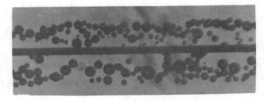

(d) $T_D = 200$ ms.

Fig.3 Process of evolution of bubbles generated from the surface of platinum wire. (setting position of the wire:2 mm below the surface of the water, d_w=30μ, T=4 ms., I_ρ=2.5 A/cm^2)

2.2.1 Diameter of the generating wire

Fig.3 shows the process of evolution of bubbles when a platinum generating wire of 30μ diameter was set 2 mm below the surface of the water. As soon as the electric potential pulse was fed, extremely small bubbles were generated densely on the surface of the wire as seen in Fig.3(a). After that, as seen in Fig.3(b) to (d), neighbouring bubbles are combined into larger ones in course of time, the number of bubbles decreased and their sizes were not uniform, being 0.5 to 1.5 times the diameter of the wire.

Fig.4 shows the behaviour of bubbles generated from a tungsten wire of 10μ diameter, which is thinner than the one mentioned above. The bubbles were generated near the wire, distributing with space between them, and they did not coalesce. Their size was fairly uniform.

Thus, in a low speed flow a sufficiently thin generating wire should be employed and the proper electric potential pulse should be applied between the electrodes. Then much sharper time-lines will be seen. A tungsten wire of 5 to 10μ diameter was used in the following experiments.

2.2.2 Electric conditions

The diameter of the bubbles can also be controlled by the electric conditions, that is, it will be smaller as the current density I_ρ, which is the current intensity per unit area of the surface of the generating wire, is smaller and the pulse width T is shorter. Thus, bubbles of smaller diameter than that of the wire can be generated under the appropriate electric condition. Fig.5 shows an example of these bubbles. Next, we tried to find the best electric conditions for each flow, considering the experiences in arriving at the results mentioned above. The range of the electric conditions thus obtained were as follows: pulse width of 1 to 4 ms., pulse voltage of 100 to 600 V, and pulse interval of 5 to 10 ms.

3. Arrangements of the generating wire

By the usual application of the short electric potential pulse to a single generating wire, the streak-lines, the path-lines and the stream-lines cannot be observed directly, even though the time-lines are observed. In order to observe these lines at the same time, some additional marks must be put on time-lines. For this purpose, a generating wire such as a kinked wire or a ladder type wire have been conventionally used(Ref. 4,5). In practice, however, it is not easy to prepare these kinds of wires.

In our experiments, a set of cross-line type wires or a single straight line type wire was used as a generating wire. A uniform flow visualized by means of these two types of wire is shown in Fig.6(a) and (b). As shown in Fig.6(a), each time-line generating from the cross-line type wire crosses the other, and divides the flow field into many quadrilateral flow elements. Corresponding to the aim of the experiment, the size of these elements can be adjusted by changing the pulse frequency for a given flow speed. Since intersecting points of these time-lines are special points on time-lines, streak-lines and path-lines can be known as well as time-lines by connecting

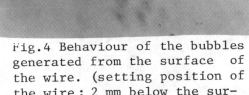

Fig.4 Behaviour of the bubbles generated from the surface of the wire. (setting position of the wire: 2 mm below the surface of the water, T_D= 20 ms., T=2 ms., I_ρ=3.5 A/cm², d_w=10μ)

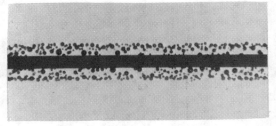

Fig.5 Behaviour of the bubbles generated from the surface of platinum wire. (setting position of the wire: 0.5 mm below the surface of the water, d_w=50μ, T=2 ms., T_D=20 ms., I_ρ=3.2 A/cm²)

the intersecting points. The new type of generating wire is very simple and easy to use as compared with a kinked wire or a ladder type wire.

Fig.6(b) shows the pattern of a uniform flow visualized by a single straight line type wire. As seen in the photograph, each path-line of bubbles can be seen clearly by using a sufficiently fine generating wire, by selecting the proper electric conditions and by taking a long exposure.

4. Observation of low speed water flows

4.1 Velocity profile

The flow between two concentric cylinders was visualized by the above mentioned hydrogen bubble technique, rotating the outer cylinder at the low peripheral speed U_0 of 5 to 20 mm/s., the inner one being at rest. Fig.7(a) shows an example of the flow pattern visualized by a single straight line type wire, and the length of the path-line is proportional to the velocity. Fig.7(b) shows the same flow visualized by a cross-line type wire, where the two generating wires form the letter V. The distance of the crossing points of each time-line is proportional to the velocity.

As will be seen in the following sections, the wire itself has much influence on the flow in low speed flows, especially in shear flows. Therefore, the single straight line type wire is preferable for measuring the velocity profile.

The velocity profile between the two cylinders was measured by a single straight line type wire. The results are shown in Fig.8, where the theoretical result(Ref.6) is also shown by a solid line. The measurements were made at a distance of at least

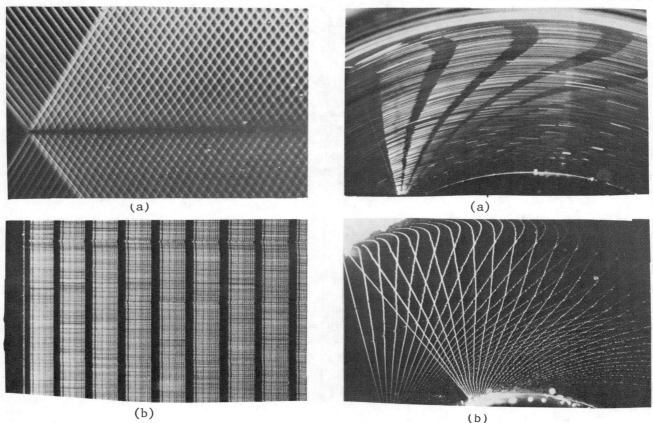

(a)

(a)

(b)

(b)

Fig.6 Uniform flow, U = 10 mm/s.
(a) Crossing time-lines by a cross line type wire. (b) Path-lines by a single straight line type wire.

Fig.7 Flow between two concentric circular cylinders, inner cylinder being stationary
(a) Single straight line type wire, U_0=5.4 mm/s. (b) Cross line type wire, U_0=11 mm/s.

more than about 1500 wire diameters downstream from the wire so that they are free from the effect of the velocity deffect in the wake of the wire itself. As shown in Fig.8, the velocity profile measured by this technique agrees well with the theoretical one, even when the peripheral velocity of the outer cylinder has the very low speed of 5.4 mm/s., which is the maximum in the flow field.

4.2 Flow elements

The wake of a flat plate parallel to a uniform flow is shown in Fig.9, being visualized by a cross-line type wire. The quadrilateral flow elements formed by crossing time-lines are seen in the figure. Thus, the rotation, extension, shear and translation of the flow elements can be directly obtained. If the rotating angle $\Delta\theta_1$ and $\Delta\theta_2$ in time Δt of the diagonals of the flow elements is measured, the value of vorticity can be determined as the sum of the angles divided by Δt, $(\Delta\theta_1+\Delta\theta_2)/\Delta t$.

4.3 Observation of stream-lines

A line connecting the diagonals of time-line quadrilaterals visualized by a cross-line type wire represents a stream-line, or a streak-line when the flow is steady. By measurement of the lines, the stream-lines of the flow between two concentric cylinders were obtained. The theoretical stream-lines are a family of concentric circles. But, measured stream-lines deviate from the theoretical ones. The same flow was visualized by the aluminium dust method and the stream-lines obtained from the trajectories of aluminium dust are shown in Fig.10 with small crosses. It is seen that the stream-lines thus obtained with no wire agree well with the theoretical line. On the other hand, when a single straight line type wire is set between the cylinders, the stream-lines past the wire, as represented by small circles in the figure, deviate toward the inner cylinder, that is, to the direction of decreasing velocity. The deviation of stream-lines increases with the increase in the diameter of the wire.

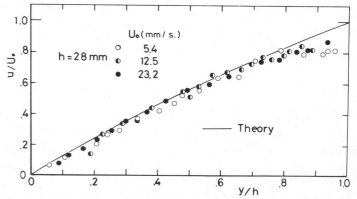

Fig.8 Velocity profile between two concentric circular cylinders determined by the hydrogen bubble technique using a single straight line type wire. (h: distance between the walls of the outer and inner cylinder, y: radial distance from the wall of the inner cylinder)

The pressure force acting on the tracer particle is the same for any particle. The centrifugal force acting on the tracer is proportional to its density. The pressure force and the centrifugal force are in equilibrium for water. For aluminium dust, the centrifugal force is greater than for water, and for hydrogen bubbles, it is smaller. However, the trajectories of aluminum dust with no wire show the same stream-lines as the theoretical ones. Thus, some other reason of the deviation of stream-lines should be considered. Thus, the effect of the density of tracers may be neglected. Therefore some other cause of the deviation of stream-lines should be considered.

The deviation of stream-lines toward the inner cylinder may be ex-

Fig.9 Wake of a flat plate parallel to a uniform flow, Re = Ul/ν = 900. (U:main flow velocity, l:length of a flat plate, ν:kinematic viscosity)

plained qualitatively as follows. A very fine generating wire in a low speed flow results in a Reynolds number flow lower than unity. In such a flow the width of the wake of the wire is fairly wide, and small hydrogen bubbles flow in the layer of lower velocity. In Fig.11 shown is a schematic drawing of the flow around the wire in a shear flow. The shear flow can be regarded as a mass of vortex filaments. The vortex filaments are deformed by being caught on the wire. By the deformation of the vortex filaments, the secondary flow is induced in the direction of the layer of lower velocity in the central layer of the wake as shown in Fig.11(b). Thus, the stream-lines deviate toward the inner cylinder at rest in the central layer of the wake of the lower flow velocity.

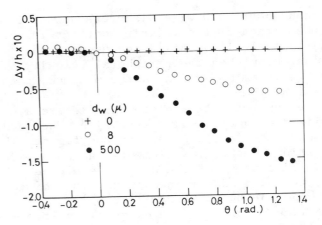

Fig.10 Deviation of the stream-line obtained from the trajectory of aluminium dust in the existence of a single straight line type wire, U_o=12 mm/s., y/h=0.52. (θ:azimuthal angle meassured from a generating wire on a radius)

5. Conclusion

The results of the present study of the hydrogen bubble technique are summarized as follows :

On application of hydrogen bubble technique, very small bubbles are generated with a fine tungsten generating wire of 5 to 10μ and very sharp time-lines are realized. Thus, the technique can be applied to a flow of lower velocity than 5 mm/s., and more information can be obtained in detail.

By using the cross-line type wire, quadrilateral flow elements can be visualized and their deformation and rotation can be measured directly.

When a fine generating wire is set across a shear flow of low velocity, stream-lines in the wake of the wire deviate toward the layer of lower velocity.

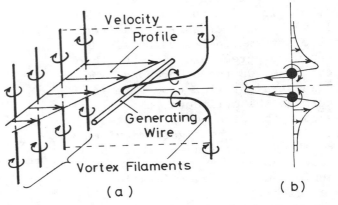

Fig.11 Explanation of inner deviation of stream-lines. (a) Defomation of vortex filaments (b) Induced velocity by a deformed filament in a wake flow.

Referances

(1) Taneda,S., KAGAKU, (in Japanese) 41-6 (1971),1714
(2) Nagata,H.,et al., Symposium on Flow Visualization (in Japanese), 1 (1973), 123, Institute of Space and Aeronautical Science, Univ. Tokyo, or Reserch Reports Gifu Univ. (in Japanese), 24 (1974), 89
(3) Nagata,H., Yasuda,H.,and Matsui,S., Turbomachinery (in Japanese), 5-2(1977), 5
(4) Clutter,D.W.,et al, Aerospace Engineering, 20 (1961), 24
(5) Schraub,F.A.,et al, Trans. ASME, Ser. D, 87 (1965), 429
(6) Schlichting,H., Boundary Layer Theory, (1968),80, McGraw-Hill.

LAGRANGIAN AND EULERIAN MEASUREMENT OF LARGE SCALE TURBULENCE BY FLOW VISUALIZING TECHNIQUES

T. UTAMI* and T. UENO*

Experimental studies are made about large scale structure of turbulent shear flow with the aim that the results can be applicable for the practical use in river hydraulics. Time sequence of flow patterns containing eddies on water surfaces and one in cross-sections are visualized by the camera moving with the mean velocity. Time variation of flow patterns containing low-speed streaks and bursts near the channel bed are also observed.

The results are used to compose the structual model of turbulent shear flows, by which developement of longitudinal spiral flows, formation of surface eddies and boils, and occurrence of bursts, sweeps and low-speed streaks are explained reasonably.

1.Introduction

During the last ten years coherent characteristics of large scale structure of turbulence have been made clear. Since these structures generally have a history of developement as they are convected downstream, observations at one or two spacially fixed stations will include realizations of a large number of structures at various stages of their life history. Time and space averages of such observations will tend to "smear out" their essential features (Ref. 1). Accordingly visual observations of Lagrangian method are supposed to be the most advantageous method to reveal these coherent characteristics.

By the way, the structual model of large scale structure of turbulent flow proposed by Kline et al (Ref. 2, 3) contains secondary flow. However, little explanation is given about the occurrence and developement of the secondary flow. On the other hand Kinoshita (Ref. 4) suggests the existence of parallel spiral flows with longitudinal axes in river flows on the basis of the aerial photogrammetry of flood waters. He also confirmed the existence of the spiral flows in experimental channels. The relation between large scale structure of turbulent flow and the secondary flow or the spiral flow have not ever mentioned in detail, but the secondary flow has rather been considered as a phase of mean flow.

The authors (Ref. 5, 6) suppose that the secondary flow is the same phenomenon with the spiral flow and this is a phase of large scale structure of turbulent shear flow. That is, they suppose that the structure of turbulent flow is composed of multiple phases which are primary vorticity, ∩-shaped vortex and the longitudinal spiral flow. They also suppose that the motion and transformation of each phase of the structure of turbulent flow can not be interpreted by statistically averaging method but can be interpreted by the dynamic. law, in which much attention is payed on the concentration and dissipation of vorticity.

* Research Assistant of Disaster Prevention Research Institute, Kyoto University, Gokanosho, Uji-City, Kyoto, JAPAN

2. Flow Patterns near the Channel Bed

Fig. 1 shows the velocity distribution visualized by hydrogen bubble method, in which electrode is settled across the flow. As observed in the figure, velocity distribution is not uniform but has some variation across the flow. The near-wall part of low-speed region is generally called low-speed streak (Ref. 2,3), where strong upward component of velocity is observed.

In order to examine the arrangement of low-speed streaks and their variatin with time, following experiment is made. Negative electrode for hydrogen bubble method is settled across the flow at the position of 0.5 cm above the channel bed. Photographs of flow patterns are taken every 1 second and resultant velocity distributions of every instant are arranged side by side as shown in Fig. 2, in which low-speed streaks are drawn with thick lines and longitudinal coodinate is time in seconds.

At a glance it seems that low-speed streaks occurs randomly and have little relation with longitudinal spiral motion whose diameter is said approximately equal to water depth. In fact, intervals between each streak become much less than water depth. However, it is noticed that there are some streaks which continue comparatively long and have intervals approximately equal to twice of water depth. These long streaks are considered to be in accordance with the longitudinal spiral motions.

3. Flow Patterns in Cross-Sections

Photographs of flow patterns in cross-sections are taken. The schematic view of the experimental facilities is shown in Fig. 3. A motor-driven camera, a mirror tilted by 45 degrees, electrodes for hydrogen bubble method, and two spotlights are all mounted on the carrying flame, which is moved downstream with the mean velocity. The spotlights are used so that only tracers in and near the cross-section passing the electrodes can be visualized. Disturbance of flow due to the mirror is considered little because it is moved with mean velocity. Examples of resultant photographs are shown in Fig. 4, in which time interval between each other is 0.6 second. In each photograph, the uppermost horizontal line is water surface and the lowermost one is channel bed. In some regions tracers occurring from two electrodes moves upward and in other regions downward, which suggests the existence of parallel spiral flows with longitudinal axes. We can observe some smaller rotating motions which have also longitudinal axes and larger velocity of rotation. These rotating motions are supposed to be some cross-sections of so-called ∩-shaped vortices. There exist some strong upward flows in narrow regions, which are considered to occur due to upward velocity component in low-speed streaks.

From these facts it is supposed that ∩-shaped vortices are lifted and stretched and streamwise parts of them drive the longitudinal spiral flow and a low-speed streak occurs due to a ∩-shaped vortex. At the same time, these ∩-shaped vortices are gathered together by the spiral motions as they move away from the channel bed. This is the reason why average interval between streak lines becomes larger as the distance from the channel bed becomes larger and at last the interval becomes approximately twice of water depth on water surface.

4. Flow Patterns on Water Surface

The method observing flow patterns with the moving camera have ever been adapted by Tietjens (Ref. 7). The authors adapt the method again to obtain Lagrangian time series of flow patterns. Some examples of photographs of flow patterns on water surface, which are taken every 6 seconds by the camera moving with mean velocity, are shown in Fig. 5. Punch trash is used as tracers. It is observed that flow patterns are composed of eddies which are arranged in longitudinal rows as schematically shown in Fig. 6. Because eddies in the same row have the same rotating direction and ones in neighboring rows have opposite directions, longitudinal velocity between the row a and the row b in Fig. 6 is small and one between b and c is large. It is also observed that in and near the longitudinal section passing TT' longitudinal component of velocity is small and there exists strong upward component of velocity and transversal distance between these low-speed regions is approximately twice of water depth. From these facts it is supposed that longitudinal spiral flows are developing as shown in Fig. 6.

Two dimensional distributions of vorticity are calculated with regard to the flow patterns of lower and left side quarter of each photograph of Fig. 5 and are shown in Fig. 7. These figures show that geometric characteristics of flow patterns have an correlation with the distribution of vorticity. That is, (1) the eddies with larger scale have larger value of vorticity, (2) the eddies arranged in the same longitudinal row have same signs of vorticity and ones in neighboring rows have opposite signs, and (3) the maximum or minimum value of vorticity is located in the neighborhood of the center of each eddy.

As for the time variation of flow patterns, the developement of eddy motions seem to have correlations with concentration or dissipation of vorticity. That is, (1) in the first step, a series of locally concentrated vorticity constitutes a band region which contains stream lines with large curvature or a chain of small eddies (e.g. upper and right part of Fig. 7(d)), (2) in the second step, this band region is transformed into arched form, and larger surface eddy with closed stream lines is formed (e.g. upper and middle part of Fig. 7(b)), (3) in the third step, the arched region developes into large mass of concentrated vorticity, and the surface eddy becomes larger and more intense (e.g. upper and right part of Fig. 7(e)), and (4) in the last step, the concentrated vorticity is diffused or dissipated and the eddy breaks into smaller, which again constitutes a link of the chain in the first step.

5. Structural Model of Turbulent Shear Flow

Turbulent shear flow is characterized by the existence of the shear layer which is located just near the channel bed and has large value of vorticity. This shear layer can be represented by vortex sheet as shown in Fig. 8(a). If a part of this layer is deformed slightly due to disturbance, vorticity is concentrated there (Ref. 8) and Ω-shaped vortex (Ref. 2,9,10) is formed as shown in Fig. 8(b),(c). This Ω-shaped vortex is lifted up according to Biot-Savart's law and at the same time stretched downstreamward due to the distribution of velocity of mean flow. In the narrow region between the legs of this vortex, as shown in Fig. 8(d),(e), upward flow is generated by the rotating motion of Ω-shaped vortex and at the same time upstreamward velocity component is also generated due to the primary vorticity which has been entrained on the Ω-shaped vortex. Thus low-speed streaks with upward flows or bursts containing in them are generated.

When the Ω-shaped vortex is lifted and stretched and upper part of it reaches near the water surface, strong upward flow due to its rotation forms a boil (Ref. 9) and a part of vortex near the water surface is stretched and dissipates. The rest part of the vortex forms surface eddy. These vortices are stretched more and more and at last the axes of them become almost longitudinal. Longitudinal spiral flows are driven by these vortices. The process mentioned above is shown schematically in Fig. 8(f).

References

1) Laufer,J.: New Trends in Experimental Turbulence Research, Annual Review of Fluid Mechanics, Vol.7, 1975, pp.307-326.
2) Kline, S.J., W.C. Reynolds, F.A. Schraub and P.W. Runstadler: The Structure of Turbulent Boundary Layer, Jour. Fluid Mech., Vol.30, 1967, pp.741-773.
3) Offen, G.R. and S.J. Kline: A Proposed Model of the Bursting Process in Turbulent Boundary Layers, Jour. Fluid Mech., Vol.70, 1975, pp.209-228.
4) Kinoshita, R.: An Analysis of the Movement of Flood Waters by Aerial Photography, Photographic Surveying, Vol.6, No.1, 1967, pp.1-17 (in Japanese).

5) Utami, T. and T. Ueno: Study on the Structure of Large Scale Turbulence by Flow Visualizing Method, Annuals, Disaster Prevention Research Institute, Kyoto University, Vol.19B, 1976, pp.267-288 (in Japanese).

6) Utami, T. and T. Ueno: Study on the Structure of Large Scale Turbulence by Flow Visualizing Method (2), Annuals, Disaster Prevention Research Institute, Kyoto University, Vol.20B, 1977 (in Japanese).

7) Prandtl, L. and O. Tietjens: Fundamentals of Hydro- and Aeromechanics, Vol.2, MacGraw-Hill, 1934.

8) Rosenhead, L.: The Formation of Vortices from a Surface of Discontinuity, Proc., Roy. Soc. London, A134, 1932, pp.170-192.

9) Ishihara, Y. and S. Yokosi: On the Structure of Turbulence in a River Flow, Annuals, Disaster Prevention Research Institute, Kyoto University, Vol.13B, 1970, pp.323-331 (in Japanese).

10) Theodorsen, Th.: The Structure of Turbulence, 50 Jahre Grenzschichtforschung, (ed. H. Görtler and W. Tollmien), Friedr. Vieweg und Sohn, 1955, p.55.

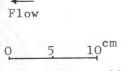

Fig. 1 Flow pattern near the channel bed. H_2 bubbles are in 10 Hz. (B=40cm, H=2.5cm, Q=0.5 l/s)

B; Channel Width
H; Water Depth
Q; Discharge

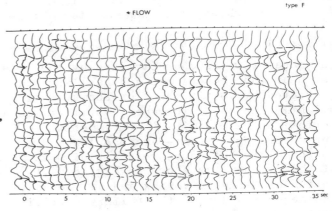

Fig. 2 Arrangement of low-speed streaks (B=40cm, H=2.5cm, Q=0.5 l/s)

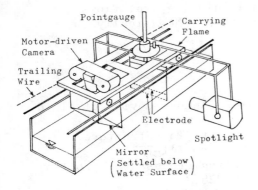

Fig. 3 Experimental facilities for the observation of flow patterns in crss-sections

224

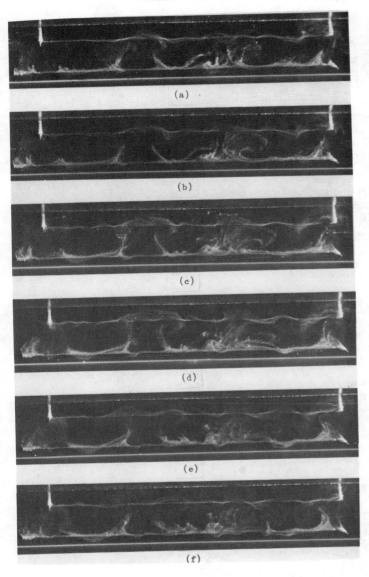

Fig. 4   Time sequence of flow patterns
         in cross-sections.
         (B=40cm, H=3.8cm, Q=0.82 l/s,
         Shutter=every 0.6sec)

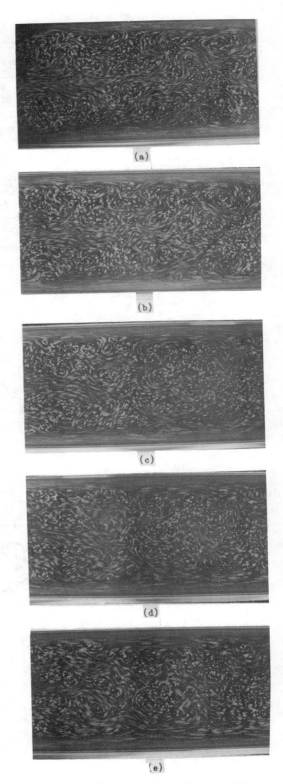

Fig. 5   Time sequence of flow patterns
         on water surface
         (B=40cm, H=5.85cm, Q=1.2 l/s,
         Shutter=every 6sec, Exposure=
         2sec)

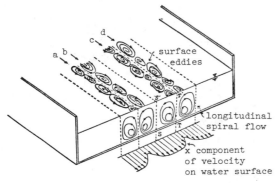

Fig. 6 Schematic view of structure
       of turbulent shear flow

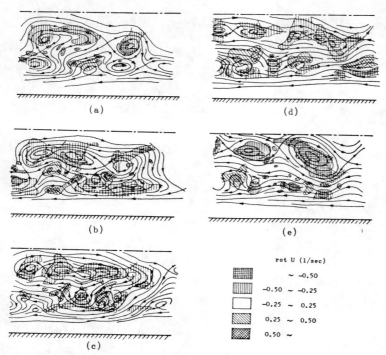

Fig. 7   Time sequence of distribution
of vorticity on water surface

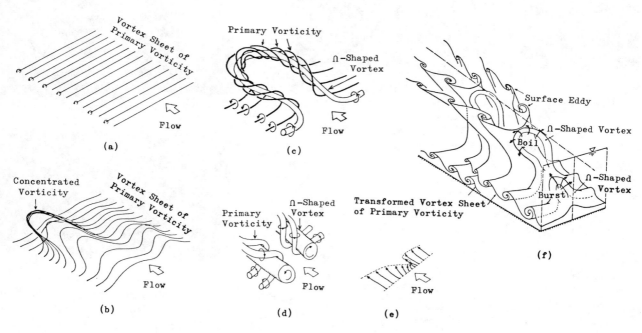

Fig. 8   Structual model of the coherent
structure of turbulent shear flow

ON THE MEASUREMENT OF FLOW VELOCITY BY MEANS OF SPARK TRACING METHOD

T. ASANUMA.* Y. TANIDA. and K. KURIHARA***

Spark tracing method has various merits. The most important one is its suitability in measuring the velocity of unsteady air flow. However, the error estimation of the method has not yet been thoroughly examined.

In the present study, some basic experiments were conducted in order to demonstrate how to estimate the errors due to diffusion effect and different density of ionized tracer in spark lines. By comparing the experimental data measured by spark tracing method with those obtained by pitot-tube or hot-wire, the various errors were discussed. Further, the spark tracing method was applied to the unsteady flow, and the flow behavior around the oscillating aerofoil was revealed quantitatively.

1. INTRODUCTION

The spark tracing method which ionizes and illuminates a portion of air by high frequency series of electric discharges suits not only for flow visualization itself but also for quantitative measurements of velocity in a wide flow range.

The application of the spark tracing method to quantitative flow analysis assumes that the ionized tracer should move with flow. There may be, however, some factors which cause measurement errors, e.g. a difference of the density between the ionized and ambient air, a diffusion between them, and so on. The errors due to these factors have not been examined thoroughly.

In the present study, the measurement errors which are peculiar to the spark tracing method will be discussed by fundamental experiments for channel flow, jet flow, and convergent flow. Further, applying this method to the case of an oscillating aerofoil, the unsteady flow will be analyzed from high-speed photographs.

2. BASIC EXPERIMENTS ON MEASUREMENT ERRORS

2.1 Experimental apparatus

In this experiment, two wind tunnels were used; that is, a low-speed wind tunnel which supplies a continuous flow of 50 m/s (max.), and a high-speed wind tunnel to which air is supplied from air tank of 15 kg/cm^2, the flow velocity being 280 m/s (max.). Test sections used were 20x40 mm rectangular duct for channel flow, circular

---

* Prof. of Faculty of Engg., Tokai Univ., Hiratsuka, JAPAN
** Prof. of Institute of Space and Aeronautical Science, Univ. of Tokyo, JAPAN
*** Associate Prof. of Metoropolitan College of Aeronautical Engg., Tokyo, JAPAN

nozzle of 20 mm dia. for jet flow, and a convergent rectangular duct with 20x50 mm exit section. Fig.1 shows the block diagram of spark discharge equipment used for this experiment (Ref.1). In the arrangement, the high-speed spark discharge apparatus "Strobokin" was used as the primary source of pulse transformer. The electric discharges can be performed by two sets of electrodes with discharge voltage $\sim$130 kV, distance of electrodes 20 to 35 mm, duration of spark $\sim$1 μs, and spark frequency 10^4 to 10^5, when the damping resistance in the spark gap circuit is fixed R_d= 10 kΩ.

2.2 Experimental results and considerations

2.2.1 Main errors in velocity measurement

When the successive spark lines of f Hz are taken on a photograph with the spark line distance δ mm (see Fig.2), a transport velocity of spark line (ionized tracer) is given by

$$U_t= \delta \cdot f/S \qquad (1)$$

where S is a scale factor on photograph. The spark frequency f is usually chosen to be much larger than 1 kHz from the view point of ion life span (Ref.2). The accuracies of S and f may be guaranteed enough, and can also be checked easily, so the accuracy of U_t is mostly dependent on the spark line distance (δ). The errors may be caused by a personal deviation in measuring due to the thickness of the spark lines and the uncertainty of the flow direction. These errors are not necessarily peculiar to the spark tracing method but rather common to any tracer method. In the following will be considered other two factors of error which are peculiar to the spark tracing method; the first is the effect of the diffusion of ionized tracer and the second is the effect of the density difference between the ionized tracer and the ambient air.

2.2.2 Effect of diffusion of ionized tracer

a) Flattening of distorted initial spark line: An initial spark line discharged between two electrodes is so unstable that it is usually not straight. We will first consider about which spark lines downstream of the electrodes should be used for obtaining the flow velocites from such a distorted initial spark line and consecutive ones in a uniform flow. Fig.2 shows a photograph of typical spark lines discharged by the needle type electrodes in the channel flow. By analyzing such ten photographs taken at the same conditions, the flow velocites on the center line of the channel were obtained at several positions downstream of the electrodes. The deviation of the flow velocity from the averaged value is shown in Fig.3, where U_0 is the averaged velocity, ΔU the root-mean-square value of the velocity deviation (U - U_0) and X the mean measurement position (A.B.C). X_A, X_B, and X_C range X = 0 to 7 mm, 10 to 17 mm, and 21 to 28 mm, respectively. Fig.3 shows that, irrespective of the flow velocity and the spark frequency, the deviation becomes greater for smaller X,e.g. the measurement error is over 2% for X < 10 mm. It is considered that the greater error in smaller X range is mainly due to the distortion of the initial spark line, and that as X increases the spark line becomes smooth and the deviation diminishes both by the repetition of the spark discharge and by the diffusion of the ionized tracer.

b) Short-circuiting of spark line: The spark line is bent strongly in the boundary layer on the wall and at the boundary of free jet. As the spark line is washed downstream, it becomes thicker due to the diffusion. So the next discharge may occur so as to choose a path of minimum electric resistance in the diffused ionized area, so that the curvature of the curved spark line becomes less by short-circuiting.
    Fig.4 shows a typical spark line obtained in free jet. By using such spark lines

the velocity distribution of free jet can be determined as shown in Fig.5, comparing with the measurements by Pitot tube. It can be seen that the short-circuiting is marked for smaller frequency, that is probably because the larger discharge interval (1/f) promotes the diffusion. As shown in Fig.5b, the short-circuiting becomes less marked for the high speed jet case, when the discharge frequency was raised according as the flow velocity increased. Fig.6 gives the velocity distribution of the flow in a rectangular channel, showing the similar short-circuiting phenomena as in the jet flow.

The short-circuiting phenomena will be considered by a simple analytical model, in which the spark line has a rectangular wave form with rounded corners of radius R, being washed downstream by a uniform flow. It is assumed that the diffusion factor is given constant, $\mathcal{D}$ = 12.5 cm²/s (Ref.3) and the electric resistance of the ionized path is inversely proportional to the ion concentration. The radius R will be determined so that the spark will be discharged successively on a path of a minimum resistance in the ionized region. Fig.7 gives some calculated results of the corner radius variation comparing with the experimental results obtained from Fig.6. There can be seen some dependence of the diffusion on the spark frequency in the experiment, whereas the effect of the spark frequency does not appear definitely in the calculation. Both the experiment and calculation show the similar variation of radius R with the distance X ; that is, at a given spark frequency the corner radius becomes larger as the velocity decreases and as the distance X increases.

2.2.3 Effect of density of ionized tracer

The ionized tracer may be of much higher temperature and therefore of lower density than the ambient air, so the tracer can not follow the fluid motion accurately when an external force such as gravity and centrifugal forces works. We will estimate the density of the spark tracer from the velocity discrepancy between the tracer and the ambient air in an accelerating channel flow.

a) Accelerating flow: In Fig.8, it shows the photograph of an accelerating flow in a convergent rectangular channel with $\theta = 10°$, and also gives the variation of the acceleration α of the main flow along the channel axis, where g is acceleration of gravity , and A,B and C indicate the locations of the electrodes. Fig.9 gives the velocity deviation of the tracer (ΔU) versus the mean acceleration ($\bar{\alpha}$). From the figure, it can be seen that the velocity deviation increases proportionally with the increase of acceleration. This relation is expressed by

$$\Delta U = 0.016 \, (\bar{\alpha}/g)^{0.94} \qquad \text{m/s} \qquad (2)$$

for spark frequency f = 20 and 30 kHz. The smaller velocity deviation for lower frequency (f = 10 kHz) may be caused by the addition of less energy per unit time, leading to relatively higher density of the ionized tracer.

b) Estimation of ionized tracer density: Now we assume that the ionized tracer is a cylinder of diameter d_t with the density ρ_t moving with a velocity U_t in the air of density ρ_f with the velocity U_f, and that the drag on the tracer is proportional to the square of the relative velocity with the coefficient ranging C_D = 1.1 to 1.4, according to their Reynolds number Re = 55 to 400. The velocity deviation between the tracer and the ambient air is given as

$$\Delta U = U_t - U_f = \Delta U_\infty (1 - \exp(-\beta t))/(1 + \exp(-\beta t)) \qquad (3)$$

where $\beta = 2[\, 2 \, C_D \cdot \bar{\alpha} / (\pi \cdot d_t) \cdot (\rho_f / \rho_t) \cdot (\rho_f / \rho_t - 1)]^{\frac{1}{2}}$

$\Delta U_\infty = [\, \bar{\alpha} \cdot \pi \cdot d_t / (2 \, C_D) \cdot (1 - \rho_t / \rho_f)]^{\frac{1}{2}}$

229

Putting tracer diameter d_t = 0.8 mm, some calculated results of Δ U are given in Fig.9 for several density ratio ρ_t / ρ_f . By comparing with the experimental results, the density ratio probably ranges between 1/5 to 1/10. Using this density ratio and considering C_D for Reynolds number less than 1.0, the buoyancy velocity of the tracer is estimated as   V = 0.022 $\sim$ 0.024 m/s.   Finally we will consider the response of the ionized tracer to an oscillating flow.   Assuming that the spark tracer is a sphere, and putting its density ratio to be 1/5 $\sim$ 1/10, the oscillating frequency should be less than 20 Hz so as to make the tracer follow the flow within 3% error (Ref.4).

3.   VISUALIZATION OF FLOW AROUND AN OSCILLATING AEROFOIL

3.1 Experimental apparatus

Fig.10 shows the aerofoil model made of bakelite (NACA-0015,150mm chord and 300 mm span).The experiment were carried out under the following conditions; wind velocity U = 20 m/s, the angle of attack for stalling α_s = 14°, the oscillatory angle of attack $\Delta\alpha$ = $\pm$ 1.5°, the oscillating frequency F = 0 to 20 Hz.

3.2 Experimental results

a) Flow direction: To measure the flow velocity over the oscillating aerofoil, it needs to know in advance the time-dependent flow direction at each measuring positions. Fig.11 shows the flow direction measured at the point I and the point V, where ϕ is the angle measured counterclockwise from the tangent of the aerofoil surface, and y is the distance of the measuring position from the surface.   It can be seen that the flow direction measured for F = 20 Hz at the point V is completely reversed when the angle of attack increases from 13° to 15°, while there can not be seen marked difference in flow direction for F = 5 Hz or at the point I.

b) Flow velocity: A typical series of photographs of the flow around an oscillating aerofoil were indicated in Fig.12. It is obvious that there occurs the flow separation or the reverse flow on the upper surface, in particular near the trailing edge. Fig.13 shows the results of velocity variation analysed from Fig.12. As is seen from the figure, at a lower oscillating frequency the flow along the aerofoil varies similarly as in the case of steady flow (F = 0)but with a slight phase lagging, while at higher frequency the phase lag increases and the amplitude of oscillation decreases.

4   CONCLUSION

The principal results are summarized as follows;
(1) For the quantitative measurements of the flow velocity, the spark line downstream of the electrodes at least by 10 mm should be used irrespective of the flow velocity.   (2) The apparent radius of curvature in a curved spark line increases as the flow velocity and the spark frequency decrease; that is mainly due to the diffusion of ionization.   (3) In the accelerating flow the velocity deviation of the spark line from the ambient flow is given by Δ U = 0.016 ($\bar{\alpha}$/g)$^{0.94}$ m/s. (4) From the above results, it can be estimated that the density of the spark line is about 1/5 to 1/10 as much as that of the ambient air and its diameter $\sim$ 0.8 mm. Then the buoyancy velocity of the ionized tracer is also estimated as $\sim$ 0.023 m/s. (5) The spark tracing method was applied to the flow around an oscillating aerofoil, and the results has proved it to be available for quantitative analysis of both steady and unsteady flows as well as qualitative observation of the flow pattern as a whole.

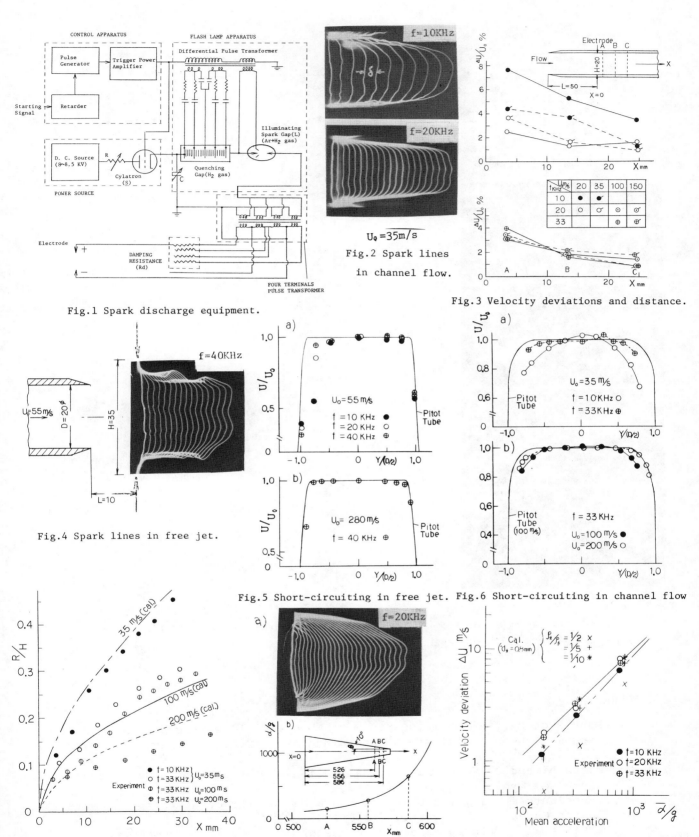

Fig.1 Spark discharge equipment.

$U_0 = 35\ m/s$

Fig.2 Spark lines in channel flow.

Fig.3 Velocity deviations and distance.

Fig.4 Spark lines in free jet.

Fig.5 Short-circuiting in free jet.

Fig.6 Short-circuiting in channel flow

Fig.7 Calculated and measured results of corner radius variation.

Fig.8 a) Spark lines in accelerating flow.
b) Variation of acceleration along the convergent channel axis.

Fig.9 Relation between velocity deviation and mean acceleration.

231

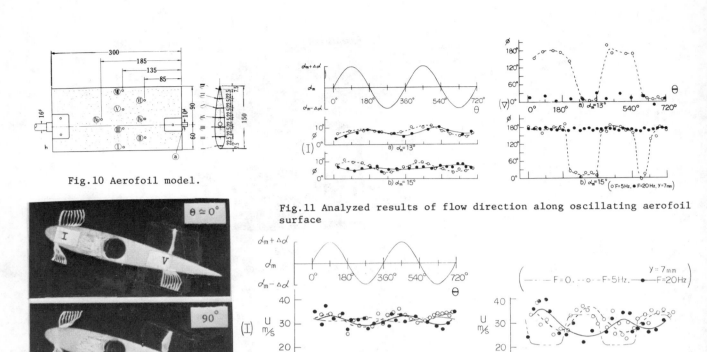

Fig.10 Aerofoil model.

Fig.11 Analyzed results of flow direction along oscillating aerofoil surface

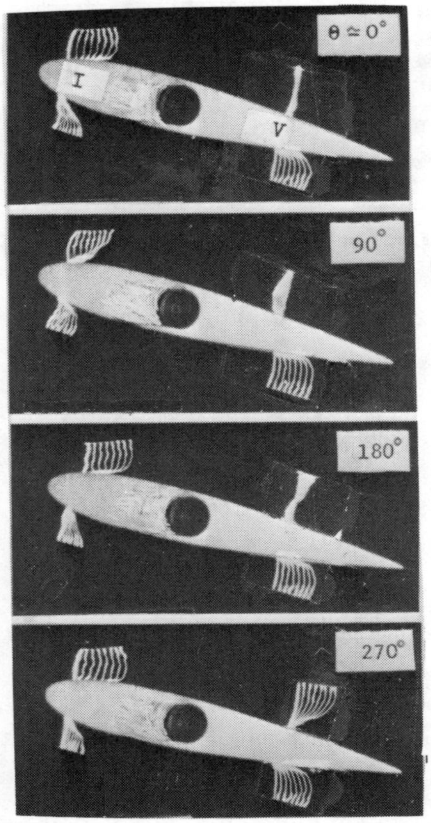

Fig.12 Series of high speed photographs of flow around oscillating aerofoil.

(U =20 m/s, α_m =15°, F=5 Hz, f=12.5 KHz.)

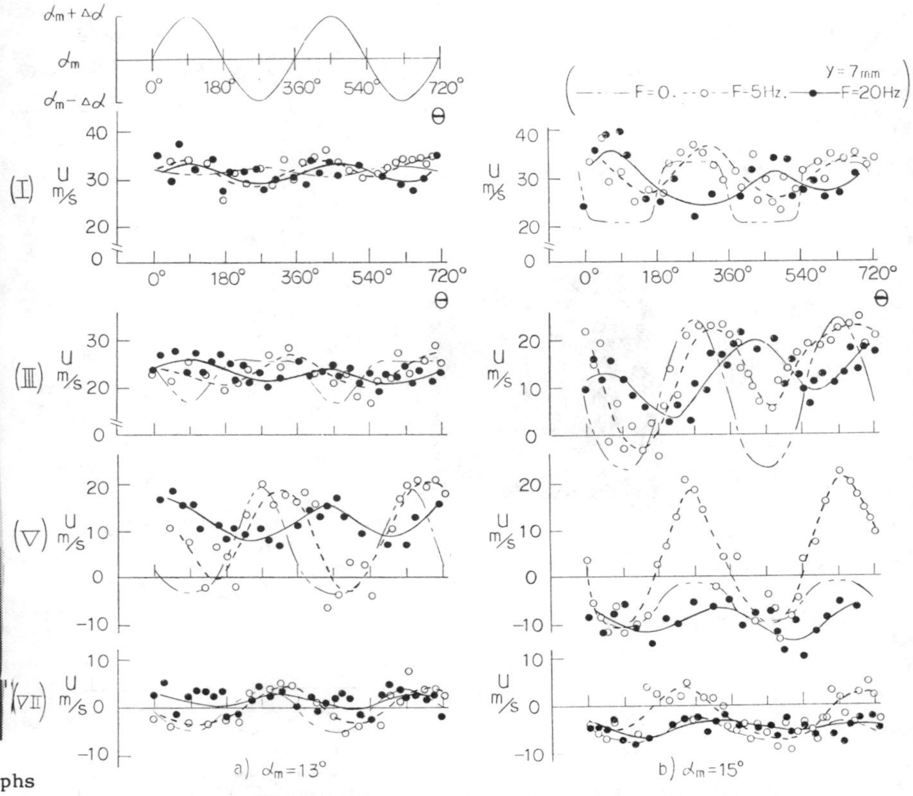

Fig.13 Analyzed results of velocity distribution on oscillating aerofoil surface.

REFERENCES

1) T. Asanuma et al., Visualizing Study on Nonsteady Flow around an Aerofoil, (in Japanese) Report of Inst. of Space and Aero. Science, Univ. of Tokyo, 7-2 (1972), 491.

2) N. Takahashi and T. Kato, Flow Visualization by the Electric-Spark Method, (in Japanese) Oyo-Buturi, 38-7 (1969), 709.

3) H.J. Bomelburg et al., The Electric Spark Method for Quantitative Measurements in Flowing Gases, Z. Flugwiss., 7-11 (1959), 39.

4) F. Durst et al., Principles and Practice of Laser-Doppler Anemometry, Academic Press (1976), 319.

MEASUREMENTS OF SHOCK TUBE FLOWS USING A SPARK TRACER METHOD

K. MATSUO,* T. IKUI,** Y. YAMAMOTO,*** and T. SETOGUCHI****

A spark tracer method, together with pressure measurements, has been employed to investigate the flows in a shock tube. The gas in the driver section expands through an unsteady expansion fan generated by the breaking of a diaphragm, and a condensation and an adiabatic shocks have occurred in this experiment. In the driven section, a shock front is formed which is followed by a hot flow of the driven gas and a cold flow of the driver gas. The shape of spark paths in such flow fields has been discussed, and the velocities obtained by the spark tracer method have been compared with the theoretical values and the results calculated from the pressure measurements.

1. INTRODUCTION

Since Bomelburg et al. (Ref.1) proposed a method for the measurement of the flow of air by means of series of electric sparks, the spark tracer method has been developed and applied to various aerodynamic investigations. Kyser (Ref. 2) employed this method for measuring velocity fields in a hypersonic wind tunnel, Lahaye et al. (Ref.3) investigated the wake behind projectiles in a ballistic range, and Nagao et al. (Ref.4) demonstrated the applicability of this method to visualize the flow in internal combustion engines.

In this paper, the spark tracer method was employed to investigate the flows in a shock tube, and the flow velocity measured by this method was compared with the theoretical values and the results calculated from pressure measurements.

2. EXPERIMENTAL APPARATUS

The shock tube used in this work consists of six tube elements which are interchangeable each other and each cross section is 38mm square. An example of the tube arrangement for the test of the driver section is shown schematically in Fig.1. A pair of electrodes located at 180mm upstream of the diaphragm are tapering needles with a diameter of 0.76mm and the spark gap is 10mm. When the diaphragm is ruptured and the head of rarefaction fan arrives at the pressure transducer for trigger, the signal from the transducer triggers the pulse generator after a prescribed time interval, and a prescribed number of sparks is discharged between the electrodes.

The static pressure is measured using Kistler piezoelectric transducers mounted flush in the sidewall at various stations along the tube. Air is used as the

---

*Associate Professor of Mechanical Engineering, Kyushu University, Hakozaki, Fukuoka, Japan
**Professor of Mechanical Engineering, Kyushu University, Hakozaki, Fukuoka, Japan
***Assistant of Mechanical Engineering, Kyushu University, Hakozaki, Fukuoka, Japan
****Graduate Student of Mechanical Engineering, Kyushu University, Hakozaki, Fukuoka, Japan

working gas. The initial pressure in the driver section P_4 is varied from one to four atmospheric pressures and that in the driven section P_1 is made vacuous so that the initial pressure ratio across the diaphragm P_{41} (=P_4/P_1) ranges from 2 to 80.

3. FLOW IN DRIVER SECTION

3.1 FLOW IN UNSTEADY EXPANSION FAN

A typical pressure-time record is shown in Fig.2, where P/P_4 is the ratio of the pressure P at the position of electrodes to P_4, and t is the time measured from the moment when the head of the expansion fan arrives at that position. The dashed line represents the results calculated by the theory of ideal shock tube flow. The difference between the experimental and theoretical values in the range of $0 < t < 0.4$ ms may be due to the fact that the expansion fan generated by diaphragm rupture is not centered (Ref.5). The pressure rise at about 0.5 ms is caused by a condensation shock (Ref.6) generated by the rapid cooling due to the expansion fan and an adiabatic shock generated by the choking of the flow at the diaphragm location. These shocks will be explained later. Eight pulse signals in Fig.2 indicate the time of electric spark discharge and the corresponding spark photograph is shown in Fig.3, where each spark path has a trapezoid shape and it is quite different from that of a spark photograph of steady flow such as Fig.9. Similar spark shapes to Fig.3 were always obtained for accelerated flows in the expansion fan.

In order to obtain velocity variation with time at the position of electrodes, it is necessary to correct the velocity measured from Fig.3, because the velocity in the expansion fan changes with both time and place. Let x and t be the distance from the diaphragm location along the tube and the time measured from the moment of diaphragm rupture, and a and u be sound velocity and flow velocity, then the following two equations hold to any point in the centered expansion fan in x-t diaphram.

$$x = (u - a)t \qquad (1) \qquad\qquad 2a_4/(\kappa - 1) = u + 2a/(\kappa - 1) \qquad (2)$$

where κ is the ratio of specific heats of the working gas. Elimination of a from Eqs. (1) and (2) yields

$$u = 2(a_4 + x/t)/(\kappa + 1) \qquad\qquad (3)$$

Using the above equation, the velocity u_{is} measured from spark photographs can be transformed into the velocity u_i at the same time and at the position of electrodes by the following equation.

$$u_i = u_{is} - 2\xi_i/\{(\kappa + 1)(t_0 + \tau_i)\} \qquad\qquad (4)$$

where t_0 is the time when the first spark is struck, ξ_i is the distance from the first spark path to the center of the i th and $(i + 1)$th spark paths along the tube, and τ_i is the time interval required for a gas particle to traverse the distance ξ_i.

Fig.4 shows an example of the variation of velocity versus time at the position of electrodes in the expansion fan. The experimental values measured from spark photographs are corrected using Eq.(4) and the corrected results agree well with the values calculated from the pressure measurements using the relation of an isentropic flow. The difference between the experimental and theoretical values is due to the same reason as in Fig.2.

3.2 CONDENSATION SHOCK AND ADIABATIC SHOCK

Fig.5 shows a spark photograph taken at $t \approx 0.5$ ms in Fig.2. As the interval between adjacent two spark paths decreases at first and then increases, it is clear that the flow is decelerated and then accelerated. This deceleration is due to a condensation and an adiabatic shocks mentioned before. These shocks occur near the diaphragm location and propagate upstream in the driver section.

Fig.6 shows the variation of pressure versus time at 2142mm upstream of the diaphragm, and a schlieren photograph taken at the instant indicated in Fig.6 is shown in Fig.7. From these figures, it may be interpreted that the pressure rise at the points b and c in Fig.6 is caused by the condensation shock and the adiabatic shock, respectively. When the head of the reflected expansion fan arrives (point d in Fig.6), the pressure decreases again.

In the present experiment, a spring-loaded needle by which a diaphragm is ruptured is attached to the shock tube wall by a strut with a thickness of 8.8mm, and the brockage ratio, i.e., the ratio of the cross sectional area of the strut to that of the shock tube is $8.8/38 = 0.232$. Therefore, the flow area is reduced appreciably at the diaphragm portion, and this will be the cause for the choking of the flow and the occurrence of the adiabatic shock. From this point of view, an experiment was made removing the strut and mounting the needle in the shock tube wall at an angle to the diaphragm. Fig.8 shows a comparison of the variation of pressure versus time with and without the strut at 100mm upstream of the diaphragm. The pressure history in the case without the strut coincides with that in the case with the strut as far as the first rise in pressure which is, mentioned before, due to the condensation shock. After that, the pressure without the strut decreases and approaches a certain value asymptotically, although the pressure with the strut experiences a short rise due to the adiabatic shock. From the above mentioned, it is evident that the adiabatic shock has occurred by the reduction of area by inserting the strut in the tube.

3.3 STEADY FLOW BEHIND UNSTEADY EXPANSION FAN

An example of spark photographs taken in the steady and uniform flow behind the unsteady expansion fan is shown in Fig.9. The ratio of the pressures behind and in front of the expansion fan P_3/P_4 is calculated using the velocity u_3 measured from Fig.9 and the relation of the centered expansion fan. The results are shown in Fig. 10 as a function of P_{41}, together with the results of pressure measurements and the experimental values of Ref.7. The deviation of the experimental values from those predicted by the theory is negligible for low values of P_{41} but becomes very marked for the higher values of P_{41}. The difference between the values predicted by the theory and the experimental values with the strut arises mainly from the condensation shock and the adiabatic shock, and the large cause for the difference in the case without the strut is the condensation shock.

4. FLOW IN DRIVEN SECTION

In order to measure the flows in the driven section, a pair of electrodes and a pressure transducer were set at 2060mm downstream of the diaphragm, which is by far downstream of the shock formation distance. The measured flow velocity just behind the shock front is plotted against P_{41} in Fig.11. At any P_{41}, the experimental values are a little smaller than those predicted by the thoery, and the difference mainly results from the viscous effect of the gas. The values obtained by the spark method agree well with those calculated from the pressure measurements. According to Ref.8, when a spark is discharged shortly before the arrival of a shock front, an ionized gas particle created by the spark has a velocity which differs much from the theoretical gas velocity after travelling the shock front, because the density of the particle is different from that of the surrounding gas. In the present experiment, the first spark was discharged immediately after the passage of a shock front, and the deviation produced by the difference between the densities of the ionized gas and of the surrounding gas was not observed.

Fig.12 shows examples of spark photographs in the hot flow and cold flow regions. As the density of the cold flow is larger than that of the hot flow, and the cold flow is more eddying than the hot flow, the spark path is uneven in Fig.(b), while it is smooth in Fig.(a). The velocity measured by the spark method in the hot flow and cold flow regions is plotted in Fig.13 against the time measured from the arrival of the shock front at the location of the electrodes. The theoretical velocity and time of the arrival of the contact surface based on the ideal shock tube theory are also shown as dash-dotted lines and arrows, respectively, in Fig.13, together with the time of the arrival of the front of contact zone estimated from the spark photographs. When $P_{41} = 5.5$, theoretical hot flow duration is 5.72 ms. It is clearly observed from this figure that the velocity increases and decreases with time in the hot flow and cold flow regions, respectively, and the maximum velocity is achieved at the front of contact zone separating the hot flow from the cold flow. This variation of velocity with time is associated with the development of the wall boundary layer whose thickness becomes maximum near the front of the contact zone. As the contact zone is accelerated due to the development of the boundary layer (Ref. 9), its front arrives at the measuring station earlier than the time predicted by the theory.

A typical pressure variation with time in the hot flow and cold flow regions is shown in Fig.14. After a sharp rise in pressure by the incident shock, the pressure decreases slightly up to the front of contact zone. Taking the development of the boundary layer into account, the pressure in the hot flow region should increase gradually rather than decrease. Therefore, the slight fall in pressure observed in Fig.14 may be caused by the characteristics of the Kistler pressure transducer, as is pointed out in Ref.10 where the pressure variations behind the incident shock were measured with a Kistler transducer.

5. CONCLUDING REMARKS

A spark tracer method has been employed to measure the flows in a shock tube. The velocities measured by this method in and behind the unsteady expansion fan and just behind the shock front agree well with the values calculated from pressure measurements. Thus the applicability of this method to measure shock tube flows has been demonstrated. The shape of spark paths is apparently unlike in each flow region in the shock tube, namely, the accelerated flow induced by the unsteady expansion fan and the steady flow behind the fan, or the hot flow and the cold flow. Based on this characteristic, it is easy to detect the arrival of the front of contact zone to a measuring station by the spark method.

REFERENCES

(1) H.J. Bomelburg et al., The Electric Spark Method for Quantitative Measurements in Flowing Gases. Zeit. Flugw., 7-11 (1959-11), 322.
(2) J.B. Kyser, Tracer-Spark Technique for Velocity Mapping of Hypersonic Flow Fields, AIAA J., 2-2 (1964-2), 393.
(3) C. Lahaye et al., Wake Velocity Measurements Using a Sequence of Sparks, AIAA J., 5-12 (1967-12), 2274.
(4) F. Nagao et al., Measurement of Air Flows by Means of Repetitive Spark Method, Bulletin of JSME, 14-78 (1971-12), 1340.
(5) J.G. Hall et al., Unsteady Expansion Waveforms Generated by Diaphragm Rupture, AIAA J., 12-5 (1974-5), 724.
(6) S.P. Kalra, Experiments on Nonequilibrium, Nonstationary Expansion of Water Vapour/Carrier Gas Mixture in a Shock Tube, UTIAS Report No. 195 (1975).
(7) I.B. Billington, An Experimental Study of the One-Dimensional Refraction of a Rarefaction Wave at a Contact Surface, J. Aeron. Sci., 23-11 (1956-11), 997.
(8) G. Rudinger and L.M. Somers, Behaviour of small regions of different gases carried in accelerated gas flows, J. Fluid Mech., 7-2 (1960-2), 161.
(9) A.G. Gaydon and I.R. Hurle, The Shock Tube in High-Temperature Chemical Physics (1963), 70, Chapmann and Hall.
(10) M.J. Lewis and L. Bernstein, Measurements of Temperature and Pressure behind the Incident and Reflected Shocks in a Shock Tube, ARC, C.P. No. 1239 (1973).

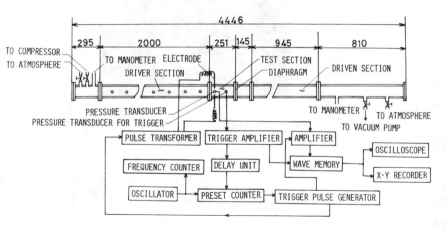

Fig. 1 Shock tube facility and its measuring system.

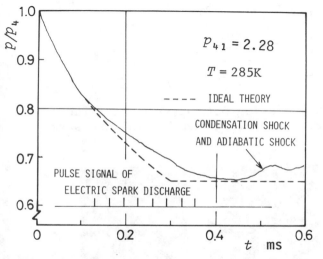

Fig. 2 Pressure variation with time at 180mm upstream of diaphragm.

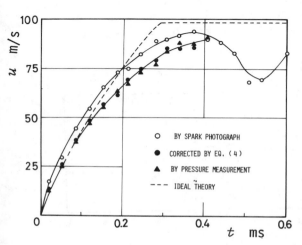

Fig. 4 Velocity variation with time in expansion fan, $p_{41} = 2.28$, $T = 283$K.

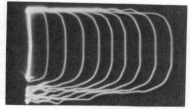

Fig. 3 Example of spark photograph in unsteady expansion fan, $p_{41} = 2.28$, $f = 30$kHz, $E_s = 14.4$kV, $n = 8$.

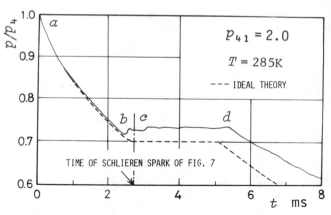

Fig. 5 Spark photograph showing deceleration of velocity due to condensation shock and adiabatic shock, $p_{41} = 2.28$, $n = 10$, $f = 40$kHz, $E_s = 14.4$kV.

Fig. 6 Pressure variation with time at 2142mm upstream of diaphragm.

Position of pressure transducer

Condensation shock   Adiabatic shock

Fig. 7 Schlieren photograph showing condensation shock and adiabatic shock, $p_{41} = 2.0$. (Flow, left to right.)

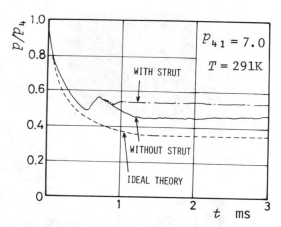

Fig. 8 Comparison of pressure variation versus time with and without strut at 100mm upstream of diaphragm.

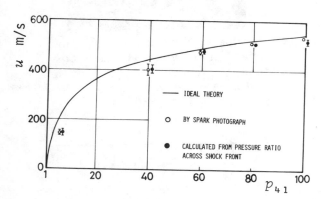

Fig. 11 Flow velocity just behind shock front, $T = 286$K.

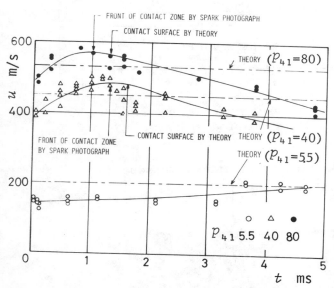

Fig. 13 Comparison of measured flow velocity behind shock front with theory, $T = 286$K.

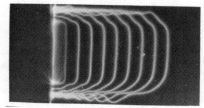

Fig. 9 Spark photograph showing steady flow behind expansion fan, $p_{41} = 8.0$, $u_3 = 160$m/s, $f = 80$kHz, $E_s = 16.5$kV, $n = 10$.

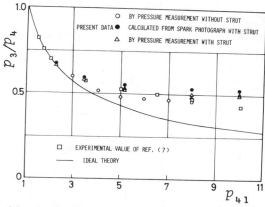

Fig. 10 Relation between expansion fan pressure ratio p_3/p_4 and initial pressure ratio p_4/p_1.

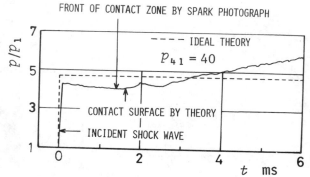

(a) Hot flow   (b) Cold flow

Fig. 12 Example of spark photograph in hot flow and cold flow, $p_{41} = 40$, $E_s = 19.5$kV, $f = 100$kHz, $n = 3$.

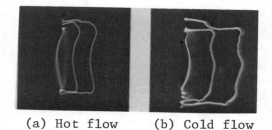

Fig. 14 Pressure variation behind shock wave at 2060 from diaphragm, $T = 286$K.

238

FLOW DIRECTION DETECTABLE SPARK METHOD

YASUKI NAKAYAMA,* SHIRO OKITSU,* KATSUMI AOKI,**
and HIROAKI OHTA***

Using the spark method,the flow velocity of a simple flow with previously known flow direction can be measured accurately. Since a complex flow, however, must be measured assuming the flow direction, it is difficult to get the accurate velocity of such a flow. The method to be reported has been developed to remove this shortcoming.

AlN, Mg_3N_2, BN etc. particles are dispersed in the gas flow , radiate in the spark and give radiant tails. By them the flow direction can be determined. The kinds and concentrations of particles are examined. Good results are obtained when applied to divergent duct flow and swirl flow.

1.Introduction

The spark method is one of the techniques of flow visualization, where suitable electrodes are arranged in the gas flow to be measured and high voltage pulses are repeatedly applied on them in adequate time intervals,and by spark discharges ionized paths are generated to visualize the flow pattern.

In recent years this method has a wide application in the varieties of flow , since in this method fluid itself is used as tracer, resulting in lack of disturbance in the flow, quantitative measurements are possible and high speed flow as far as super sonic flow can be observed (Ref.1).

In accordance with these situations some papers have been published on the various ways observing electric spark and fundamental characteristics such as the form of electrodes and its materials (Ref.2), required discharge voltage (Ref.3) and so forth as well as on the application of this method.

In this method,however,only time lines are obtained. Although in the case when the flow directions are known correct velocities can be ascertained, in the case of unknown flow directions, other methods are needed to identify its directions. This is the great disadvantage of this method.

This paper aimes to get rid of this shortcoming and presents a development of a new spark and radiant particle combination method, in which particles are dispersed in the flow to be measured and in the spark discharge they generate radiant trails that show the flow direction.

For the purpose of detecting flow directions by the use of this combination method, some experiments have been carried out using various particles and the dispersed particles optimum for these observations have been searched for.

*Professor, **Assistant Professor, ***Instructor,
Department of Mechanical Engineering, Faculty of Engineering,
Tokai University, Kitakaname, Hiratsuka, Kanagawa, 259-12, Japan.

2.Experimental apparatus and method

Fig.1 shows the rough view of the approximate two dimentional divergent duct which is used for this experiment and the method of supplying particles. The divergent angle of this duct is $30°$ and the area-ratio of outlet to inlet is 2.34. The particles used for the second tracer are of 10 kinds: $Ca(OH)_2$, $NaCl$, K_2CO_3, Al_2O_3, Al, AlN, Mg_3N_2, BN, SiO_2 and C.

Fig.2 shows the whole view of this experimental apparatus and Fig.3 shows the block diagram of high voltage high frequency pulse generator. The characteristics of this pulse generator are shown in Table 1. In the pulse generating circuit,arbitrary pulse width varying from 1 μs to 1 s are generated through the frequency division circuit of 1 MHz pulses input from crystal oscillator. By presetting the counter,the pulses are given to the trigger pulse circuit through the gate circuit up to the preset number when the start signal is given to the preset counter. These trigger pulses are amplified and they trigger the high voltage switching circuit. Its output is given to the pulse transformer and the boosted output from this transformer is applied to the electrodes. The needle-shaped electrodes (diameter 0.4 mm) are set on the wall of the divergent duct at a right angle as shown in Fig.1.

The particles are accumulated in the supplying hopper and the wind is slowly blown by the air compressor. The particles induced out from the hopper are distributed in the divergent duct according to the flow so the particles on the spark line make the radiant trails to the flow direction. This experiments are performed on the condition that output velocity of the duct is 10 m/s , supply voltage-10 kV, pulse width-150 μs and number of pulses-20.

3.Experimental results and considerations

Fig.4 shows the observed examples of flow pattern and radiant trails when the particles are dispersed in the flow, varying kinds of particles in each experiment, and electric sparks are made.

Fig.4(a) shows the flow pattern using $Ca(OH)_2$ particles (size, 80~120 μm) as tracer. The velocity distribution in the divergent duct is shown clearly and the particles on the spark lines make orange light and generate strong radiant trails without disturbing the velocity distribution made by spark lines. In the case of higher concentration of dispersed particles, however, the radiant trails overlap and the spark lines are shaded off. So it is difficult to catch precisely the flow pattern near the wall. In this experiment, good results are obtained with volume concentration of particles of about 1%. Orange colour of radiant trails can be improved to some degree by using filter.

Fig.4(b) shows the case using $NaCl$ particles. As these particles burn in pale and make radiant trails, they enlarge, overlap easily with each other and the spark lines are shaded off. By holding the concentration of particles a little lower and using a filter, however, the direction of the main flow can be checked sufficiently. Good flow patterns are observed with the volume concentration of particles of the order of 1%. Fig.4(c) shows the case using K_2CO_3 particles (size, 100~300 μm). The radiant trails seem as burning and the spark lines are shaded off like the $NaCl$ ones.

Fig.4(d) shows the case using Al particles (size, 1~30 μm), where in higher concentration of dispersed particles, as shown in the figure, the spark seems to be convected with Al particles which results in spark lines whose directions are different from the velocity distribution. That is why Al particles are not conveniant for tracer. Fig.4(e) shows the case using AlN particles (size, 1.5~6 μm), where the radiant trails draw sharp tails and without disturbing the spark lines grasp accurately not only the direction of the main flow, but also the reverse flow due to separation of flow in the neighbourhood of wall surface. Moreover, they have a wide application range from 10^4 to 10^8 particles /cm³ to take precise flow pattern. The other nitrides such as Mg_3N_2, BN used in this experiment have brought comparatively good results.

Fig.4(f) is the case using flaky graphite,which shows the directions of both the main stream and the local one. Channel wall surfases, however, are coated with black particles, so that the clearness of the spark line decreases a little.

As for the other particles such as SiO_2, Al_2O_3 etc., they yield no radiant trails and proved to be unadequate as tracer.   From these the results shown in Table 1 are obtained.

4.Example of application

The flow in swirler, as shown in the Fig.5, is measured as an example of application.   The swirler is of axial-flow type with 16 swirl vanes whose angle is 45°. Its swirl rate S (Ref.4) is 0.722.

Fig.6 shows one example of flow pattern in swirler when the mean velocity of its outlet is about 13 m/s.   By this photograph,the front and side view flow patterns of the swirler are found to have charactristic shapes.   Namely , as flow patterns approach swirling to the wall , so the intervals of spark lines increase little by little.   The radiant trails generated by secondary tracer make the flow direction clear.   The side view photograph shows reverse flow in central zone,whose directions are made clear from the radiant trails.

Fig.7 shows the method to calculate the flow velocity from the intervals of spark lines in photographs.   The method of calculation is as follows.

$$v_t = (\Delta l / a) \, f \qquad , \qquad v_a = (\Delta l' / a) \, f$$

v_t : tangential velocity
v_a : axial velocity
$\Delta l, \Delta l'$: interval of spark lines on the photograph
f : pulse frequency
a : enlargement of photograph
$F(s)$: directional curves of gas flow that are obtained from radiant trails

Fig.8 shows tangential and axial velocity distributions in the radial direction to each section of swirler when its outlet velocity is 10 m/s.

5.Conclusions

This study has made the following items clear :
(1) when particles are dispersed in gas flow and spark is dischaged , the flow directions can be detected by the radiant trails of particles on the spark lines.
(2) As material for the dispersed particles , nitrides , in particular AlN , are convenient producing sharp radiant lines and indicating the flow direction clearly.
(3) The optimum size of particles is of the order of several μm.
(4) The volume concentration of particles,yielding good results,is suitable to be 1%.
(5) As for the gas flow, containing solid particles, the measurement of velocity, the visualization of particles and their flow direction have also become possible.
(6) Good results are obtained when this combination method applied to the divergent duct flow and the swirl flow.

Reference
(1) T.Asanuma : Flow Visualization and its Applications,Journal of the J.S.M.E., vol. 77, No.666 (1974-5), 567.
(2) Y.Nakayama and M.Hayashi : Flow Visualization Using Spark or Hydrogen Bubble Method , Pro.2th International JSME Symp.   Fluid Machinary and Fluidics , Vol.3 , Tokyo Japan, (1972-9), 131.
(3) N.Takahashi and T.Kato : Flow Visualization by Electric Spark Method, Appl. Phys.

, Vol.38, No.7 (1969), 709.
(4) J.M.Beer and N.A.Chiger : Combustion Aerodynamics, Applied Sci., pub.,(1972),112.

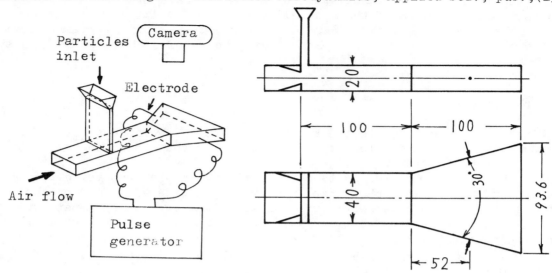

Fig.1 Two dimensional diviergent duct

Fig.2 Experimental apparatus

Table 1 Specifications of the pulse generator

Output voltage	(kV)	10～100
Pulse width	(μs)	0.5
Pulse frequency	(kHz)	1～75
Output energy	(J/pulse)	0.05～0.5

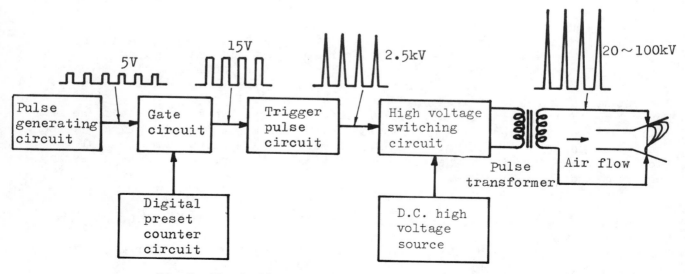

Fig.3 Block diagram of high voltage pulse generator

Table 2 Performance of particles

article	Specific gravity	Melting point °C	Particle size μm	Concentration particles cm³	State of radiant trail	Proprety Main flow	Proprety Local flow
$Ca(OH)_2$	2.24	580	80~120	$5 \times 10^3 \sim 2.5 \times 10^4$	orange light	◯	◯
NaCl	2.164	800.4	100~250	$8 \times 10^3 \sim 2 \times 10^4$	pale light	◯	△
K_2CO_3	2.428	891	100~300	$5 \times 10^2 \sim 2 \times 10^4$	pale light	△	△
Al_2O_3	3.5~4.0	2050	1~3.5			✕	✕
Al	3.93	191	1~30		pale light	✕	✕
AlN	3.05	2150	1.5~6	$10^4 \sim 10^8$	white light	◎	◎
Mg_3N_2	2.712	1500	1.5~8	$10^4 \sim 10^7$	white light	◯	◯
B N	2.34	3000	1.5~8	$10^4 \sim 10^7$	white light	◯	◯
SiO_2	2.20	2230	80~125			✕	✕
C	2.25	3700	1~5	$10^4 \sim 10^8$	orange light	◯	◯

◎ very good , ◯ good , △ usable , ✕ not usable

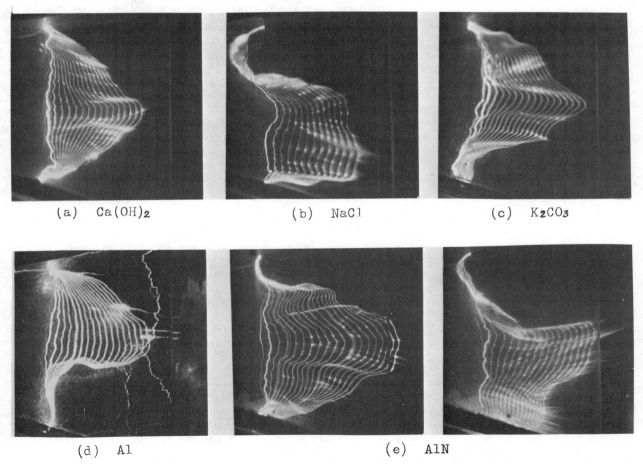

(a) $Ca(OH)_2$ (b) NaCl   (c) K_2CO_3

(d) Al   (e) AlN

Fig.4 Observed examples of particles

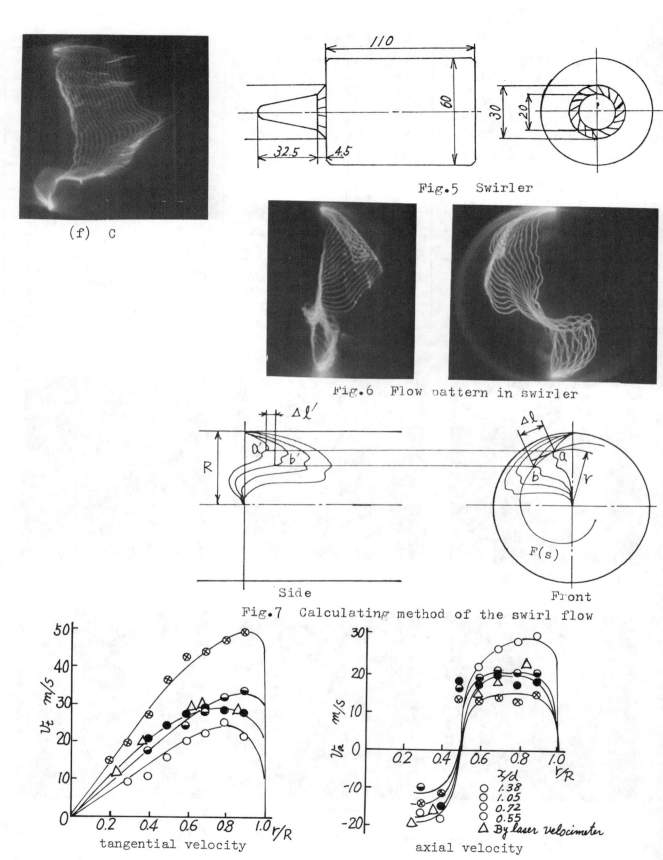

(f) c

Fig.5 Swirler

Fig.6 Flow pattern in swirler

Fig.7 Calculating method of the swirl flow

tangential velocity

axial velocity

x/d
1.38
1.05
0.72
0.55
By laser velocimeter

Fig.8 Radial and axial velocity distributions to each section of swirler.

VISUAL STUDIES OF LARGE EDDY STRUCTURES IN TURBULENT SHEAR FLOWS BY MEANS OF SMOKE-WIRE METHOD

NOBUHIDE KASAGI,* MASARU HIRATA,**
and SEIICHI YOKOBORI***

So-called well-ordered structures in turbulent shear flows are becoming of much interest to an increasing number of researchers. In the present study, the Smoke-Wire method, which is adequate to the flow-visualization of the turbulence three-dimensional behaviours, is applied to the several turbulent shear flows, namely, the turbulent boundary layers, the plane impinging jet, the separated and reattaching flow, where the coherent structures in large scales such as bursting phenomena and turbulent longitudinal vortex streets are found and confirmed. In addition, the combined technique of the Smoke-Wire method and the hot-wire measurement is developed and the typical example of its application is reported.

INTRODUCTION

Many turbulent flows can be easily observed in natural and industrial systems and the study of turbulence is evidently an interdisciplinary activity. Inherent randomness and nonlinearity of turbulence, however, make any theoretical treatment of the turbulence problem nearly intractable and any kind of prediction method has to rely on the available empirical relations obtained experimentally. Therefore, the experimental findings have been of much importance in many research works of turbulent flows conducted up to date (Ref.1).

Recent experimental evidence, in addition, indicates that there exist the well-ordered structures in the turbulent shear flows and leads to a better understanding of turbulence phenomena that turbulent flows are not as chaotic as has been previously assumed (Ref.2). Since such structures have the relatively strong coherence and play predominant roles on the turbulence production, the Reynolds stress generation as well as the heat and mass transfer mechanism, much attention has been paid to their behaviours in the flow fields. It is, however, rather difficult to detect and analyze the unsteady three-dimensional phenomena of these structures which exist with a considerable randomness in space and time, if one depends only upon the conventional measuring devices fixed in space. It is noted with a particular emphasis that the important findings in the studies of the turbulent boundary layer (Ref.3) and the plane mixing layer (Ref.4) were obtained by the flow-visualization techniques which made it possible to observe the characteristic motions of turbulence of comparatively low wave-numbers in the Lagrangian frame. Although the significant attempts such as conditional sampling and pattern recognition in turbulence signals have been carried with fruitful results after the discovery of these structures, it can be said that the applications of flow-visualization techniques should give great support to what is understood by the quantitative measurements and might bring further new findings

---

  * Associate Professor, Department of Mechanical Engineering, University of Tokyo, Bunkyo-ku, Tokyo, Japan
 ** Professor, Department of Mechanical Engineering, University of Tokyo
*** Graduate Student, Department of Mechanical Engineering, University of Tokyo

in turbulence phenomena.

In the present study, the Smoke-Wire method, which is adequate to the flow-visualization of the turbulence three-dimensional motions in air, is applied to the several turbulent shear flows, namely, the turbulent boundary layer along a flat plate, the turbulent boundary layer around a rotating cylinder, the two-dimensional impinging jet and the separated and reattaching flow downstream of a rearward facing step. Then further applicability of the Smoke-Wire method is discussed concerning the combined technique with quantitative measurements of turbulence.

GENERAL DESCRIPTION OF SMOKE-WIRE METHOD

The principle of the Smoke-Wire technique has been proposed in the early 1950s (Ref.5) and the quantitative measurement of velocity distribution in the laminar boundary layer has been performed by Yamada (Ref.6), while the varied appearance of smoke generation due to the electric charge on the heated wire has been reported by Torii (Ref.7). This technique has been further applied to the phenomenological studies of various turbulent shear flows by Kasagi (Ref.8).

In the Smoke-Wire method, a fine nichrome wire is located in the desired region to be visualized and a mixture of liquid paraffin and machine oil is painted on the wire initially. This wire is impulsively heated by a D.C. charge which is accumulated by capacitor banks, then the liquid mixture evaporates and condenses into white smoke of high visibility in the gas flow. This smoke becomes fluid tracer without any considerable disturbances on the flow field, if the gas flow velocity is in adequate range which is generally estimated as 1 to 20 m/s. After a fixed time interval from the smoke generation on the wire, it is possible to take a photographic record of the streak lines of smoke by flashing a stroboscope. It is also possible to obtain stereoscopic visual records by using several cameras simultaneously in different directions. These programmed sequence can be automatically carried out by an electronic circuit which is triggered by a shutter of camera.

Diameters of the wires used up to the present are usually 50 to 100 μm selected according to the dimensions of experimental apparatus. When liquid paraffin is painted on a fine wire, small droplets are formed along a wire due to its surface tension. Therefore, in order to avoid a formation of unwanted large drop, some additives, e.g. machine oils, are mixed with liquid paraffin to control the surface tension of a mixture.

The electric capacity of capacitor banks and the D.C. charge for wire heating should be varied according to factors such as the wire resistance and the flow velocity. Although the smoke can be generated by a D.C. charge of comparatively wide range, in the case the charge is too high the droplets scatter and the smoke generation ends in a short time. On the other hand, the smoke produced is small in quantity in the case of lower D.C. charge. In addition, the timespan of smoke generation on the wire can be also controlled by the value of capacity of electric condensers.

Several practical notes are mentioned as follows:

(1) There is a time delay in the smoke generation after the wire heating is begun. This time delay can be estimated by operating the Smoke-Wire method in a uniform flow with changing the fixed time constant of the delayed pulse circuit.

(2) At a low flow velocity of the order of 0.1 m/s, the effects of the natural convection by heating a wire and the initial velocity of the smoke from the wire cannot be neglected.

(3) The upper limit of flow velocity for the Smoke-Wire method may be usually about 20 m/s, because of a rapid diffusion of smoke.

(4) When this method is applied to gas flows at a high temperature above 400 K, the visibility of smoke tends to fade due to the inactive condensation of evaporated liquid.

(5) Care must be taken so that the wire initially tensioned in the flow field should not vibrate violently due to its thermal expansion.

APPLICATION OF SMOKE-WIRE METHOD TO TURBULENT SHEAR FLOWS

Fig.1 shows the experimental setup of the Smoke-Wire method for the visual investigation of the wall region in the fully-developed turbulent boundary layer on a flat plate without pressure gradient (Ref.9). The free stream velocity is 4.69 m/s and the boundary layer thickness is 48.7 mm, while the Reynolds number based on the momentum thickness is 1620. A couple of photographs have been taken simultaneously from the normal two directions as shown in Fig.1 and the typical series of photos in the various regions are represented in Fig.2, where the distance of the wire from the wall, y, is non-dimensionalized by the wall parameter, $\nu/\sqrt{\tau_w/\rho}$. As has been indicated by the recent visual study (Ref.3,10), so-called low-speed streaks, of which spacing is about $100\nu/\sqrt{\tau_w/\rho}$, can be observed near the viscous sublayer and the lift-up of a low-speed fluid lump (bursting) as well as the wallward motion of a higher momentum fluid (sweep) are clearly visualized respectively.

The turbulent boundary layer around a circular cylinder rotating at a constant speed in a stationary surrounding air has been also visually investigated as shown in Figs. 3 and 4 (Ref.11), where the diameter of the cylinder is 82.14 mm and the Reynolds number, Re_d, is based on the peripheral velocity and the diameter of cylinder. There exist also the well-ordered motions such as the bursting and the sweep, although they are remarkably augmented by the centrifugal forces, and this results in the high transport rate of heat and momentum in this type of boundary layer flows.

In the stagnation region of the two-dimensional submerged jet impinging normally on a flat plate, the large scale vortex-like structures (Ref.12) have been confirmed as shown in Figs. 5 and 6, where the nozzle width is denoted as B and the distance of the impinging plate from the nozzle exit, H. From the accumulated experimental evidence through the flow-visualizations, it has been found that these unsteady vortex-like structures are amplified and produced by the three-dimensional distortions of the shear vortices convected from the upstream free shear layers of the jet. It is inferred straightforward that these large eddies must play a predominant role on the enhancements of heat and mass transfer rate in the stagnation region which is generally realized in impinging jet systems.

In Figs. 7 and 8, the Smoke-Wire technique is applied to the turbulent separated and reattaching flow downstream of a rearward facing step, the height of which, H, is 20 mm (Ref.19). The free stream velocity is 16.9 m/s, and the upstream turbulent boundary layer near the separation line is almost in the developed state and is about 10 mm in thickness. Similarly to the case of the impinging jet mentioned above, there also be found the disturbance amplification phenomena in the reattaching region where the streamwise vortex-like structures in the scales of the order of a step height are clearly visualized.

As described above, the further confirmations and also new findings of the existence of well organized structures in turbulence have been obtained by making use of the Smoke-Wire method in the several kinds of turbulent shear flows and this indicates the possibility that the application of this technique will give important clues on more complicated flow systems.

COMBINED TECHNIQUE OF SMOKE-WIRE METHOD AND TURBULENCE SIGNAL MEASUREMENTS

The Smoke-Wire method can be combined with simultaneous quantitative measurements of velocity or pressure fluctuations in the turbulent flow field. In this paper, the continuous measurements by the X-probe of hot wire anemometer are integrated with the Smoke-Wire technique and the instance of the strobo-flash is marked on the time axis along with the velocity fluctuations, u', v', and the Reynolds stress, u'v', from the anemometer with the aid of a data recorder. The typical examples of this application to the bursting phenomena are represented in Figs. 9 and 10, which are the flat plate turbulent boundary layer and the turbulent boundary layer around a rotating cylinder respectively, explained in the previous paragraph. The effect of the wake of a fine wire on the X-wire measurements has been confirmed to be negligible, while the thermal disturbance by heating the wire has been appreciable to some extent, but it is also satisfatorily small compared with the fluctuations in the near-wall region of turbulent boundary layers. In both figures, the photographs of flow-visualization have been taken at the instance when the outward or wallward motions take place just upstream of the X-probe. In the wall region of the turbulent boundary layer in Fig. 9, the outward ejection of low-speed fluid from the viscous sublayer as well as the wallward inrush of high speed fluid are recognized to be predominant in the Reynolds stress generation, i.e. the turbulence production, strongly in an intermittent manner. In the case of Fig. 10, the similar features are observed, although the outward ejection of higher momentum fluid lump has a distinguished role on the transport process due to the unstable centrifugal effect.

CONCLUSION

The Smoke-Wire method has been applied to the several kinds of turbulent shear flows, where the well-ordered eddy structures in large scales are found and confirmed successfully. The combined technique of the Smoke-Wire method and the quantitative measurements of turbulence is also developed and the typical examples are demonstrated. These flow-visualization techniques are indicated to be effective for the research works even of the unsteady three-dimensional flow systems.

ACKOWLEDGEMENT

The first author of the present paper would like to express his appreciation to the financial promotion for this work by the Sakkokai Foundation.

REFERENCES

(1)   J.O. Hinze, Turbulence(2nd Edition), (1975), McGraw-Hill.
(2)   J. Laufer, New Trends in Experimental Turbulence Research, Ann. Rev. Fluid Mech., 7 (1975), 307.
(3)   S.J. Kline, W.C. Reynlods, F.A. Schraub and P.W. Runstadler, The Structure of Turbulent Boundary Layers, J. Fluid Mech., 30-4 (1967), 741.
(4)   G. Brown and A. Roshko, On Density Effects and Large Structure in Turbulent Mixing Layers, J. Fluid Mech., 64-4 (1974), 775.
(5)   T. Asanuma, Handbook of Flow-Visualization (in Japanese), (1976), Asakura Publ. Corp.
(6)   H. Yamada, Instantaneous Measurement of Air Flows by Smoke-Wire Technique, Trans. Japan Mech. Engrs., 39 (1973), 726.
(7)   K. Torii, Flow Visualization by Smoke-Wire Method, Proc. 3rd Symp. on Flow Visualization, ISAS Univ. of Tokyo, (1975-6), 5.
(8)   N. Kasagi, Study on Transport Mechanisms in Turbulent Boundary Layers With Body Forces (in Japanese), Dr. Engrg. Thesis, (1974-12), Univ. of Tokyo.
(9)   N. Kasagi and M. Hirata, 'Bursting Phenomena' in Turbulent Boundary Layer on a Horizontal Flat Plate Heated From Below, Proc. 1976 ICHMT Seminar on Turbulent Buoyant Convection, Dubrovnik, (1976-8), 27.
(10) e.g., S.G. Nychas, H.C. Hershey and R.S. Brodkey, A Visual Study of Turbulent Shear Flow., J. Fluid Mech., 61-3, (1973), 513.
(11) N. Kasagi and M. Hirata, Transport Phenomena in Near-Wall Region of Turbulent Boundary layer Around a Rotating Cylinder, ASME Paper 75-WA/HT-58, (1975).
(12) S. Yokobori, N. Kasagi and M. Hirata, Characteristic Behaviour of Turbulence in the Stagnation Region of a Two-Dimensional Submerged Jet Impinging Normally on a Flat Plate, Proc. Symp. on Turbulent Shear Flows, Penn State, (1977-4), 3.17.
(13) N. Kasagi, M. Hirata and H. Hiraoka, Large-Eddy Structures in Turbulent Separated Flow Downstream of a Rearward Facing Step, ibid., 16.14.

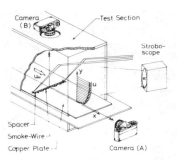

Fig.1 Schematic of Smoke-Wire method
in flat plate turbulent boundary layer

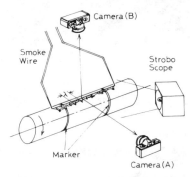

Fig.3 Schematic of Smoke-Wire method
in turbulent boundary layer around a
rotating cylinder

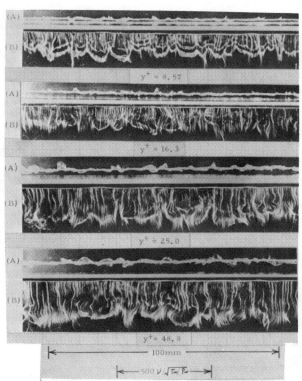

Fig.2 Flow-visualization in wall-region
of flat plate turbulent boundary layer

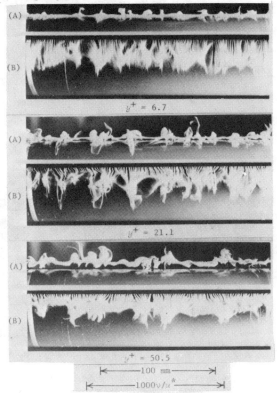

$Re_d = 1.00 \times 10^4$

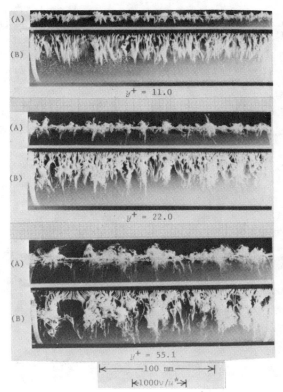

$Re_d = 3.00 \times 10^4$

Fig.4 Flow-visualization in near-wall region of turbulent boundary layer
around a rotating cylinder

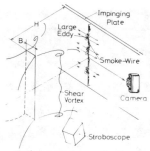

Fig.5 Schematic of Smoke-Wire method in 2-D impinging jet

Fig.7 Schematic of Smoke-Wire method in separated flow

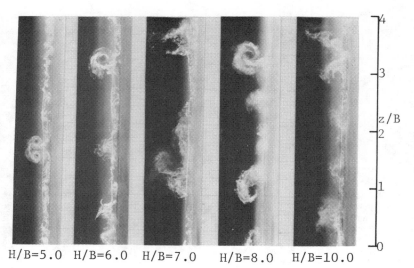

H/B=5.0 H/B=6.0 H/B=7.0 H/B=8.0 H/B=10.0

Fig.6 Flow-visualization in stagnation region of 2-D impinging jet (B=120mm, Re_B=8.12x10^4)

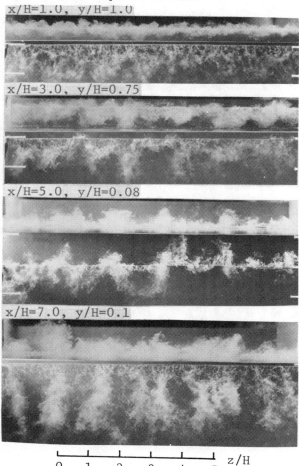

x/H=1.0, y/H=1.0

x/H=3.0, y/H=0.75

x/H=5.0, y/H=0.08

x/H=7.0, y/H=0.1

0 1 2 3 4 5 z/H

Fig.8 Flow-visualization in separated and reattaching flow over a step

Bursting   Sweeep

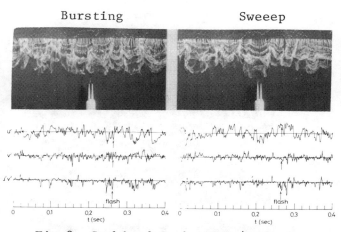

Fig.9 Combined Smoke-Wire/hot-wire technique in flat plate turbulent boundary layer

Bursting   Sweep

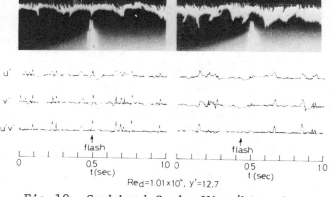

Re_d=1.01x10^4, y^+=12.7

Fig.10 Combined Smoke-Wire/hot-wire technique in turbulent boundary layer around a rotating cylinder

FLOW VISUALIZATION
BY SMOKE-WIRE TECHNIQUE

K. TORII*

The essential data and technique of a smoke-wire method
visualizing air-flow with a velocity up to tens of meter per
second are presented. A nicrome wire of 0.1 mm in diameter is
coated with oil which is vaporized by passage of an electrical
pulse through the wire. This electrically controlled smoke can
trace either the time-lines or streak lines of a flow at a
very fine pitch. The technique is successfully applied to visualize the laminar and turbulent boundary layers on a flat plate
and the wake behind a cylinder. The technique is also applied
to the flow around a sphere situated in a turbulent stream to
find the mechanism that the heat transfer is augmented by the
free-stream turbulence.

INTRODUCTION

Many methods of visualizing air flow by means of tracer have been proposed, but
failed to prove their feasibility for quantitative analysis and are rather difficult
to be applied to a flow faster than several meters per second. The smoke-wire method
being proposed here has its excellent feature in principle as it can electrically
control the smoke to make quantitative measurements and can visualize a flow without
disturbing the flow so much. However, it has not been used so often partly because
of the poor technique of smoking, though the idea itself was introduced in the early
fifties (Ref.1).   By considering the mechanism of smoking, the author improved the
technique to visualize the time-lines of smoke for a wide range of flow velocities in
a turbulent flow as well as in a laminar one. The technique was also developed to
visualize streak-lines of a flow at a very fine pitch.   The essential data and
technique such as the time-lag and duration-time of smoking, the buoyant effect and
the control of smoking are obtained by visualizing the laminar and turbulent boundary
layers on a flat plate and Karman's vortex streets behind a cylinder. The flow
around a sphere is visualized to find the mechanism that the heat transfer is augmented by the free-stream turbulence.

PRINCIPLE AND METHOD

A fine wire in air flow is coated with liquid paraffin which is vaporized by
passage of an electrical pulse through the wire and then condenses to form photogenic
white fog. The resultant line source of smoke is then photographed at a certain
time-delay, t_f after the pulse passes. The distance traveled by a point in the smoke-
trace downwind from the wire is proportional to the local mean flow velocity.
Figure 1 shows the smoke wire system as utilized during velocity profile experiments. The circuit diagram is shown in Figure 1(b). The intensity and duration-time

---

* Associate Professor of Mechanical Engineering, Yokohama National University,
Yokohama, Japan.

of the pulse is varied by the charged voltage, E and capacitance, C of a condenser respectively. As this stable delay-circuit has two channels, it can opperate either two strobes independently in IND position of Switch, S10 or one strobe flashing twice in the COM position. To prevent the strobe from flashing due to induced pulses, condensers are inserted in the relays and switches and the smoke-wire section is separated from an alternate power source by a transformer.

Mechanism of Smoking: Joule energy generated in the smoke wire is partially transfered to the oil film and the rest of all is stored as internal energy in the wire, which results in the rise of the wire temperature according to the heat capacity. The heat transfered to the oil film is partially transfered to the air flow and the rest is absorbed as the temperature rise and evaporation of the oil. As compared with several kilowatts of the joule energy, the heat transfered to the flow is only a few watts. Therefore, the smoking rate depends almost solely on the factors determining the heat transfer between the wire and the oil film, such as joule heating, wire temperature, oil-film thickness, boiling point and surface tension. This is very important to find the ways of the control of the smoke wire.

Control of Smoke: In order to obtain a straight sharp line of smoke in which the time-lag, t_s and duration-time, Δt of smoking are uniformly short, it is essential to make uniformly thin oil film on the wire. This is to be obtained naturally and reproducibly only if we carefully wipe off the oil-droplets left on the wire, since the thickness of the film is determined by the diameter of the wire and the physical properties of the film. Then, either (E=250 V, C=47μF) or (E=150 V, C=220 μF) can generate a good smoke line, because it can be realized only if the input power (=CE^2 /2) is supplied enough to keep the wire temperature appropreate for the evaporation of the oil-film. The adjustment should be done by E rather than by C. Once we find the optimal values of E and C, we need not change it according to the flow velocity. When the flow gets faster, the smoke-line gets wider and blurs more for a constant duration-time of smoking. In this case, a sharp smoke-line can be obtained by splashing out the oil from the wire by means of raising the wire temperature.   When the oil is coated on the wire, it naturally draws up into very small beads, from which smoke can be emitted for a considerably long time (up to 300 ms) if the wire is controlled to moderate temperature not to splash them. The streak-lines of a flow can be visualized at a very fine pitch (≈1 mm) in this way. Assumed that the evaporation rate is constant for simplicity, the maximum temperature of the wire is approximately given as follows,

$$(T_w - T)_{max} = \frac{2}{\pi} \frac{C}{\rho c} \frac{E^2}{\ell d^2} (1 - \frac{\tau}{2\Delta t} (1 + \ln \frac{2\Delta t}{\tau}))$$

$$\simeq \frac{2}{\pi} \frac{C}{\rho c} \frac{E^2}{\ell d^2} \qquad \text{for } \Delta t >> \tau \equiv RC \tag{1}$$

where ρ, c, l, d and R represent the density, heat capacity, length, diameter and electrical resistance of the wire respectively. When we use a different wire, we may adjust E and C according to Eq. (1) to hold the same optimal temperature. The duration-time of smoking at the optimal condition is found to be approximately proportional to the diameter of the wire, d in case of $\Delta t >> \tau$.

Wire: The smaller is the diameter of the smoke wire, the sharper the smoke line becomes in a low speed flow. In a flow faster than several meters per second, where the high rate and short duration-time of smoking are necessary, the diameter must be large enough to get the high smoking rate and to withstand the tension at so high temperature as to splash the oil. The optimal size may be about 0.1 mm. According to Eq. (1), the wire is not necessarily to be a material of high specific resistance. As the appropriate resistance of the wire is ten ohms or so for reasonable capacity of the power supply, such material is chosen as the wire has this appropriate resistance in its length necessary for the measurement of the flow visualization.

Oil: In a high speed flow liquid paraffin is preferable but is apt to be splashed, so machine oil or kerosene is mixed at the sacrifice of the density of the smoke. Silicon oil of a little viscosity and surface-tension can be coated very thin on the wire to result in a uniformly fine smoke-line, though the density of the smoke is less than that of liquid paraffin.

RESULTS AND DISCUSSIONS

Laminar Boundary Layer: The laminar boundary layer on a flat plate is visualized and compared with the theory by Blasius so that the precision of the velocity-profile measured by the smoke wire technique may be found. Vertical wind tunnel in Fig.1(a) is employed to eliminate the buoyant effect. A flat plate (650x100x2mm) is placed vertically in the duct and a nichrome wire (d=0.1mm, l=50mm) connected with copper lead-wires of 0.2 mm is horizontally stretched through a hole at the distance, x_o of 500 mm from the sharp leading edge of the plate. Figure 2 shows that two time-lines of smoke are photographed by flashing one strobe two times (t_{f1}, t_{f2}). The distance along the plate $(x_2(y)-x_1(y))$ which the first time line (I) travels to the second time line (II) is measured at the distance from the wall, y and the velocity profile is obtained by $U(y)=(x_2-x_1)/(t_{f2}-t_{f1})$, which is plotted by the open circles (II-I) in Fig.3 and shows an excellent agreement with the theoretical one. The velocity profile obtained by a single time-line is less accurate as plotted by squares or triangles in Fig.3. A single time-line cannot always express the exact velocity profile due to the non-uniformity of smoking, the buoyant effect and the disturbance of the wire. Two time-lines or more are necessary for a turbulent or unsteady flow. The duration-time of smoking is determined by the width of the smoke line in the free stream, b as $\Delta t=b/U_\infty$. The time-lag of smoking is obtained by $t_s=t_f-(x-x_o)/U_\infty$ in the free stream.

Buoyant Effect: The buoyant effect may be much stronger in a vertical smoke-wire than in a horizontal one, since it is proportional to the square root of the vertical length of the wire. The rising velocity of the smoke-line, U_b is determined to be $U_b=$ 5cm/s by the vertical shifts of two smoke lines from the vertical smoke wire situated in a horizontal wind tunnel as shown in Fig.4. The smoke line seems to curve more at the lower part due to the combined effects of the buoyancy and the boundary layer than at the upper part.

Turbulent Boundary Layer: The two time-lines of smoke of the turbulent boundary layer on a flat plate are photographed by two strobe-lights. The apparatus employed in Fig.5 (a) is shown in Fig.1 (a). A piece of sand paper is stuck on the entrance-wall of the duct to trip the laminar boundary layer on the duct wall. A flat plate (700x150x14.25mm) with a round leading edge is situated in a wind tunnel (300x150mm) for the measurement in Fig.5 (b). (Ref.2) The large scales of the intermittent eddies in the turbulent boundary layer are observed in Fig.5. The time-lag and duration-time of smoking are found to be controlled well to 1~2ms and 1~1.5ms respectively, independent of the flow velocity. The velocity profiles measured by the smoke wire method are compared with those measured by a hot-wire anemometer and show good agreements in Fig.6.

Wake Behind Cylinder: Figures 7(a), (b), (c) and (d) visualize Karman's vortex streets behind a circular cylinder of 6 mm in diameter,D in the wind tunnel shown in Fig.1(a) by the streak lines of the smoke wire situated at x_o=+6mm downstream of the rear stagnation-point. A constantan wire of 0.2 mm is used for the longer duration-time of smoking in Fig.7(a) and (b). The difference due to the diameter of the wire is shown in Fig.7(b) and (c). A cylinder of 11 mm is situated in the duct (800x300x 150mm) in Fig.7(e).

Flow Around Sphere: The author has been studying the free stream turbulence effects on the heat and mass transfer around a sphere. Heat transfer augments more with the increase of the free stream turbulence in the rear half of the sphere than in the front. The distribution of heat transfer around a sphere has a peak, which becomes higher with the increase of Reynolds number and turbulence intensity at the angular position of 110 or 120 degrees from the front stagnation point. Flow visualization by the smoke wire technique is expected to enhance physical interpretation of the mechanism of the heat transfer augmentation described above. A sphere of 50 mm in diameter, D is suspended in a 300x300 mm duct by eight wires of 0.3 mm in diameter. The flow around the sphere without free stream turbulence is visualized by the two smoke-wires in Fig.8(a). One wire is 120 mm upstream of the rear stagnation point $(x_0=-120$ mm) and the other is 5 mm downstream $(x_0=+5$ mm). As shown in Figs.8(a) and (b), the flow separates near 80 degrees, the free shear layer has transition from laminar to turbulent layer at x=0, which might be transition from laminar flow to ring vortex street, and the boundary of the wake becomes minimum in width 2D downstream (x =100 mm), which might mean that the vortex begins shedding spirally. The velocity distribution in recirculating zone which cannot be measured by a hot-wire anemometer, is obtained by the two time-lines of smoke in Fig.8(c). It has considerably flat backward velocity $(U=0.4U_\infty)$ covering 0.8D wide at x=0.5D as shown in Fig.8(d). The absolute value of the velocity at the axis of symmetry(r=0) coincides with the one by a single hot-wire probe. The flow around the sphere with free-stream turbulence of 3.4 % in intensity, Tu is visualized in Fig.9. As compared with the flow without free-stream turbulence, it seems that its separation point moves downstream to 90 degrees and that the boundary of the wake gets slender a little but the recirculation gets stronger. However it is still uncertain, since the Reynolds number is so low that the effects of free-stream turbulence may not be strong enough to be observed. Further investigations are being conducted. An attempt will be made to visualize the flow three-dimensionally with two cameras or high speed cinecamera in future experiments.

CONCLUSIONS

1) The smoke wire is a simple but powerful tool visualizing airflow quantitatively.
2) The smoke wire is an excellent anemometer of a low velocity free from calibration.
3) The applicable range of velocity without losing accuracy may be from 0.5 to 20 m/s.
4) The duration-time of smoking can be controlled from 1 ms to 300 ms.

ACKNOWLEDGEMENT

The author wishes to express his appreciation to Mr. T. Horikoshi of Yokohama National University for valuable assistance, discussions and suggestions.

REFERENCES

(1) J.J. Cornish, A Device for the Direct Measurement of Unsteady Air Flows and Some Characteristics of Boundary layer Transition, Aerophysics Res. Note, 24 (1964),1.
(2) T. Asanuma ed., Flow Visualization Hand Book, (1977), 305, Asakura.

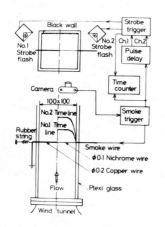

    (a) Apparatus.   (b) Circuit diagram.
Fig. 1. Smoke wire system.

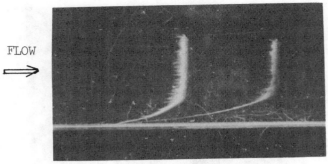

FLOW ⟹

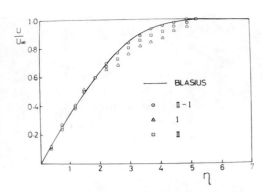

U_∞=1.28m/s,Re_x=45100,x=557mm,x_0=500mm,
d=0.1mm,l=50mm,t_{f1}=30.7ms,t_{f2}=62.8ms,
E=120V,C=100μF,Δt=2.3ms,t_s=2.4ms.

Fig.2 Laminar boundary layer on flat plate.

Fig.3 Comparison between the visualized and theoretical laminar velocity profiles.

FLOW ⟹

WIRE U_∞=0.87m/s,d=0.1mm,l=50mm,
t_{f1}=89.7ms,t_{f2}=182ms.

Fig.4 Rising Velocity due to buoyancy.

(a)   (b)

(a) U_∞=4.57m/s,Re=145000,x_0=500mm,d=0.1mm,
    l=50mm,t_{f1}=8.77ms,t_{f2}=17.42ms,E=110V,
    C=300μF, t=1.1ms,t_s=0.9ms.

(b) U_∞=17.3m/s,Re=343000,x_0=300mm,d=0.1mm,
    l=80mm,t_{f1}=4.51ms,t_{f2}=7.47ms,E=120V,C=220
    μF,Δt=1.1ms,t_s=1.29ms.

Fig.5 Turbulent boundary layer on flat plate.

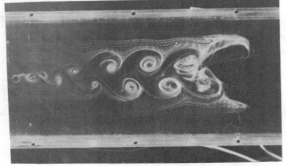

Smoke wire probe
○ U_∞=17.3 m/s δ=14mm
△ U_∞=4.6 m/s δ=14mm
—— Hot wire probe

Fig.6 Comparison between smoke wire and hot wire probes in turbulent boundary layer.

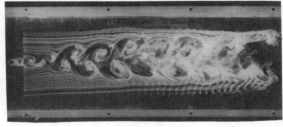

Fig.7(b) U_∞=1.6m/s,Re=596,D=6mm,x_0=6mm,d=
0.2mm,l=70mm,t_f=161.2ms,E=110V,C=550μF.

(a) U_∞=0.9m/s,Re=335,D=6mm,x_0=6mm,d=0.2
    mm,l=70mm,t_f=193.7ms,E=105V,C=550μF.

Fig.7(c) U_∞=1.6m/s,Re=596,D=6mm,x_0=6mm,d=
0.1mm,l=70mm,t_f=1o4.5ms,E=50V,C=550μF.

Fig.7 Flow behind a circular cylinder (Karman's vortex street).

$U_\infty=4.7m/s, Re=1760, D=6mm, x_o=6mm, d=0.1$
$mm, 1=70mm, t_f=55.2ms, E=60V, C=550\mu F.$

Fig. 7 (d) Flow behind circular cylinder.

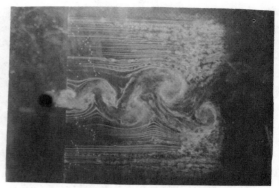

$U_\infty=8.6m/s, Re=6670, D=11mm, x_o=11mm, d=0.1$
$mm, 1=150mm, t_f=17.6ms, E=120V, C=220\mu F.$

Fig. 7 (e) Flow behind circular cylinder.

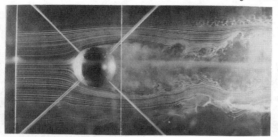

(a) $x_o=(-120mm, 5mm), 1=(120mm, 90mm),$
$t_f=70.4ms, E=140V, C=330\mu F.$

(b) $x_o=20mm, 1=90mm, t_f=77.7ms, E=100V, C=330\mu F.$

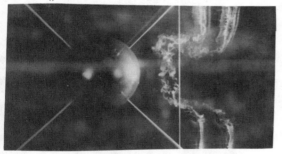

(c) $x_o=25mm, 1=90mm, t_{f1}=4.09ms, t_{f2}=9.25$
$ms, \Delta t=1.8ms, t_s=2.0ms, E=150, C=300\mu F.$

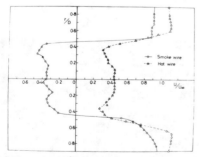

(d) velocity profile obtained by Fig.8(c).

Fig. 8 Flow around sphere without free-
stream turbulence; $U_\infty=6.5m/s, Re=21000, D=50$
$mm, d=0.1mm.$

(a) $x_o=-110mm, 1=110mm, t_f=50.9ms, E=100V.$

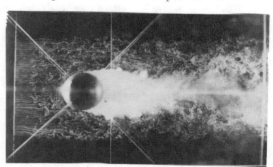

(b) $x_o=(-110mm, 10mm), 1=(110mm, 90mm), t_f$
$=64.1ms, E=130V.$

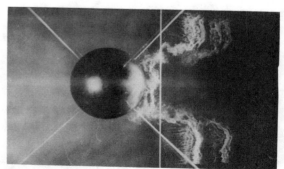

(c) $x_0=10mm, 1=90mm, t_{f1}=5.77ms, t_{f2}=$
$10.74ms, t_s=2.0ms, E=140V.$

Fig. 9 Flow around sphere with free-stream
turbulence; $U_\infty=6.5m/s, Re=21000, Tu=3.4\%, D=$
$50mm, d=0.1mm, C=330\mu F.$

VISUALIZATION OF TURBULENT AND COMPLEX FLOWS USING CONTROLLED SHEETS OF SMOKE STREAKLINES*

H. M. NAGIB**

The "smoke-wire" technique is used to visualize the flow
past single and multiple bluff bodies, and flows in thick turbulent boundary layers simulating the atmospheric surface layer.
A synchronization circuit controls the duration of time the wire
is supplied with the heating current and triggers the camera and
strobe lights after an adjustable delay. Operation of vertical
or horizontal "smoke wires" and the supporting electronics is
controlled from outside the tunnel even while experiments are
in progress. Instantaneous and time-averaged records of the
flow are used to highlight the various scales of motion.

Introduction

Until recently, smoke visualization in wind tunnels has been limited by one or
more of the following: maximum useful velocities and turbulence levels of the flow;
means of local introduction and ease of repositioning of the smoke streaks; generation of flow disturbances by the technique; contamination of tunnels, especially in
recirculating systems; and density and quality of the smoke. A new visualization
technique, incorporating a vertically oriented "smoke wire" mounted on a portable
probe, was developed (Ref. 1) to minimize these limitations and to maximize its adapt-
ability over a wide range of flow conditions and test facilities. Based on our experience in water (e.g., Ref. 2, 3), the objective was to arrive at a method in air
which would be comparable in its quality and simplicity to the hydrogen-bubble technique. Therefore, we required that the technique does not result in flow disturbances
in the field of interest, and is capable of generating dense sheets of controllable
smoke streaklines, which could be readily photographed.

As pointed out and demonstrated by Hunt et al. (Ref. 4), topology is a very helpful tool in studying complex flow fields with the aid of flow visualization. At IIT
we have believed for some time in the utilization of several visualization techniques
to highlight the various aspects of such flow fields (e.g., Ref. 5). In addition, the
"smoke-wire" and the hydrogen bubble methods permit the utilization of the same technique in several regions of the flow as well as observation from various angles.

In presence of turbulence or unsteadiness in the flow, the use of several time-
exposures of different durations, including very short or instantaneous ones, leads
to enhanced appreciation of the details of the flow. With the aid of a control system
like the one introduced by Corke et al. (Ref. 1) and used in this paper, the "smoke-

* Supported under Grant ENG76-04112 from the National Science Foundation, U.S.A.

** Associate Professor of Mechanical and Aerospace Engineering, Armour College of
Engineering, Illinois Institute of Technology, Chicago, Illinois 60616, U.S.A.

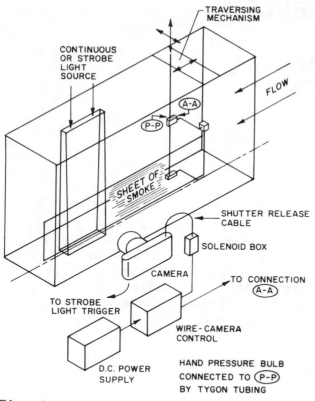

Fig. 1 Set-up of "Smoke-Wire" Visualization Technique

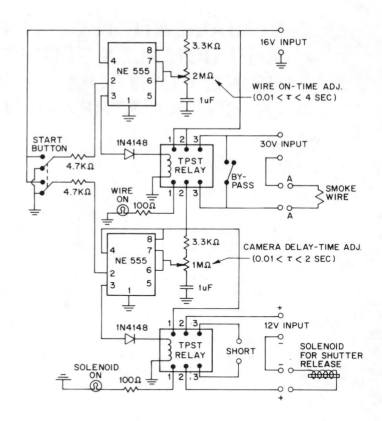

Fig. 2 Electronic System for Controlling Wire Energizing Time and Camera Delay

Fig. 3 Variation of Flow Past a Complex Bluff Body, Composed of a Cylinder and a Flat Plate, With Cylinder-Plate Separation Distance

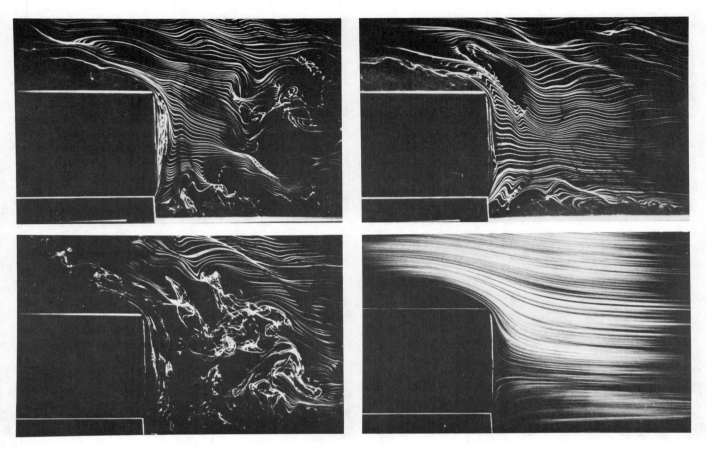

Fig. 4   Three Instantaneous Records, and One Time-Exposure, of Flow Past a
Rectangular Bluff Body Placed With a Ground Gap on a Flat Plate in a Turbulent Flow

wire" method can be used to provide controlled and reliable records of different time
exposures. Further developments of this control circuit are presently in progress
and some of them will soon be presented by Corke et al. (Ref. 6). These developments
will permit the triggering of the circuit by an external signal, such as a condition-
ed anemometer output, the use of the smoke wire at any orientation (including horizontal) and the automatic re-coating of the smoke-wire for the next photographic record.
The external triggering approach is being used to capture particular flow structures,
such as the shedding from a bluff body, in several records at the same phase of the
motion, i.e., state of the development of the flow structure. The application of this
technique to study the role of large scale and coherent motions in turbulent flows is
planned.

Finally, with the aid of advanced photographic techniques, fine resolution and
high quality images of the flow can be obtained from the smoke wire method as demonstrated by Corke et al. (Ref. 1) and in this paper. Digital image processing techniques, which have been developed over the past few years, in particular for medical
applications, can be employed in the processing of the photographic records of complex flow fields. The digital records can be generated by direct processing of a
video output signal or with the aid of an arrangement for scanning the photographic
records. Such an arrangement would incorporate stepping motors and position indicators as well as a light source and a light intensity detector. The digitized images
can then be processed by a data aquisition and processing system or on a large computer (Ref. 7). The same images may then be processed digitally by various methods
including conventional or conditional averaging techniques.

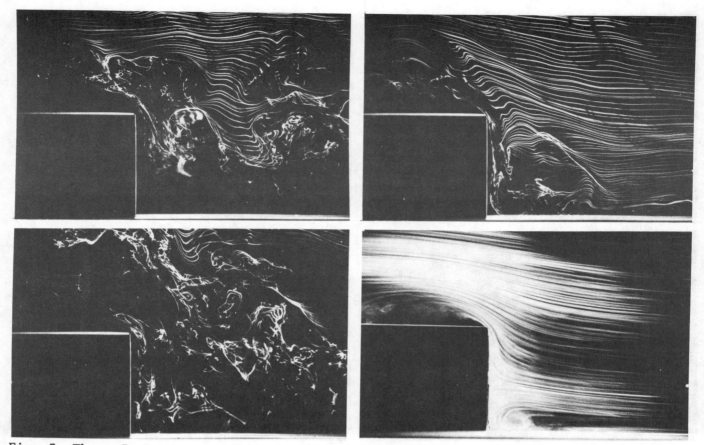

Fig. 5   Three Instantaneous Records, and One Time-Exposure, of Flow Past a Rectangular Bluff Body Placed on a Flat Plate in a Turbulent Flow

Smoke-Wire Technique

The "smoke wire" consists of a 0.1mm diameter wire onto which regulated droplets of mineral oil are allowed to fall. The oil coats the wire surface forming minute droplets along its length. The spacing and density of these droplets are a function of the oil and wire surface characteristics. By passing an electronic current through the wire, the oil is burned off to produce a sheet of smoke consisting of droplet-produced, discrete streaklines. The wire is mounted on a portable probe which can be supported through the wall of the wind tunnel or in a traversing mechanism. A typical arrangement for the application of the technique is schematically represented in Fig.1.

A synchronization circuit, which is activated by triggering the wire, controls the duration of time the wire is supplied by the heating current and triggers the camera and strobe lights after an adjustable time delay. Thus, only small amounts of smoke are required to obtain quality photographs, thereby reducing the contamination of the flow in recirculating wind tunnels caused by the accumulation of smoke. A full range of flow speeds and turbulence levels can be met by adjusting the synchronous delay and the duration of the heating of the wire. The circuit diagram is shown in Fig. 2. For further details on the technique the reader is referred to Corke et al. (Ref. 1).

Applications

In the first of the two applications presented here, the flow is visualized past a complex two-dimensional bluff body consisting of a thin flat plate of width D perpendicular to the flow, and a cylinder of diameter D placed parallel to the plate with a gap s between them. The distance along the lateral direction between the center-

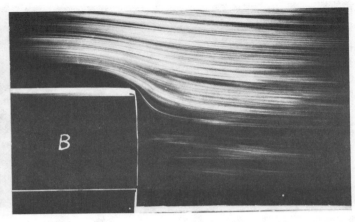

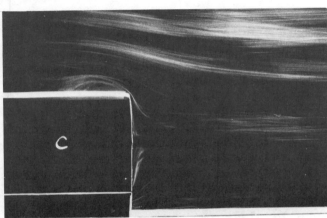

Fig. 6 Effect of Increasing (From A to C) Turbulence Intensity, Velocity Gradient and Thickness of Boundary Layer on Flow Past a Rectangular Body Placed With a Ground Gap on the Floor of Simulated Atmospheric Surface Layers

lines of the two bodies is d; i.e., s = d - D. While various downstream separations between the cylinder and the plate have been investigated, the examples presented here are limited to cases where the centerlines of the two models are located along a plane perpendicular to the direction of the flow. By changing the gap s, various regimes of flow can exist in the wake of such a composite bluff body. Figure 3 displays some of these regimes for D = 1.27cm and Re_D = 3200. The gap s between the two bodies in photographs a, b and c is 3.58, 1.42 and 0.25cm, respectively. The circular light-color body encompassing the two models is the support system outside the test section.

While most of the details of the potential part of the flow past the bodies are revealed by these instantaneous photographs (exposed by strobe flashes of a few microseconds), only some of the global features of the flow in the wake can be ascertained. Positioning the wire in the wake of the body leads to the enhancement of the details of the wake. By changing the spacing between the streaklines (Ref. 1) various levels of details are captured in the photographs. Some of the observations that can be made from the photographs of Fig. 3 are: 1) the difference in the angle of the separation streakline for the cylinder and the flat plate and the resulting larger size of the part of the wake associated with the flat plate, e.g., see photograph a; 2) the difference in the shedding frequency on the two sides of the composite wake, see photograph b, in particular; 3) the coexistence of various scales of semi-periodic motions on the two sides of the composite wake, e.g., see top side of the wake of photograph c; and 4) the bi-stable separation pattern in the middle part of the composite wake, e.g., see wake area between two models in photograph b.

In the second application, turbulent flows past a bluff body with and without a ground gap are visualized. The height and width of the parallelopiped body are 10cm each and its length along the flow direction is 20cm. The 2.5cm gap between the body and the floor of the boundary layer in which it is immersed is provided by four corner cubical blocks. The Reynolds number based on the width of the body and the free stream velocity is of the order of 10^4. Three cases are presented here and designated as A, B and C. The boundary layers are intended to simulate the atmospheric surface

Fig. 7 Effect of Increasing (From A to C) Turbulence Intensity, Velocity Gradient and Thickness of Boundary Layer on Flow Past a Rectangular Body Placed on the Floor of Simulated Atmospheric Surface Layers

layer, and are developed, with the aid of the counter-jet technique and surface roughness (Ref. 8) along the 7m floor of a 1.2 x 1.8m test section of a wind tunnel. A series of grids at the entrance to the test section maintain the turbulence level at 1% in the free-stream above the boundary layer. The bluff body was placed on a flat-plate located in the free stream, and on the floor of the test section in two boundary layers of increasing velocity gradient and turbulence intensities and tested in Flow Cases A, B and C, respectively. All of the photographs shown here depict the condition along the centerline of the body.

The three realizations of an instantaneous visual record in Fig. 4 demonstrate the continuous changes in the details of the flowfield in presence of only 1% turbulence, i.e., Case A. In particular, the conditions in the stagnation region appear to vary considerably and the flow through the gap seems to display some unsteadiness. While in the top two visualization records, as well as in the time-mean record, some flow is shown passing under the bluff body, the flow through the gap appears to be momentarily "choked" in the bottom left photograph. The exposure duration for the bottom right picture of this figure, as well as for others shown in the following figures, is typically from 0.25 to 0.5 seconds; i.e., the time required for a fluid particle to travel a distance of approximately one to two heights of the bluff body.

Eliminating the ground gap leads to substantial changes in the flow past the bluff body as can be deduced from comparing Figs. 4 and 5. A horseshoe vortex is developed even for Case A where the approach boundary layer thickness is approximately 1.5cm; i.e., 1/8 of the height of the body. As many as ten substantially different patterns of the flow past the model, like those of Figs. 4 and 5, may be obtained by repeated instantaneous records of the flow. These records reveal the variability and the intermittency of the turbulent flow. On the other hand, time exposure photographs of sufficiently long duration were found to be quite repeatable.

The effect of changing the conditions of the incoming flow, by using Cases A, B and C, are demonstrated, with the aid of time-exposure records, for the bluff body with and without the ground gap in Figs. 6 and 7, respectively. As demonstrated by the photographs and during the course of the study, the smoke wire is a useful tool which may be utilized quantitatively and can handle the turbulence in the flow. Some of the key observations made from the second application are: 1) the secondary flow, in the form of a horseshoe vortex, is very unsteady and, hence can only be visualized in time-exposure records; 2) this secondary flow can be rather complex and composed of several cells, as shown in Fig. 7; 3) the horseshoe vortex disappears in the presence of a gap, producing a flow under the bluff body, and grows larger and diffuses as the velocity gradient is increased and the turbulence level raised, respectively; 4) as displayed in Figs. 6 and 7, increasing the upstream-flow velocity gradient and turbulence for a bluff body with or without a gap leads to the upward shift of the stagnation "point" and to the faster reattachment of the flow separated along the top leading edge; 5) introducing a gap which allows flow under the body leads to the downward shift of the stagnation region; 6) even with only a relatively short segment of the ground plate upstream of the body in Case A, a boundary layer develops quickly and the rapid change in its displacement thickness leads to some inclination of the average streamlines, as depicted by the smoke streaklines at a negative angle of attack with respect to the body; 7) the ground plate effects may be limited to bodies with small gaps, small being measured by the absence of significant flow through the gap.

Acknowledgment

The author expresses his appreciation to Marsha G. Faulkner for her editing and expert typing of the manuscript, and to Jimmy Tan-atichat and Tom Corke for assistance in preparing the figures. This research was supported under NSF Grant ENG76-04112.

References

1. Corke, T., Koga, D., Drubka, R., and Nagib, H., "A New Technique for Introducing Controlled Sheets of Smoke Streaklines in Wind Tunnels," Proceedings of International Congress on Instrumentation in Aerospace Simulation Facilities, IEEE Publication 77CH1251-8 AES, 1977, p. 74.

2. Loehrke, R.I. and Nagib, H.M., "Experiments on Management of Free-Stream Turbulence," AGARD Report, R-598, 1972, AD-749-891.

3. Fejer, A.A., "Flow Visualization Techniques for the Study of the Aerodynamics of Bluff Bodies and of Unsteady Flows," Proceedings of International Congress on Instrumentation in Aerospace Simulation Facilities, 1973, p. 175.

4. Hunt, J.C.R., Abell, C.J., Peterka, J.A. and Woo, H., "Kinematical Studies of the Flows Around Free or Surface Mounted Obstacles; Applying Topology to Flow Visualization," to appear in J. Fluid Mech., 1978.

5. Nagib, H.M. and Hodson, P.R., "Vortices Induced in a Stagnation Region By Wakes-- Their Incipient Formation and Effects on Heat Transfer from Cylinders," AIAA Paper 77-790, 1977.

6. Corke, T.C., Crawford, A.C. and Nagib, H.M., "Visualization of Turbulent Flows Using Controlled Sheets of Smoke Streaklines," Bulletin Am. Phys. Soc., Nov. 1977.

7. Way, J.L., "Applications in Fluid Mechanis Research of a Portable Data Acquisition and Processing System," Proceedings of International Congress on Instrumentation in Aerospace Simulation Facilities, 1975, p. 50.

8. Nagib, H.M., Morkovin, M.V., Yung, J.T. and Tan-atichat, J., "On Modeling of Atmospheric Surface Layers by the Counter-Jet Technique," AIAA J., 14, 2, 1976, p. 185.

USE OF SMOKE WIRE TECHNIQUE IN MEASURING VELOCITY PROFILES OF OSCILLATING LAMINAR AIR FLOWS

HIDEO YAMADA*

Smoke wire technique was applied to measure velocity profiles of oscillating laminar air flows in the following cases: (1) flows on an oscillating flat plate immersed in a fluid at rest as well as in a uniform flow, and (2) simple periodic flow as well as transient one from rest to an oscillating flow in a rectangular channel. The evaluations of the smoke wire technique were made primarily by comparing experimental results to the theoretical velocity profiles. The smoke wire technique was found to be effective for measuring instantaneous velocity profiles in the arbitrary phases of oscillating flows as described above, although some restrictions in experiments are involved.

Introduction

The smoke wire technique is simple and useful in measuring velocity profile of laminar air flow and in visualizing the flow field. In the smoke wire method a time delay circuit is actuated at the discharge of the electric current through the wire; after a predetermined delay time a strobe light fires to illuminate a smoke profile and simultaneously photographs are taken. The delay time favorably used for measuring velocity profiles ranges usually between about 5 to 100 msec. It may be able to measure approximate, instantaneous velocity profiles of an unsteady flow if the substantial derivative of the velocity of the flow field is relatively small during the delay time used.

The present paper aims to apply the technique to the measurement of velocity pofiles of the incompressible, two-dimensional, oscillating laminar air flows, and evaluate the usefulness of the technique by comparing experimental results to the theoretical ones. In this paper experimental results for two contrastive cases are described, namely (I) the case that a solid boundary performs harmonic motion, and (II) the case that a main flow itself oscillates harmonically while solid boundaries being kept stationary. In the case of (I), the following two cases are considered: (1) when the oscillating plate is immersed in a still fluid, and (2) when it is immersed and placed parallel to a uniform flow. On the other hand, in the case of (II), harmonic flows in a rectangular channel are discussed for both simple harmonic flow and transient one.

---

*Lecturer of Fine Measurement and Instrumentation Engineering, Nagoya Institute of Technology, Gokiso-cho, Showa-ku, Nagoya, Japan

Experimental Apparatus and Methods

Case (I)*:

The oscillating plate used in the tests was 0.4 cm thick aluminium plate with a polished and lacquered surface and was 40 cm long and 21 cm wide. The flat plate was driven by gears and an eccentric harmonic drive system. The amplitude of the oscillation was 2.54 cm. The oscillating plate was mounted horizontally at the middle of the test section of a low speed wind tunnel. The frequency of the oscillating plate was adjusted by a variable power supply. A nichrome wire of 40μ diameter and 4.6 cm length was used as a smoke wire. The probe of the smoke wire was mounted perpendicular to the plate. The oil coated on the smoke wire was liquid paraffin. The delay time used in the tests was 50 msec. It is important that the delay time be very much less than the period of the oscillation of the plate so that the measured velocity can be considered to give the value of the instantaneous velocity. The frequencies of the oscillation of the plate in the tests were 100 and 120 cpm. Thus, the delay time was less than 10 percent of one period of the oscillating motion. The test section of the wind tunnel was enclosed in a dark room, in which a camera was set up.

In order to synchronize the measuring apparatus with the oscillating plate, a piece of metal, fixed to the side of the plate, and a microswitch, fixed to the supporting frame of the oscillating apparatus, were used. When the metal pushed the microswitch, the delay circuit started to operate. The symbols for phases of the oscillating motion, which were tested, are defined as follows. Phase angle $\pi/2$ corresponds to the position of the oscillating plate when the plate has its maximum velocity to the right, i.e., toward the main flow direction. Phase angle π corresponds to the position of the plate when it stops instantaneously at the end point to the right. Other phases, for instance, $3\pi/2$ and 2π are similarly defined.

Case (II):

The rectangular channel used in the tests was 5 cm wide, 15 cm high, and 430 cm long. The test section of 50 cm length was placed at the middle of the channel; it was constructed entirely of plexi-glas for photography and lighting. The channel was open to the atmosphere at one end. A harmonic flow was generated by a piston at another end of the channel. The amplitude was 3.8 cm. A nichrome wire with diameter of 54μ and of 5 cm length was used. It was horizontally strained across the channel at the middle of the test section.

In order to synchronize the phase of the oscillating flow to the delay circuit, the circular motion of the piston was utilized. When a piece of metal fixed on the circumference of the disc pushed the microswitch, the delay circuit started to operate. The phase of the oscillating flow was defined in an exactly similar manner as that of the previous case.

Observations and Results

Case (I):

In Figs.1(a) through (d), the smoke profiles of the flow near the surface of the oscillating plate, when the main flow is at rest, are shown in a series for the four phases in the case of 100 cpm frequency and of 50 msec delay time. The experimental velocity profiles are plotted for the phases $\pi/2$ and π together with the theoretical curves in Figs.2(a) and (b) respectively. The disagreement with the theoretical curves is due primarily to the effect of the buoyancy of the smoke and the inaccurate synchronization of the phases.

In Figs.3(a) through (d), the smoke profiles of the flow are shown when the plate oscillates in a uniform flow of 43 cm/s. It is interesting to note that the

*The experiments of case (I) were conducted during the author's stay in Mississippi State University.

266

smoke profile has three inflection points in the boundary layer for the cases 3(a) and 3(b). The corresponding theoretical velocity profiles could not be obtained for this flow.

Case (II):

In Fig.4(a) through (e), the smoke profiles of the oscillating flow in the rectangular channel are shown in a series every π/4 radians. Fig.4(c) shows the smoke profile corresponding to the phase π/2. The frequency of the piston was 1 cycle/s; the delay time was 60 msec. The flow velocity takes its maximum in the neighborhood of the solid boundary; this is well known as "Richardson annular effect"(Ref.1). This phenomenon is clearly observed in the pictures. In order to compare the experiment with the theory (Ref.2), the following procedures are assumed: (1) to measure velocity at the center of the channel from the smoke profile obtained, (2) to find theoretically the phase θ corresponding to the measured velocity, and (3) to compare the theoretical velocity profile with that of the experiment for the phase of θ. The experimental velocity profile are compared with the theoretical ones in Fig.5(a) through (d), where the solid lines show the theoretical results. The rigorous evaluation of the argument between the theory and the experiment is not so simple; because various factors should be taken into account such as phase and frequency of the oscillating flow, and delay time, so on. In the present case the effect of buoyancy was almost insensible by straining the wire horizontal; in addition the synchronization of phase was much improved by making use of circular motion of the disc. The test was performed for the frequency of 4 cycles/s, Fig.6(a) and (b). The delay time was shortened to 15 msec. Another case of the frequency of 0.5 cycles/s was tested, shown in Fig.7(a) and (b); delay time of 100 msec had to be used in order to obtain measurable and "good shape" smoke profile. Owing to this large delay time, the effect of buoyancy appeared to be appreciable. The delay time of 100 msec is considered to be practically the upper limit in the smoke wire technique when an unsteady laminar air flow is measured. In other words, the delay time restricts not only maximum frequency of the flow but its minimum.

So far, we have discussed "steady" periodic flow. Now, we treat the "transient" case, i.e., the oscillating flow that starts from rest. It should be remarked that the oil coated on the wire does not vaporize at one time unless the electric current through it is very large. Usually it may be able to produce smoke several times for every application of electric current after oil is coated once. The present experiment makes use of this nature of the oil coated on the wire. In Fig.8 the successive transient states are shown for the phase π/2. The agreement between the theory and experiment is comparatively well, as shown in Fig.9(a) and (b).

Conclusions

smoke wire technique was applied to measure velocity profiles for some kinds of oscillating laminar air flows.

The smoke wire technique was found to be effective in measuring the instantaneous velocity profiles of the incompressible, two-dimensional, laminar air flows if following cares are taken: (1) the delay time is very much less than the period of the oscillation of a flow, and (2) a smoke wire is strained horizontally to minimize the effect of buoyancy. It is worth noticing that the smoke wire technique can also be applied to measure velocity profiles of the transient harmonic flow equally well.

The author wishes to express his sincere appreciation for the help provided by Professor T. Matsui, Gifu University, during preparation of this paper.

References

(1) E.G. Richardson and E. Tyler, The Transverse Velocity Gradient near the Mouths of Pipes in which an Alternating Continuous Flow of Air is Established, Proc.

Phys. Soc., 42(part 1)-231 (1929),1.

(2) C. Fan and B. Chao, Unsteady Laminar Incompressible Flow through Rectangular
    Ducts, ZAMP, 16-3 (1965), 351.

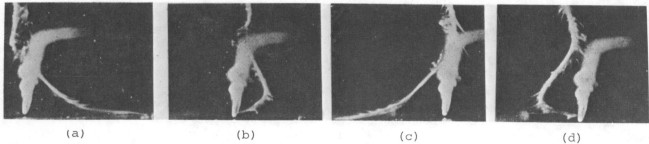

(a)　　　　　　　(b)　　　　　　　(c)　　　　　　　(d)

Fig.1 Smoke profiles of Stokes' second problem flow when phase angles of the
      oscillating flat plate are (a) π/2, (b) π , (c) 3π/2, (d) zero.

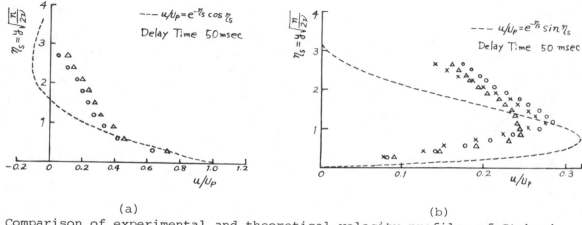

(a)　　　　　　　　　　　　　　　　(b)

Fig.2 Comparison of experimental and theoretical velocity profiles of Stokes'
      second problem flow when the phase angles of the plate are (a) π/2, and
      (b) zero.

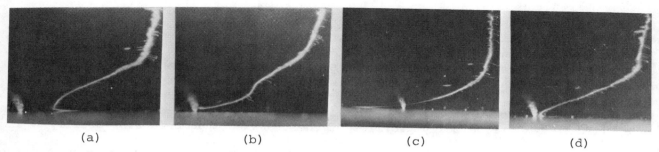

(a)　　　　　　　(b)　　　　　　　(c)　　　　　　　(d)

Fig.3 Smoke profiles of laminar boundary layer flow on an oscillating flat plate,
      phase angles being (a) π/2, (b) π, (c) 3π/2, and (d) zero.

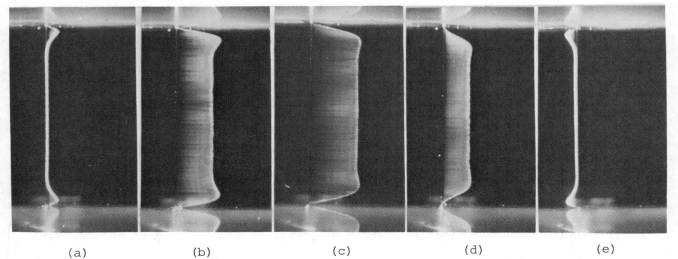

Fig.4 Smoke profiles of an oscillating flow with the frequency of ω = 1 cps in a
rectangular channel, phase angles being (a) zero, (b)π/4, (c) π/2, (d) 3π/4,
and(e) π.

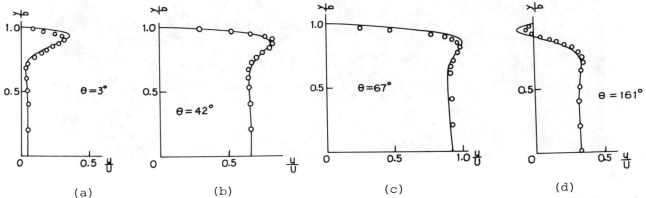

(a)　　　　　　　(b)　　　　　　　(c)　　　　　　　(d)

Fig.5 Comparison of experimental and theoretical velocity profiles of an oscillating
flow with the frequency of ω = 1 cps, phase angles being (a) zero, (b) π/4, (c)
π/2, and (d) 3π/4.

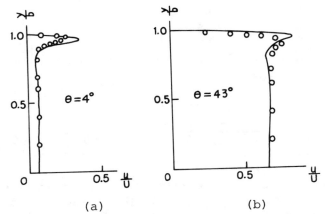

(a)　　　　　　　(b)

Fig.6 Comparison of experimental and theoretical velocity profiles of the oscillating flow with the frequency of 4 cps in a rectangular channel, phase angles
being (a) zero, and (b) π/2.

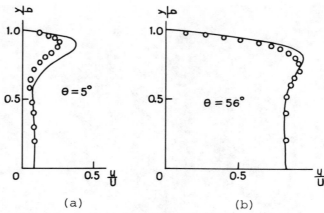

(a)   (b)

Fig.7 Comparison of experimental and theoretical velocity profiles of an oscillating flow with the frequency of ω = 0.5 cps, phase angles being (a) zero, and (b) π/2.

Fig.8 Successive smoke profiles of a transient flow with the frequency of ω = 1 cps, phase angle of π/2. The profile proceeds from left to right.

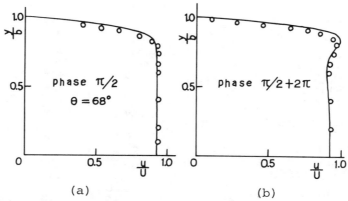

(a)   (b)

Fig.9 Comparison of experimental and theoretical velocity profiles of a transient harmonic flow with the frequency of ω = 1 cps, phase angles of (a) π/2, (b) 2 + (π/2).

OPTICAL METHODS

FLOW VISUALIZATION BY STEREO SHADOWGRAPHS OF STRATIFIED FLUID

C. W. McCUTCHEN*

The refractive index of stratified water decreases with height. This has no obvious effect until the fluid is disturbed by a passing body. Then it causes a jump in refractive index across any vortex sheets created by the body, and makes them visible in shadowgraphic projection. Two stereo views display the sheets in three dimensions. From the shape and motion of the sheets the flow everywhere in the fluid can in principle be computed. The observer seems to do this calculation in his head. The method is especially useful for visualizing time-varying flows, and being cheap and dramatic it should be well suited for teaching.

Introduction

No present method of flow visualization measures the velocity everywhere in the flow. It is not practical to collect this much information, and if it were somehow collected it could not all be displayed at once. But most of it is not needed if viscous shear stresses are insignificant throughout most of the flow; for then, if the flow is revealed in correctly chosen regions, it can be inferred everywhere else. (Turbulent flow, in which viscous shear stresses are important almost everywhere, has not yet been visualized in any way that explains it very well.)

In theoretically derived flow patterns the velocity everywhere is known, and could be displayed if one wanted to and knew how to do it. In practise these flows are illustrated by a few streamlines in one or two selected planes, by the pattern of shock waves, or by the distribution of vortex lines.

Flow visualization methods edit their data in similar ways. They reveal streak lines, path lines, shock wave patterns and so on (Ref. 1). I present here a method that makes vortex sheets visible.

The method

Suppose a fish swims through thermally stratified water. The waters parted by his nose are seldom in perfect vertical register when they meet at his tail. The surface of confluence between them is marked by a jump in temperature, and therefore in refractive index, and is visible in shadowgraphic projection (Ref. 2). Except in special cases or places, the rejoined waters are in motion relative to each other, so the surface of confluence is a vortex sheet.

---
*Research Physicist, The Laboratory of Experimental Pathology, The National Institute of Arthritis, Metabolism, and Digestive Diseases, The National Institutes of Health, Bethesda, Maryland 20014, USA

Two shadowgraphs in stereo display the vortex sheet in three dimensions. In Fig. 1, crossed polarizers polarize the shadowgraphs at right angles to each other, and crossed analysers reveal each one only to the correct eye. The method shows the shape and motion of all vortex sheets. From this the flow everywhere else can in principle be calculated (Ref. 3). The observer seems to do this calculation in his head. It is as if a vortex sheet drawing of a flow were brought to life in space and time.

Prandtl and Tietjens (Ref. 4) illustrate the vortex sheet behind a lifting wing. A model wing moved through the water tank makes the same picture, except that the sheet moves downward and rolls up at the edges before one's eyes. Also it develops waves and then becomes randomly rough and tangled as turbulence develops.

Steady flow patterns can be explored with dye or smoke streaklines, which in steady flow are streamlines and pathlines, and easy to interpret. In unsteady flow the picture presented by streaklines or pathlines is hard to understand (Ref. 5) whereas the vortex sheet representation is easy. Furthermore, in an unsteady flow experiment, the model may leave the vicinity of the streaklines, and enter unmarked fluid.

Figure 2 shows the wake left by a 3.15 cm Zebra Danio (Brachydanio Rerio) that has accelerated with a change of course, coasted in a straight line, and then braked with his pectoral fins and a puff of water out his mouth (Ref. 2). Each action has left its mark in the picture. Wherever the fish goes in the tank the stratified water records his actions. If he is reluctant to perform, the water waits patiently until he does. Compare this with making die streamers and then persuading the fish to swim through them.

After each experiment the pattern fades, as thermal conductivity smooths sharp temperature variations, and buoyancy restratifies the fluid. Experiments can follow each other at 15 second intervals, a great convenience.

If the next experiment is started before the disturbance from the last has entirely faded, the left-over inhomogeneities will mark the fluid outside the surfaces of confluence and reveal its motion. This answers certain questions more directly than can the surfaces of confluence. Stirring the water with a thin rod leaves sharp but separated inhomogeneities that do not obstruct too much of the view.

Movies

Figure 2 is from a movie taken at 44 frames/sec with the apparatus shown in Fig. 3. The large field lenses need only focus the image of the point source well enough so it fits within the aperture of the camera lens. Because shadowgraph contrast results from small-scale compressions and stretchings of the final image, small distances, such as the thickness of the fish's body, may be badly distorted. This distortion could be almost eliminated by switching to a schlieren system, but then the field lenses would have to be much freer from aberrations.

The method is quantitative. The movie shows how fast the vortex sheet moves normal to its surface. I used this information to compute the Froude propulsive efficiency of the fish shown in Fig. 2 (Ref. 6). The motion of any waves on the sheet gives the average of the tangential velocities of the waters on the two sides.

The movie was in mono. Debler and Fitzgerald (Ref. 7) say that single-view shadowgraphs of stratified fluid cannot yield quantitative data on three dimensional phenomena. This seems unduly pessimistic. Other data often resolve the ambiguities in the shadowgraphs. I am not sure from their text that Debler and Fitzgerald realized that they were observing vortex sheets. Pierce, on the other hand, (Ref. 8) identified the shadowgraph features as vortex sheets, but did not propose stable stratification as a way to get the necessary difference in refractive index across them.

Strickler (Ref. 9) has used a different consequence of stratification (by dissolved solute) to mark the wakes of swimming copepods. Because copepods climb and descend as they swim, the fluid they drag with them in their boundary layers and then shed to form the viscous wake does not match the refractive index of the fluid on either side of the wake.

Stratification has little effect on the motion of the fluid if the observed event runs its course in a time short compared to the the fluid's Vaisala-Brunt period (Ref. 10)

$$2 \pi \left[g \frac{d(\ln \rho)}{dh} \right]^{-\frac{1}{2}}$$

where g is the acceleration of gravity, ρ the density of the fluid, and h is distance measured downward from the surface. In water at 20°C the period is 14 seconds if the temperature gradient is 1°C/cm. For events lasting a second or so stratification has little effect on the flow. It lets us see the drama without itself playing a major role.

The temperature gradient does not have to be linear, and needs only be steep enough to make the surfaces of confluence visible. One or two degrees C per cm is enough.

We will now watch the movie of the Zebra Danio. Even though the speed will be 24/44 of real life you may see more than you can pay attention to. I have run it again and again at lower speeds, and analysed some of it frame by frame, without extracting all the information it contains. The marker is 2.54 cm. long. (At this point the movie was shown.)

The magic cap

For visual observations I prefer the magic cap, shown in Figure 4. It keeps the polarizers and analysers aligned with each other, and it keeps the lights spaced apart in the same direction as the separation between the observer's eyes, a necessity if the brain is to combine the shadowgraphs into a three-dimensional impression. Because the lights follow the motions of the observer's head, the scene has natural parallax, though it appears as it would if the observer's eyes were where the lights are.

Above the shadowgraph image is the fish or model seen directly. One soon learns to ignore it.

The lights should be about 10 cm apart. To increase the impression of depth move them further apart. A mathematical account of stereo projection is given in Ref. 2. As point sources I use microscope illuminating lamps (GE 1493), run in series by an isolated 12V storage battery to reduce the shock hazard.

The viewing screen must diffuse the light without upsetting its polarization. Matte finished metal is satisfactory. So is aluminum paint. A bead screen is not. A ground glass transmission screen is all right. I use the matte side of household aluminum foil stretched over a large wooden embroidery frame. The frame expands when it soaks up water and tightens the foil. The foil is laid with its along-the-strip direction parallel to the line between the observer's eyes, because it diffuses light more broadly in this direction. This keeps the brightness of the left and right eye shadowgraphs more nearly equal.

Stratification is created by adding hot water to the top of the tank, cold at the bottom, and extracting water at the level where the greatest sensitivity is desired. Putting this above the middle of the tank increases the sensitivity by increasing the distance from the disturbance to the viewing screen. Drains that draw from beneath the surface yet regulate the water level are shown in Fig. 5. Given time, an abrupt temperature step will develop at the height of the outlet. Mixing, intentional or by the experiments themselves, will turn the step into a slope.

The apparatus can display classic fluid dynamic phenomena that are usually taken on faith. The vortex sheet left by a wing accelerated from rest rolls its aft edge into a starting vortex. The Karman vortex sheet flutters from the cylinder that caused it like a flag behind a flag pole. When a sphere is accelerated from rest one can see a vortex ring develop behind it and then get left behind.

Models are conveniently manipulated through the liquid surface. Cheap, convenient and dramatic, the method should be useful in teaching. My own experience with it has been in research, where it answers more questions per hour and per dollar than any other method I have used.

Because the models need no internal plumbing they are simple to make. Transparent models are useful for looking at separating boundary layers. I know of no transparent material that matches the low refractive index (1.33) of water, but a thin-walled, water-filled model of non-matching material does not distort the light paths too severely. Vacuum forming might be used to make complicated shapes.

Channels and tunnels

Experiments done in a sink are limited to low Reynolds Numbers. To go faster and farther either the model must be towed through a long tank, or the stratified fluid must be run past the model. The later would be more convenient. Preliminary experiments, in which stratified fluid in a large settling basin has been emptied through a water channel, have revealed no problems. A recirculating channel or tunnel would give more observing time for the same energy input. The recirculating channel of Odell and Kovasznay (Ref. 11) gives too low a water velocity, but a faster one seems entirely practical.

References

1. W. Merzkirch, Flow Visualization, (1974), Academic Press.
2. C. W. McCutchen, Flow Visualization with Stereo Shadowgraphs of Stratified Fluid. J. Exp. Biol., 65-1 (1976), 11. (Versions lacking the section on movies were submitted without success to J. Fluid Mech. in Aug. 1971, to the AIAA Journal in Aug. 1972, and to Am. J. Physics in June 1973.)
3. H. Lamb, Hydrodynamics, (1945), 43, Dover.
4. L. Prandtl and O. G. Tietjens. Applied Hydro and Aeromechanics, (1957), 191, Dover.
5. National Committee for Fluid Mechanics Films, Illustrated Experiments in Fluid Mechanics, (1972), 34-35, MIT Press.
6. C. W. McCutchen, The Froude Efficiency of a Small Fish Measured by Wake Visualization. In, Scale Effects in Animal Locomotion, T. J. Pedley, Ed., (1977), 339, Academic Press.
7. W. Debler and P. Fitzgerald, Shadowgraphic Observations of the Flow Past a Sphere and a Vertical Cylinder in a Density-Stratified Liquid. University of Michigan Technical Report EM-71-3 (1971).
8. D. Pierce, Photographic Evidence of the Formation and Growth of Vorticity Behind Plates Accelerated from Rest in Still Air, J. Fluid. Mech, 11-3 (1961), 460.
9. J. R. Strickler, Swimming of planktonic Cyclops species (Copepoda crustacea). In Swimming and Flying in Nature V2. T. Y. -T. Wu, C. J. Brokaw, and C. Brennen, Eds., (1975), 599, Plenum Press.
10. M. N. Hill,. The Sea, Vol 1, (1962), 37, Interscience.
11. G. M. Odell and L. S. G. Kovasznay. A New Type of Water Channel with Density Stratification. J. Fluid Mech., 50-3 (1971), 535.

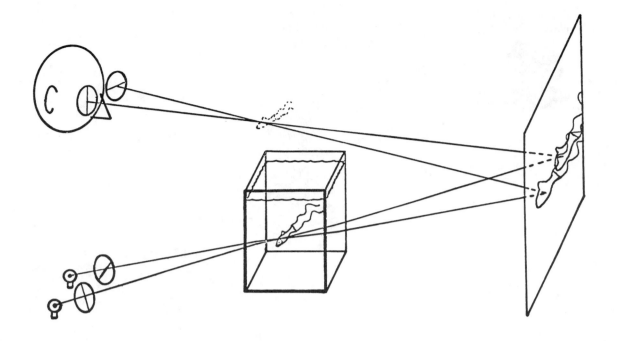

Fig. 1. (above) Two shadowgraphs in stereo are reconstructed by the brain into a three dimensional impression of the surface of confluence behind the fish.

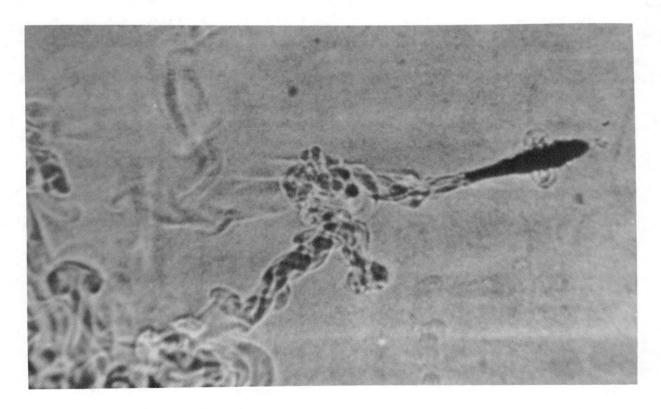

Figure 2. A 3.15 cm Zebra Danio has accelerated with a change of course, coasted, and then slowed himself down with a flick of his pectoral fins and a puff of water out of his mouth.

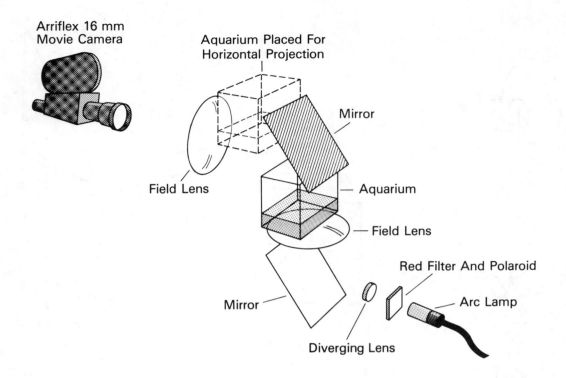

Figure 3. (above) Shadowgraph movie apparatus. The red filter reduces the effect of chromatic aberration of the field lenses. The Polaroid polarizes the light in the plane of incidence of the mirrors to reduce the intensity of unwanted reflections from their front surfaces. The point source was a zirconium arc demagnified by a negative lens. A 3mm thick sheet of clear plastic was floated on the surface with styrofoam floats, otherwise surface waves made the shadowgraphs wobble like jelly.

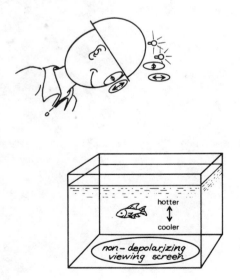

Figure 4. The magic cap that makes water visible.

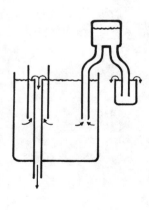

Figure 5. Drains that draw from beneath the surface yet regulate the water level.

VISUALIZATION OF SHOCK AND BLAST WAVE FLOWS

JOHN M. DEWEY* and **D. J. McMILLIN****

Two visualization techniques and their application to studies of shock and blast wave flows are described. The first is an adaptation of the shadowgraph technique, and has been used to photograph the three-dimensional curved shocks produced by large scale explosions. The second technique involves the high speed photography of smoke tracers introduced into the air just before the passage of a shock or blast wave so that the trajectories of the gas particles within the wave can be accurately recorded and used to compute the particle velocity, density and pressure throughout the wave. This technique has been applied to one and two-dimensional flows in shock tubes, and to blast waves produced by large scale explosions.

Refractive Image Photography

The large density gradient at the leading edge of a shock wave produces an extreme refraction of the light rays tangential to the front, and this phenomenon can be used as shadowgraph and schlieren photography to record the position of a shock front. In large scale experiments, using explosive sources, the resolution of the shock front image can be improved by the use of a backdrop of black and white stripes, as illustrated in Figure 1. The canvas screens used for the backdrop in this photo-

Fig. 1 Refractive image of shock front system against striped background.

---

*Professor of Physics, University of Victoria, B.C., Canada

**Research Associate, University of Victoria, B.C., Canada

graph were approximately 15 m high. This technique has been used to study the explosions of charges ranging from a few kilograms to several hundred tons of TNT. The refractive images are recorded using high speed photography at framing rates varying between 3,000 and 10,000 pictures a second, the lower speeds being used for the larger yield experiments and the higher speeds for the smaller yield experiments.

The object being photographed in such experiments is an expanding three-dimensional refractive surface. A measurement from a photograph does not give directly the position of the shock front, but only the direction of the line from the camera tangent to the refractive surface. The problem of determining the actual position of the shocks in an experiment with a complex configuration of explosions is illustrated in Figure 2. In order to calculate the radial positions of the shocks it is necessary to assume that the free-field shocks from the charges are spherical and that the Mach stem shocks produced by shock interactions and reflections are cylindrically symmetrical. As the shock waves expand, the trajectories of the points on the shock surfaces whose positions are being measured in the photographs do not lie on a plane but on a spherical surface in the case of a spherical free-field shock, and on a cylindrical surface in the case of a Mach stem shock. The charge centre and the camera are diametrically opposite points on these spherical and cylindrical surfaces.

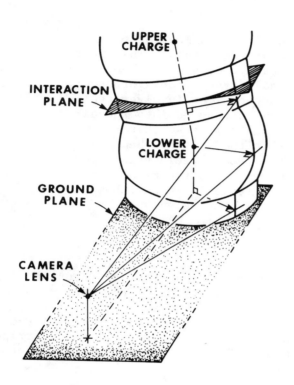

Fig. 2 Photogrammetry of an expanding three-dimensional shock surface.

After applying the appropriate geometrical corrections to allow for the non-planar measurement surface, radial shock positions at a sequence of times are obtained. The shock trajectory is then described by least squares fitting these data to an analytic function. For most explosively-produced shock waves, theory does not indicate a preferred form for the fitting function and a function is chosen that is able to describe the general characteristics of an expanding shock; namely, that at a short time after the detonation pulse the shock will have a finite radius corresponding to the radius of the charge; that the shock radius increases and the shock speed decreases monotonically with time, and that at large times the shock speed asymptotically approaches a constant value. After least squares fitting the measured data with such a function, the shock speed at any time can be obtained by differentiation. The shock speeds are then used in the well known extensions of the Rankine-Hugoniot equation to calculate the physical properties of the shock (Ref.1, 2).

Particle Tracer Photogrammetry

The technique described above provides information only about the shock front at the leading edge of a shock or blast wave. Another visualization technique has been developed to provide information about the physical properties throughout such waves. The technique involves the introduction of smoke tracers into the air ahead of the shock wave so that the subsequent movements of the tracers can be recorded by high speed photography.

Fig. 3 Smoke trails used in a 500 ton TNT experiment. The refractive image of the hemispherical shock may also be seen (upper left).

This technique, as used to study the blast wave from a hemispherical explosion, is illustrated in Figure 3. On the right hand side of this 500 ton TNT explosion a series of white smoke trails were produced just before charge detonation by firing an array of miniature mortars. The smoke trails were photographed at approximately 5000 pictures per second as they moved in front of a specially prepared background of black smoke. The background was used because the white smoke trails would not have been clearly visible against a sky background. The trails were formed in a plane containing the charge centre and parallel to the film plane of the camera. The movement of the trails were measured from the film record and the trajectories analyzed to provide a measure of the physical properties within the blast wave. For example, the particle velocity was obtained directly from the velocity of the smoke tracers which are known to move exactly with the air flow, and the gas density was determined by applying the conservation of mass equation to the spacing between adjacent trails.

As a shock front passes a smoke trail it produces an entropy change in the gas. This entropy change can be determined from the shock speed at that point. In the subsequent flow the entropy of each gas element remains constant, and is therefore known. The shock waves studied in these experiments normally are not strong enough to produce real gas effects in the air through which they are passing so that an ideal gas equation of state can be assumed. Using this equation of state, together with the entropy and density measurements just described, it is possible to calculate the other physical properties within the wave, such as temperature and pressure (Ref.2).

A similar visualization technique has been used to study flows within shock tubes. In this case the smoke tracers were laminar jets of tobacco smoke projected across the shock tube, and the movement of the smoke was recorded with a standard schlieren or shadowgraph system, as illustrated in Figure 4. Smoke jets, such as the one illustrated, have been projected without difficulty across a 25 cm diameter shock tube, and up to 15 such jets at 10 cm spacing have been used throughout the shock tube in a single experiment. In this way the particle trajectories throughout the flow regime of the shock tube have been recorded, and by using a method of analysis similar to that described for free air explosions, all of the physical properties within the flow have been determined (Ref.3).

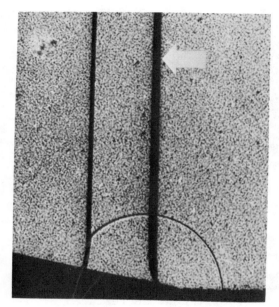

Fig. 4   Smoke jet in shock tube.

In both of the applications described above it was assumed that the particle trajectories were rectilinear; radial in the case of a symmetrical explosion and along the length of the shock tube. This assumption permitted the use of continuous trails or streams. In order to apply the technique to shock and blast flows in which the particle trajectories are not rectilinear it has been necessary to develop a discrete system of point smoke tracers.

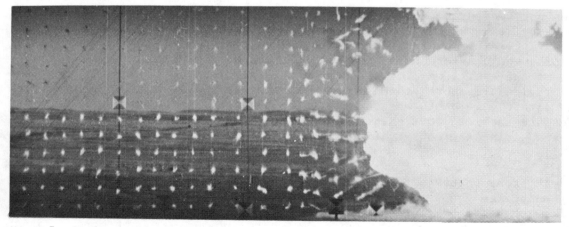

Fig. 5   Smoke puff array used in a 500 kg TNT double charge experiment.

Figure 5 is a frame a high speed film which illustrates how the technique has been applied to study the simultaneous detonation of two 500 kg charges. A two-dimensional rectangular array of 240 individual smoke puffs was formed adjacent to the charges to a height of 19 m in a plane containing the charges and parallel to the film plane of the camera. To enhance their visibility the smoke puffs seen against the sky were red and those below the horizon were white. The movement of the puffs by the blast waves was photographed at approximately 4000 pictures per second and the results were analyzed in a manner similar to that described above to provide a complete mapping of the particle velocities, densities and pressures throughout the smoke puff region (Ref.4).

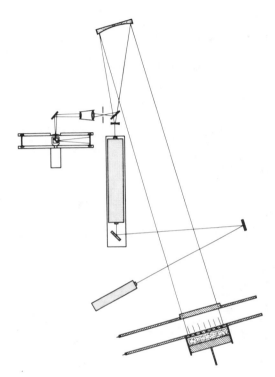

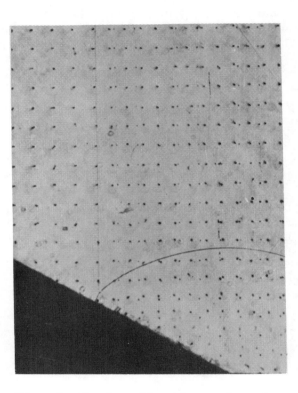

Fig. 6 Double pass multiply-
pulsed laser schlieren system.

Fig. 7 Shock reflection and smoke
tracers in a shock tube.

The study of two-dimensional flows in a shock tube has required the development of a new type of schlieren system, illustrated in Figure 6. This is a plan view of the window section of a shock tube. One of the windows has been replaced by a plane front surface metal mirror of schlieren quality. An array of 0.3 mm diameter holes has been drilled through this mirror at a spacing of 1 cm. Behind the mirror there is a smoke chamber, and when an appropriate pressure is applied laminar jets of smoke are formed by the holes and traverse the 7.5 cm width of the tube. These are viewed end-on by the schlieren system and appear in the resulting photographs as small points of smoke. The shock tube window is 25 cm by 25 cm and an array of over 600 smoke tracers can be seen in the schlieren pictures. The light source for the system is a pulsed laser with a pulse duration of approximately 30 nanoseconds, and the images are recorded on a simple rotating mirror camera at a rate of up to 50,000 pictures per second (Ref.5).

Figure 7 is a photograph taken using this system showing the reflection of a plane shock at a wedge. Ahead of the shock the smoke tracers and the holes in the mirror coincide. Behind the shocks the displacement of the smoke relative to the holes in the mirror can be seen. The holes in the mirror form an accurate calibration grid for measurements made from the photographs. The trajectories of the smoke puffs can be analyzed to map the physical properties within the smoke puff array.

Conclusions

Two flow visualization techniques have been developed for studying shock and blast wave flows in laboratory shock tubes and from large scale explosions. These techniques produce a minimal disturbance of the flow being measured and provide a considerable amount of information, much of which is not available by other means. Densities and pressures measured using these techniques compare well with those measured using other techniques such as electronic transducers, and the use of flow visualization permits the study of large areas which would not be feasible using other methods. The spatial resolution of the particle trajectory analysis is limited by the minimum spacing possible between adjacent smoke tracers.

References

(1) J.M. Dewey, D.F. Classen and D.J. McMillin, Univ. of Victoria, Phys. Fluids Rept., UVIC-PF 1-75, 1975.

(2) J.M. Dewey, Proc. Roy. Soc. Lond. A324, 275-299, 1971.

(3) J.M. Dewey and B.T. Whitten, Phys. of Fluids, 18, 4, 1975.

(4) J.M. Dewey, D.J. McMillin and D. Trill, Univ. of Victoria, Phys. Fluids Rept., UVIC-PF 1-77, 1977.

(5) J.M. Dewey and D.K. Walker, J. Appl. Phys., 46, 8, 1975.

VISUAL STUDIES OF RESONANCE TUBE PHENOMENA

C. E. G. PRZIREMBEL,* R. H. PAGE,** and D. E. WOLF***

Several different flow visualization techniques were used to provide a more detailed physical understanding of the various flow phenomena which characterize gas dynamic resonance tubes. The thermal aspects of the resonance phenomenon were demonstrated by the ignition of a tube consisting of a blind hole in a piece of wood. The hydraulic flow analogy in conjunction with dye injection was used to investigate the effect of various geometric changes on a resonance tube. Color schlieren and shadowgraph techniques were combined with high speed motion picture photography to study the flow field between a nozzle and a resonance tube. Good correlation between the water table and gas dynamic resonance tube results was obtained.

Introduction

A resonance tube (Fig. 1), in its simplest form, is a cylindrical or rectangular tube closed at one end and opposed by a coaxial fluid jet at the open end. If appropriate geometric and fluid flow test conditions are satisfied by the nozzle/tube combination, the flow field in and around the tube is excited to violent, periodic motion. If the resonance tube is sufficiently long, the intense harmonic oscillations give rise to strong thermal effects, which result in resonance tube endwall temperatures significantly higher than the corresponding jet stagnation temperatures. The gas heating effect has been attributed to the periodic motion of compression or shock waves in and out of the resonance tube. For a subsonic and low Mach number supersonic gas jet, the fundamental frequency is independent of the jet Mach number and can be approximated by $f = c/4L$, where c is the speed of sound of the jet, and L is the length of the resonance tube. From time-dependent pressure measurements and flow visualization studies, the idealized characteristic phases of one period during resonance can be described as follows:

(1) At the beginning of a cycle, most of the jet mass flow enters the resonance tube. An accompanying train of compression waves or a shock wave moves into the fluid trapped in the tube. The shock wave is then reflected from the endwall, and moves toward the tube exit.

(2) The shock wave begins to advance into the jet, as simultaneously an expansion wave moves into the resonance tube. During this process, some of the fluid in the resonance tube starts to spill out.

*Associate Dean of Engineering, Rutgers University, New Brunswick, New Jersey, USA
**Professor of Mechanical and Aerospace Engineering, Rutgers University, New Brunswick, New Jersey, USA
***Instructor of Engineering Science, Somerset County College, North Branch, New Jersey, USA

(3) The resonance tube outflow penetrates into the nozzle jet region. In order for this flow to be established, the local total pressure of the resonance tube outflow must be equal to or larger than the local total pressure of the primary jet. During this outflow phase, the average stagnation pressure in the resonance tube continuously decreases.

(4) As the outflow decreases, the local axial stagnation region begins to move toward the resonance tube, and the cycle is repeated.

Brief Literature Review

The resonance tube was first described by the Danish physicist, Julius Hartmann (Ref. 1). While investigating the axial stagnation pressure distribution of an underexpanded sonic jet, he observed high amplitude pressure oscillations in selected portions of the jet. Hartmann and his co-workers became interested in this phenomenon primarily as a means of generating acoustic signals. The intense acoustic radiation has been used in sirens, fog dissipators, ultrasonic cleaning apparatus, and other engineering applications. Sprenger (Ref. 2) was the first investigator to report the thermal aspects of the resonance tube. This experimental study was concerned primarily with the measurement of average endwall pressures, average endwall temperatures, and sound intensities as a function of tube geometry and stagnation pressures. Maximum temperatures of 425°C (797°F) were obtained for a resonance tube with a length-to-diameter ratio (L/D) of 34. Because of the obvious practical applications of this thermal phenomenon, there has been a considerable experimental effort to establish the dependence of the heating effect upon the fluid flow, operating parameters and various geometric configurations of the resonance tube. Most of the experimental measurements have been obtained for underexpanded, sonic jets. Recently, Przirembel and Fletcher (Ref. 3) have investigated the use of subsonic jets for resonance tubes. A typical endwall temperature variation with nozzle/tube separation distance is shown in Fig. 2. As can be observed from these measurements, strong resonance with a subsonic jet, (M = 0.73), occurs only when a trip or small wire is placed across the nozzle exit.

In an effort to visualize the complex flow interactions, two techniques have been utilized, namely, schlieren or shadowgraph still photography using short duration exposures and the application of the hydraulic analogy. Hall and Berry (Ref. 4) obtained short duration (about 1 microsecond) shadowgraphs, which showed the existence of shock waves in the resonance tube. Thompson (Ref. 5) showed from a series of timed, short duration shadowgraphs, that the flow field between the nozzle and the resonance tube is temporarily steady during the tube inflow and outflow. Vebralovich (Ref. 6) proposed a cyclic process within the resonance tube, based on time spark schlieren photographs of the flow processes in and near the tube. Kang (Ref. 7) obtained time shadowgraphs of both the internal and external flow fields associated with a correctly expanded supersonic nozzle (M_{exit} = 1.86) and a square resonance tube.

Hydraulic analog studies of resonance tubes have been reported by Hartenbaum (Ref. 8), Solomon (Ref. 9), and more recently by Skok and Page (Ref. 10). Although it is recognized that there are limitations to the validity of quantitative data for certain aspects of the hydraulic analogy, it remains a very useful technique for visualizing high speed flow phenomena.

The current paper improves the utility and quality of these two fundamental flow visualization techniques by using high speed photography and color schlieren methods for the optical studies, and by combining dye injection and motion picture photography with the hydraulic analog. The actual techniques used in this investigation will be briefly described. The principal results of these studies are presented in a 16mm color motion picture film.

Flow Visualization Techniques

The analogy between the flow on a water table and the supersonic flow of a gas was first pointed out by Riabouchinsky (Ref. 11). In this apparatus, water flows through an open channel with a converging nozzle, and then discharges freely into the working section. The flow in this section is above the critical velocity ($v > \sqrt{gh}$), and is called either superundal or shooting. The depth of the liquid (h) is the analog of the density of the gas. From the mathematical analysis, strict quantitative correspondence of the phenomena requires the ratio of the specific heats to have a value of 2. Shock waves in the gas are represented by hydraulic jumps on the water table. The latter are sudden increases in the depth of the flowing layer of water. The apparatus used for the current investigation is shown in Fig. 3. The resonator channel was placed in the working test section facing the nozzle exit. Quantitative measurements of the time-history of the hydraulic jumps were reported by Skok and Page (Ref. 10). Standard motion pictures (16 or 24 frames per second) were used to record the resonant flow phenomena. These relatively low framing rates were possible because the hydraulic analog provides a time scale reduction on the order of 1000 (Ref. 10).

The optical schlieren studies used various modifications of the standard Toepler type system, which is shown in Fig. 4. The underlying principle for these optical techniques is that the local refraction index of light is a function of the density of the gas through which the light is being propagated. Hence, with proper optical techniques (Ref. 12), fluid density gradients may be investigated without introducing instrumentation into the actual flow phenomena. Furthermore, the entire flow field can be photographically recorded at any instant or over an extended period of time.

The schlieren system consisted of a continuous D.C. light source, two parabolic mirrors, a front-surfaced mirror and a camera stand. The light source was a Xenon filled arc bulb operating at a power of 1000 watts. A system of lenses focussed the light at the slit source. A heat reflecting filter eliminated approximately 20% of the infrared region of the spectrum.

For the conventional schlieren system, the light from a uniformly illuminated slit source is collimated by the first parabolic mirror, and then passes through the test section. It is then focussed by the second parabolic mirror. At the focal point, where an image of the source exists, a knife-edge is introduced in order to intercept about one-half of the source image. With constant density in the test section, the image at the camera plane is uniformly grey. When flow is established, light rays passing through density gradients normal to the light path will be deflected. Depending on the sign of the density gradient, more or less of the light passing through the test section will escape the knife-edge and illuminate the film. Thus, this system makes density gradients visible in terms of intensity of illumination.

The color schlieren system was obtained by replacing the knife-edge with a tricolor filter. The filter consisted of a blue center section, and the two adjacent colors were red and yellow. The width of the blue filter was determined by the width of the light source slit. Hence, the undisturbed light passed only through this section. When a density gradient, such as generated by a shock wave, was present in the test section, light was refracted either through the yellow or red filter depending on the algebraic sign of the gradient.

The direct shadowgraph method is obtained by removing completely the knife-edge from a conventional Toepler system. It can be shown that the variation of the intensity of illumination at the film plane is proportional to the gradient of the density gradient.

A 16mm Hycam camera (Red Lake Model K1001) was used to obtain high speed motion pictures of the optical images. The camera, which utilized a rotating prism for optical compensation, had a capacity of 100 feet (30.48m) of film. The frame rate was variable from 150 to 8000 frames per second.

Discussion of Results

The essence of this study is a 16mm film demonstrating various aspects of the resonance tube phenomena using several flow visualization techniques. The three major portions of the film will be described briefly.

In the first sequence of the film, the thermal aspects of the resonance tube phenomena are vividly demonstrated by the rapid ignition of a piece of wood. The piece of wood has a blind hole drilled along its axis (L/D $\approx$ 30), and is aligned with a subsonic/sonic jet. The closed end of the tube suddenly burns through, and sparks are observed. The exothermic-reaction temperature for wood is close to 273°C (523°F); it is independent of moisture content, and scarcely varies with kind of wood. Ordinarily, ignition takes place at temperatures considerably in excess of this value.

The second series of sequences depict several geometric variations of the resonance tube in the presence of a supercritical, underexpanded jet. This investigation utilized the water table as a test facility. Using dye injection in the resonance tube, it was possible to observe the approximate residence time for the fluid initially trapped in the cavity. During each cycle, some of the dye mixed with the jet fluid that penetrated into the resonance tube. This portion of the dye was then removed from the cavity during the outflow stage. By using two different colors of dye (one in the jet, and one in the tube), an estimate of the location of the so-called contact surface was observed. One other important feature observed during these studies was that a blunt leading edge on the resonance tube may result in the absence of a traveling shock wave in the tube. For the same length-to-diameter ratio (L/D), a hydraulic jump was clearly present for a sharp leading edge, and not present for a blunt leading edge.

The last sequences of the film demonstrate the above result with the use of various optical flow visualization techniques. For relatively short gas dynamic resonance tubes, no evidence of strong shock waves was observed. However, endwall temperature measurements had indicated the existence of thermal effects.

Acknowledgment

This work was supported in part by the Rutgers Research Council and by Picatinny Arsenal under contract number DAAA21-73-C-0781.

References

1. Hartmann, J., "On a New Method for the Generation of Sound-Waves," Physical Review, Vol. 20, No. 6, 1922, p. 719.

2. Sprenger, H. S., "Uber Thermische Effekte Bei Rezonanzrohren," Mitteilungen Aus Dem Institut Fur Aerodynamik An Der E.T.H., Zurich, Nr. 21, 1954, p. 18.

3. Przirembel, C. E. G. and Fletcher, L. S., "Aerothermodynamic Characteristics of a Resonance Tube Driven by a Subsonic Jet," AIAA Paper No. 77-236, AIAA 15th Aerospace Sciences Meeting, Los Angeles, California, January 1977.

4. Hall, I. M. and Berry, C. J., "On the Heating Effect in a Resonance Tube," J. Aero/Space Sci., Vol. 26, No. 4, 1959, p. 253.

5. Thompson, P. A., "Jet Driven Resonance Tubes," AIAA J., Vol. 7, No. 7, 1964, p. 1230.

6. Vrebalovich, T., "Resonance Tubes in a Supersonic Flow Field," Jet Propulsion Lab., California Institute of Technology, Report No. 32-378, 1962.

7. Kang, S. W., "Resonance Tubes," Ph.D. Thesis, Rennssalaer Polytechnic Institute, 1964.

8. Hartenbaum, B., "Hydraulic Analogue Investigation of the Resonance Tubes," M.S. Thesis, M.I.T., 1960.

9. Solomon, L., "A Hydraulic Analog Study of the Hartmann Oscillator Phenomenon," J. Fluid Mechanics, Vol. 28, 1967, p. 261.

10. Skok, M. W. and Page, R. H., "An Analog Investigation of the Gas Jet Resonance Tube," Proceedings of Fifth Cranfield Fluidics Conference, Upsala, Sweden, 1977, p. 37.

11. Riabouchinsky, D., "On the Hydraulic Analogy to Flow of a Compressible Fluid," C. R. Acad. Sci., Vol. 195, 1938, p. 998.

12. Schardin, H., "Toepler's Schlieren Method--Basic Principles for Its Use and Quantitative Evaluation," David Taylor Model Basis, U.S.N., Translation 156, July 1947.

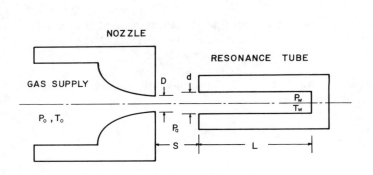

Fig. 1.   Schematic of a Resonance Tube

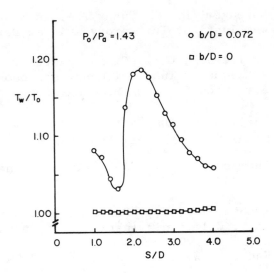

Fig. 2.   Subsonic Jet – Variation
          of Endwall Temperature
          with S/D

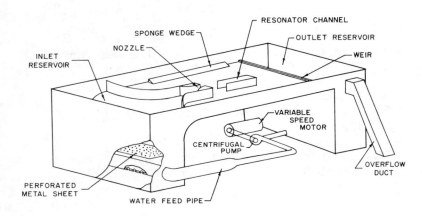

Fig. 3.   Schematic of Water
          Table for Hydraulic
          Analog Studies

Fig. 4.   Conventional Toepler
          Schlieren System

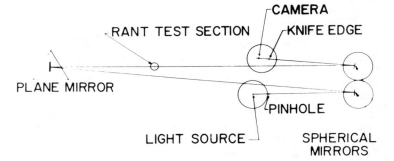

OPTICAL INVESTIGATIONS IN THE PHASE INTERACTION OF A STEADY STATE TWO-PHASE FLOW AND DURING DEPRESSURIZATION

F. MAYINGER,* H.-J. VIECENZ, and H. LANGNER***

The mutual influence of vapour and liquid in a boiling two-phase flow is investigated by the aid of optical methods. The processes at the phase separation during the excaping of vapour from a horizontal liquid surface, the flashing, resulting from decompression evaporation at a sudden pressure decrease, and the entrainment behaviour in an upwards co-current angular flow are discussed as examples for application. The application of high speed cinematography and of specific test devices, adapted to the optical measuring technique, makes possible the direct observation of the occuring phenomena.

Introduction

Investigating thermohydraulic processes in a two-phase flow, difficulties frequently arise in the metrological recording of the local values. The flow pattern, phase distribution and all phase changes can already at steady state not be described only by measurements of pressure and temperature. At transient conditions, occurring for example at a sudden depressurization of the system pressure, these difficulties even increase since the fluid behaviour can no longer be decribed by the thermodynamic equilibrium. These difficulties make the application of a method describing the fluid dynamics in addition to the indirect measurements of pressure, temperature and void fraction indispensable. Optical measuring procedures have proved particularly qualified since they allow a direct local investigation of the occurring phenomena and the resulting fluid behaviour qualitatively as well as quantitatively.

Optical Equipment and Devices

Due to the fast processes in the fluid, as given by the investigated mutual influences of vapour and liquid with regard to the phase separation and liquid entrainment of a free surface or the flash evaporation in the fluid the high speed cinematography is particularly qualified as a measuring technique. With the help of an appropriate high

---

*Professor and Director of the Institut für Verfahrenstechnik, Technical University of Hannover, Hannover, FRG

**Co-worker of Prof.Dr.-Ing. F. Mayinger, Institut für Verfahrenstechnik, Technical University of Hannover, Hannover, FRG

***Co-worker of Prof.Dr.-Ing. F. Mayinger, Institut für Verfahrenstechnik, Technical University of Hannover, Hannover, FRG

speed camera and the necessary high frequency flash device for short light pulses exposure frequencies up to 20.000 f.p.s. can be obtained. For the often used range of approximately 4 - 5000 f.p.s. as applied for our investigations the original process is extended by a factor of about 200, supposing a normal reproduction frequency of 25 f.p.s. and thereby shifted into a well evaluable dimension. A Hitachi 16 mm rotating prism camera and a drum camera are at our disposal. For illumination a high frequency Strobolight is used which can be applied for onlight as well as for light penetration techniques. The optical equipment is presented in fig. 1. For the recording of the investigated processes the fluid is illuminated by the high frequency flash device. The flash light is fed by an external current supply and controlled by an electronic unit or directly by the applied high speed camera. The pulse generator, integrated in the controlling unit, allows flash frequencies of 10 cycles to 20 kilo-cycles. The length of a single flash is 1 μs at 20 kilo-cycles, while 1 WS is produced.

Due to this very short flash time, each photo of a high speed film is uniformly illuminated and the velocity unsharpness of the dynamic processes are reduced to a minimum. Since the light flashes are generated in an inert gas atmosphere the spark light is extremely short-waved whereby the optically and thermally favourable blue-white wave lengths predominate over the long-waved infra-red range. For investigations of phase change or evaporation processes this spectrum guarantees that the test medium is only slightly influenced by the measuring technique.

For the recording two types of high speed cameras are at our disposal as shown in fig. 1. The drum camera does not have a shutter, separating single pictures but exposes the whole film length during one rotation of the drum. Therefore, the camera can only operate in combination with the flash device. The film material is thereby fixed directly on the drum and illuminated when passing the objective.

The maximum rotation speed is 4.000 r.p.m. and can continously be regulated. The advantage of this camera consists in the possibility of obtaining large single negatives of a test process and in the application of normal (36 x 24 mm) film sizes which are available in many phototechnical specifications. As a disadvantage the limited film length of only 1.5 m has to be mentioned, which does not allow an observation of dynamical processes over a longer period of time. Therefore the rotating prism camera is frequently used in its place. The principle of the exposure process, using a rotating prism, is shown in fig. 2. Applying a square prism (fig. 2), exposure frequencies up to 10.000 f.p.s. can be obtained. This frequency can be increased up to 20.000 f.p.s. as a maximum if an 8 plane prism is installed.

From the frequency range of 4.000 - 5.000 f.p.s., sufficient in many cases of application, and the maximum possible film length of 120 m results a recording time of about 5 sec.

As an additional advantage in comparison to the drum camera the rotating prism camera has two internal markers which make use of the edges of the film for the recording. On the one hand a constant high frequency signal can be stored as a time scale for the single negatives and on the other hand an external signal, for example the beginning of the investigation period, can be marked on the film. With the help of this additional information the films can qualitatively as well a quantitatively be used for the evaluation of a temporally variable process.

Examples of Application of the Optical Measuring Method

For the investigation of the separation behaviour of vapour from a stationary liquid surface and the thereby occurring entrainment of liquid particles a glass-made test section was constructed which is shown in fig. 3. It serves the investigation of steady state and transient tests.

In steady state tests vapour is injected in the fluid through the nozzles at the bottom of the vessel to study the separation processes at the liquid surface. In transient tests the mutual influence of the phases and also the phase change behaviour during the flash evaporation is observed. Therefore the system pressure in the glass-made test section is decreased all of a sudden.

The measuring devices, illustrated in fig. 3 serve the recording of the thermodynamic state of the fluid. Rectangular to the vessel the hydraulic fluid behaviour is optically investigated. In all cases of application for the investigation of the thermohydraulic fluid behaviour and the mutual phase influence the principle of light penetration photography was applied. It has the decisive advantage of very satisfactory contour sharpness whereby entrained droplets at phase separation or bubble growth and flashing effects can clearly be recognized and evaluated even during the flash evaporation. Fig. 4 shows a series of pictures, put together from a film recorded with the drum camera at 4.000 f.p.s. By regarding only every fourth of the pictures 1/1000 sec lies between the illustrated scenes. Thereby the phenomenological process of the phase separation at the penetration of a stationary liquid surface and the mechanism of droplet formation, resulting in the entrainment of liquid in the upstreaming vapour, can be visualized very well. The upstreaming bubble reaches the surface, penetrates it and produces a number of droplets of small diameters which are carried away by the upstreaming vapour. At the location of the escaping bubble a liquid jet is formed, due to the equalizing of the liquid surface. From the top of this jet droplets of different diameters are ejected. Depending on their diameter, they are either carried away with the upstreaming vapour or falling back to the surface, due to the gravity force.

For the quantitative investigation of the processes at the phase separation tests were carried out, in which the main influencing parameters (liquid level above the nozzle, injected vapour rate, system pressure) were varied systematically. As an example the influence of the liquid level above the injector nozzle is presented in fig. 5. With the help of these pictures which were taken from several films it is possible to make statements about the droplet entrainment behaviour. By measuring and counting the droplets, the diameter spectrum can be calculated and by summarizing the droplets in a known depth of field, the liquid hold-up of the droplets, carried along by the upstreaming vapour, can be obtained. Furthermore the dependence of the produced droplet diameters on the experimental parameters could be determined (Ref. 1).

Comparing the entrance velocity of the vapour and calculating the extension of the vapour jet and the separation area of the vapour in the fluid from the photographs, the separation velocity of the vapour is known and the produced droplet size can be described by the dimensionless Weber-number.

In order to investigate the phase interaction of vapour and liquid in the two-phase mixture during a flash evaporation and to clarify the question whether separation effects occur in such cases also tests were carried out with a sudden decrease of the system pressure. As the main influencing parameter the initial void fraction in the mixture was varied.

Fig. 6 shows a series of pictures which were taken with the rotating prism camera at an exposure frequency of 4.000 f.p.s.

The presented series of pictures shows that at a very low initial void fraction in the mixture - a single-bubble was injected - the flashing process mainly starts at that bubble, at the tips of the injector nozzle and at the liquid surface. These zones represent areas of high boiling nucleii concentration. In the fifth picture a second flashing-front can be seen, rolling upwards from the bottom of the test section. In the last picture these fronts merge and fill the entire test section.

In comparison to those pictures fig. 7 shows the flashing behaviour at a higher initial void fraction in the mixture. Here the system pressure above a two-phase mixture in the state of bubble boiling is suddenly decreased. The optical arrangement and the exposure frequency of the rotating prism camera maintained.

This combination of scenes shows the beginning of the flashing process and the formation of a homogeneous mixture in the vessel. Due to the beginning flash evaporation in the fluid, the at first stationary volumetric void fraction strongly increases, recognizable at the growing and multiplying bubbles in the mixture. Phototechnically - due to the enlarged number of bubbles and the resulting increase of phase boundary layers - the incoming light is more reflected which causes increasing darkness with increasing void fraction visible on the pictures. Fluidhydraulically the increase of vapour in the mixture results in a volume increase whereby the fluid in the vessel flashes and rises to the outlet. This test yielded that the vapour bubbles, existing in the mixture in the beginning, clearly represent the dominating boiling nucleii. Consequently a minimum void fraction exists due to bubbles. In this case the flash evaporation by bubble growth starts throughout the mixture and not at favoured zones in the vessel or at the liquid surface.

Analyzing such photos quantitatively with regard to the temporal process, essential informations can be obtained which allow in combination with the usual thermodynamic measuring values the physical and mathematical description of such flash evaporation processes (Ref. 2).

Phase Distribution and Liquid Entrainment in a Co-Current Upwards Flow

The liquid entrainment behaviour and the mutual influence of the phases in a stationary two-phase mixture with a horizontal surface as well as in a tube flow still need a more exact investigation. Due to the still not final clarification of the hydraulic fluid behaviour, i.e. the phase distribution and the liquid entrainment burnout damage of the tube walls occurs repeatedly. This damage is particularly important in the technically very frequent case of an annular flow at a heated tube because the usual calculation methods predict a liquid film at the tube wall and thereby a better heat transfer only regarding the decrease of the liquid film by evaporation and without considering the droplet entrainment.

With regard to the clarification of such problems a test section was constructed, consisting of a heated metal tube and a glass-made test section at one end. The upper part of this test section and the arrangement of the camera are shown in fig. 8.

Constructing the outlet as a slit device, the liquid film wetting the tube well, can stream out without influencing the flow profile in the investigation area. Consequently it is possible to look and to photograph directly into the tube through a glass plate built in the top of the slit device.

The glass tube in the upper part of the test section simultaneously allows a view rectangular to the flow direction and serves for the illumination of the fluid. Fig. 9 shows two pictures which were taken simultaneously from these two different views. Normal miniature cameras and a particularly contrasty film were used.

It can be recognized that the axial view technique with direct view into the tube gives essentially more information about the droplet distribution, the droplet size, the thickness of the angular film and its extension at the wall than the recording rectangular to the flow (Ref. 3). Applying a high speed camera instead of the miniature camera shown in fig. 8, the dynamic processes can very well be investigated and statements about the motion of the liquid film and the droplets can be made. The velocity and direction of the droplets then represents a direct measure for the entrainment respectively deposition behaviour of the droplets. Fig. 10 shows two pictures, taken according to the axial view method and clarifying the different flow forms qualitatively.

At low mass fluxes, i.e. 300 kg/m² sec to approximately 500 kg/m² sec, and high vapour qualities a clear angular flow with a relatively thin liquid film at the tube wall and with entrainment in the gas core of the flow occurs (fig. 10b). Increasing the mass flux at constant quality, according to Bennet (Ref. 4) the regime of the so-called "whispy annular flow" is reached, in which a clear separation of the phases cannot be observed any longer. In this regime liquid bridges and large circumferential waves which partially reach the centre of the tube are formed in the liquid film, as shown in fig. 10a. With the help of pictures obtained in that manner, the flow pattern in a two-phase tube flow can satisfactorily be explained. In order to get a clear impression the flow pattern map of Bennet (Ref. 4) was illustrated mainly for the large regime of the annular flow. This presentation is given in fig. 11. The change of the hydrodynamic phase distribution is clearly recognizable, observing a fixed value for the mass flow and increasing the void fraction. Starting from a bubble flow with very high liquid fraction, in the regime of the annular flow the film thickness decreases and causing an increasing entrainment rate in the gas core. The dryout line in Bennet's flow pattern map can clearly be identified by the complete disappearance of the liquid film. By this line the regime of spray flow is indicated.

Summary

Optical investigations with the help of high speed cinematography offer the possibility of making important statements in a two-phase flow about the hydrodynamic behaviour of the single phases. The large temporal extension makes the phenomenological investigation of processes at the phase boundaries possible. Analyzing the movies quantitatively, the processes yield the informations required in addition to the thermodynamic test results. Thus a basis for the mathematical description of the entire process can be gained.

References

(1) F. Mayinger and H.-J. Viecenz, Comparison of phase separation models by use of own experimental data, Paper presented at the NATO advanced study institute on two-phase flow and heat transfer, Aug. 16-27, 1976, Istanbul, Turkey

(2) F. Mayinger, H. Langner and H.-J. Viecenz, Phase interaction and entrainment tests to evaluate theoretical models, Paper presented at the Thermal reactor safety meeting, July 31 - Aug. 4, 1977 Sun Valley, USA

(3) F. Mayinger and H. Langner, Steady state entrainment investigations at inside cooled tubes, Paper presented at the NATO advanced study institute on two-phase flow and heat transfer, Aug. 16-27, 1976, Istanbul, Turkey

(4) Bennet, A.W. et al., Flow visualization studies of boiling at high pressure, (1965), AERE Report 4874, Harwell, UK

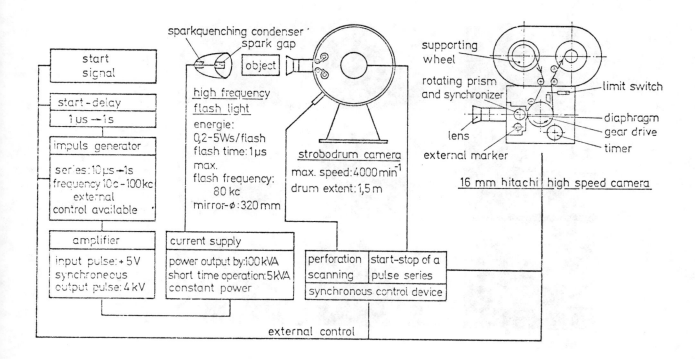

start signal

start-delay
1 us → 1s

impuls generator
series:10 μs → 1s
frequency 10c - 100kc
external
control available

amplifier
input pulse:+5V
synchronous
output pulse: 4 kV

sparkquenching condenser
spark gap
object

high frequency
flash light
energie:
0,2-5Ws/flash
flash time:1μs
max.
flash frequency:
80 kc
mirror-ϕ:320 mm

strobodrum camera
max. speed:4000 min⁻¹
drum extent:1,5 m

current supply
power output by:100 kVA
short time operation:5kVA
constant power

supporting
wheel
rotating prism
and synchronizer
lens
external marker
limit switch
diaphragm
gear drive
timer

16 mm hitachi high speed camera

perforation scanning / start-stop of a pulse series
synchronous control device

external control

Fig. 1 : High frequency flashing device and strobodrum camera

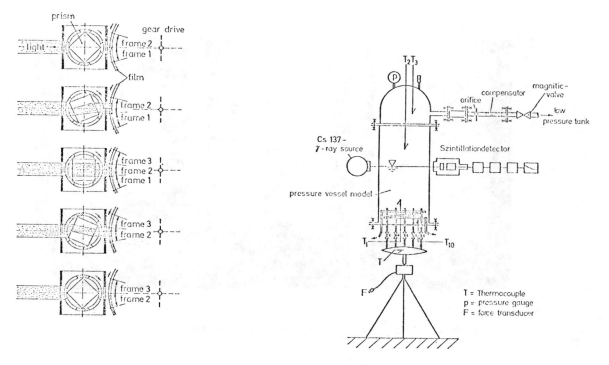

Fig. 2 : Exposing mechanism of the square prism

Fig. 3 : Instrumentation of the pressure vessel model

T = Thermocouple
p = pressure gauge
F = force transducer

297

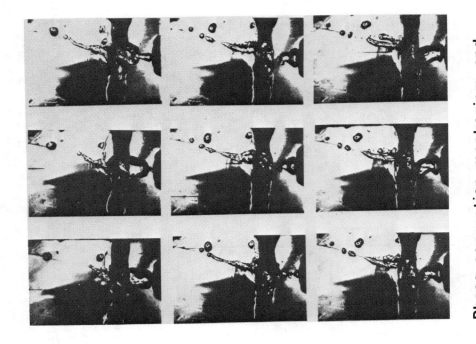

Phase separation mechanism and droplet entrainment

Fig. 4:

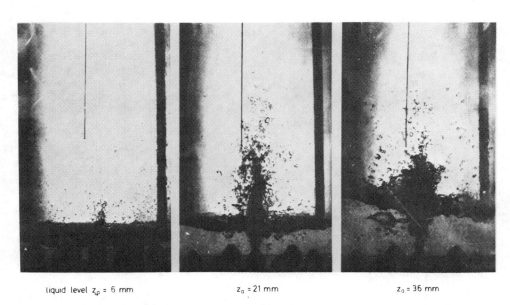

liquid level z_{fl} = 6 mm z_a = 21 mm z_a = 36 mm

Influence of varied liquid level on phase separation and droplet entrainment behaviour

Fig. 5:

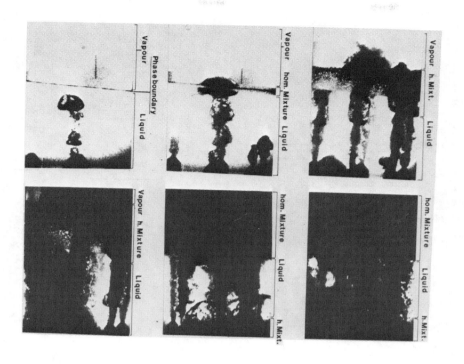

Fig. 6: Fluid behaviour during blowdown
(ε = 0,001)

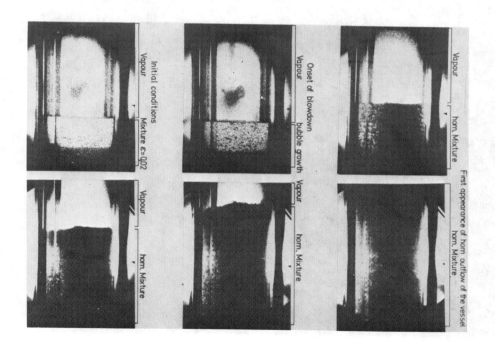

Fig. 7: Vessel blowdown of a mixture of low void
fraction (ε = 0,02)

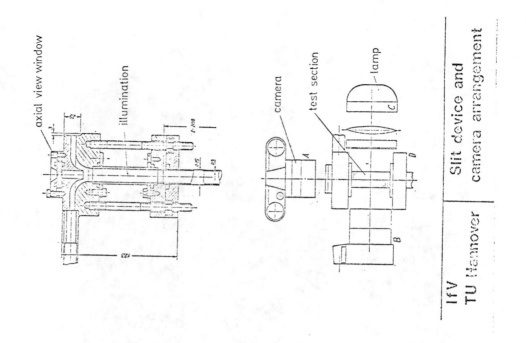

Fig. 8:

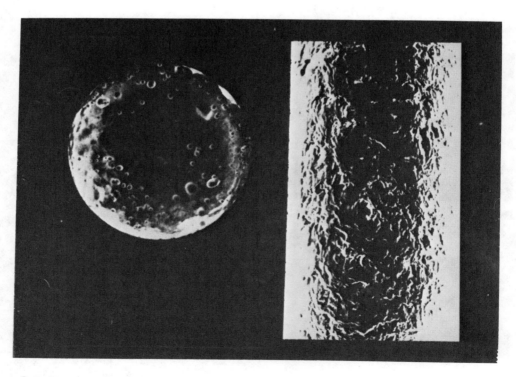

Fig. 9: Comparison of rectangular view and axial view photos

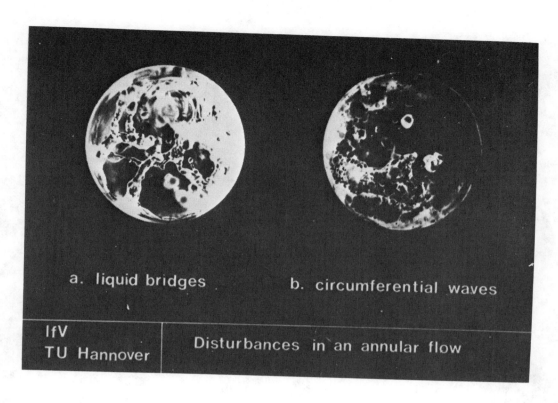

a. liquid bridges   b. circumferential waves

| IfV TU Hannover | Disturbances in an annular flow |

Fig. 10 :

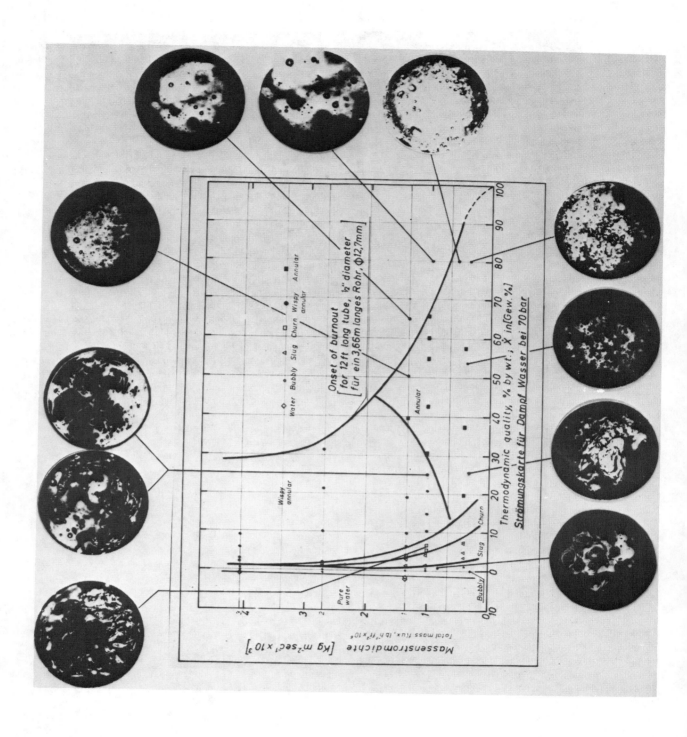

Fig. 11: Flow Pattern Map of Bennet, illustrated with axial
view photos

STUDY OF THE DISRUPTION PATTERN OF A LIQUID JET IN A SWIRLING AIR FLOW USING FLOW VISUALIZATION TECHNIQUES

A. N. RAO, V. GANESAN, R. NATARAJAN, K. V. GOPALAKRISHNAN, and B. S. MURTHY*

This paper reports the study of disruption pattern of a water jet issuing from a swirl plate atomiser into a surrounding swirling flow. The study has been carried out for different flow rates of water and different outer swirl intensities. From the present investigation it is found that below a critical flow rate of water the surrounding air flow plays an important role in the disruption pattern. Back lighted photographic technique was used to study the flow pattern.

INTRODUCTION

Swirling flows are found to improve the combustion intensity and flame stability in Combustion Chambers by creating a recirculation zone. In most combustion systems like Gas Turbines, high inten-sity furnaces etc., a liquid fuel is sprayed into the recirculation zone with the objective of achieving complete combustion. However, to understand the details of the combustion process the mechanism of interaction of a swirling liquid jet with swirling ambient air is to be understood first. In the present work, interaction of a water spray from a swirl plate atomiser with a recirculating swirling air flow issuing from a vane swirler has been reported. Back light pho-tographic Technique has been adopted to study the water spray flow pattern.

PREVIOUS WORK

Because of the complexities of the flow process, it is not sur-prising that investigations involving even a qualitative assessment of the interaction between a swirling liquid jet and the swirling ambient atmosphere is very limited. However, there have been a few investigations on the interaction between a non-swirling liquid jet and a non-swirling air stream, either coaxially or transversely (Ref. 1-4).

+ Department of Mechanical Engineering
Indian Institute of Technology
Madras-600 036
INDIA

PRESENT CONTRIBUTION

The present work is a preliminary study of the disruption pattern of a water jet from a swirl atomiser into the swirling outer air jet. Studies have been carried out for different swirlers with different swirl velocities and for different discharge rates of water. The disruption pattern has been obtained for the various condition using a back lighted photographic Technique. A 35 mm camera with an electronic flash (about 150 μs flash duration) has been used for this purpose.

EXPERIMENTAL SET UP AND MEASUREMENT

The experimental set-up is shown in Fig.1. A vane swirler of 100 mm dia. with a 34 mm dia. hub and having a central hole of 17 mm is fixed at the end of a low turbulence wind tunnel to produce annular swirling air jet. A swirl-chamber nozzle having a 7 mm dia. swirl chamber with four numbers of 1 mm tangential holes and a discharge orifice of 1 mm dia. designed to give a nominal spray cone angle of 80° is fixed at the centre of the hub to produce the central water jet. The intensity of swirl of the outer jet is changed by using swirlers with different vane angles (0,30,45 and 60 degrees). These vanes were of constant width and length. The mean velocity of the outer-jet was changed from 3 to 15 m/sec. The supply pressure for the water jet was varied from 0.2 Kgf/sq.cm to 6 Kgf/sq.cm(7.5 to 44 Kg/hr flow rates). The outer swirl was in the anti-clock-wise direction. Initially, measurements were made with a 5 hole pitot sphere to study the effect of vane angle, on recirculation zone Geometry in the absence of the central water jet. Back-lighted photographs of the jets were taken with a short duration (15 μs) strobe flash and a longer duration (150 μs) electronic flash. In the first case it was possible to 'freeze' the droplets, whereas in the second streaks were obtained due to the longer duration of the flash. These streaks indicate the instantaneous direction of motion of the droplets.

RESULTS AND DISCUSSION

From Fig.2 it is evident that the intensity of inlet swirl has considerable effect on the geometry of the recirculation zone. From Figs.5 and 6 it is quite clear that the disruption pattern is dependent on the outer air velocity, the intensity of the outer swirl when the discharge rate of water is below 24 Kg/hr. Above this discharge rate the outer swirling air does not impart any significant effect on the disruption pattern. Fig.6e gives the details of the disruption pattern when a short duration (15 μs) strobe was used for taking the photograph.

From Fig.3 it is evident that the spray cone angle, measured at an axial distance of 30 mm from the nozzle (measurements made by projecting the negatives over a screen), increases with the increase of the intensity of the swirl. From Fig.4 it is clear that the trajectory angle increases with the increase of outer swirl intensity and its velocity; but, it decreases with the increase of water flow rate and axial distance.

When the water flow rate is high, since the radial moments of the water particles and the tarbulance in the water jet are greatly in excess of the viscosity and surface tension, they cause a quick disruption of the jet. On the other hand if the water flow rate is small, the air forces have considerable influence on the disruption pattern. Due to the recirculation created by the outer swirl the water sheet coming out of the nozzle deflects backward giving out a flower like shape. After disruption the water particles travel along the direction of the air flow.

CONCLUSIONS

When a swirling air flow and a central injection of water interact, the spray cone angle of the water jet increases with increase in the intensity of outer swirl. Above a certain discharge rate of water (24 Kg/hr), the outer swirling flow has a negligible effect on the disruption pattern. hence it can be concluded the pressure corresponding to this flow rate, viz., 2 Kg/sq.cm is the critical discharge pressure.

NOMENCLATURE

QW Rate of water discharge through the swirl plate nozzle Kg/hr
Q   Vane angle of the Swirler
UR Average reference velocity through the swirler, determined from the total flow and cross sectional flow area of corresponding $0°$ swirler in m/sec.

TRAJECTORY ANGLE: Inclination of the flow path of the droplets with respect to x-axis.

REFERENCES

1. A.M.Rothrock, Effect of High Air Velocities on the distribution and penetration of fuel sprays, NAGA Tech. Note No.376, 10 April 1931.

2. L.Strazhewski, The spray range of liquid fuel in an opposing air flow, Tech. Phys. U.S.S.R. Vol.4, No.6, 1937, p.438.

3. Cigier,N.A., Aerodynamic interaction between Burning spray and Recirculation zones. Comb.Inst. Eur. Symp. 1973, pag.96, p.517-582.

4. Clarke,A.E., Gerrard,A.J. and Holiday, Some experience in gas Turbine combustion chamber practices using water flow visualisation techniques, N.Ninth Symposium (International) on Combustion, Academic press, New York, 1963, p.878.

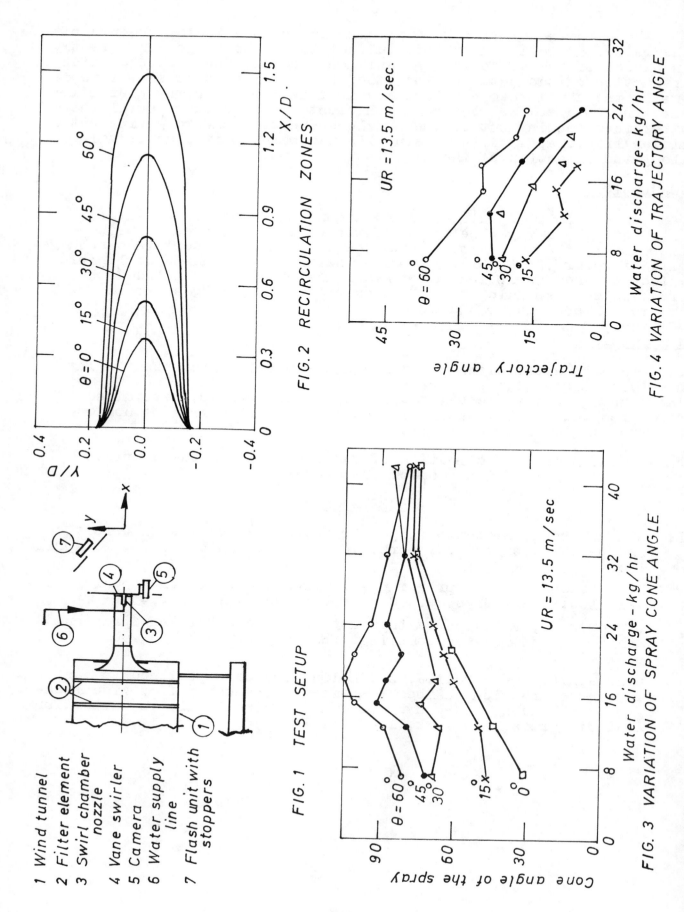

FIG.2 RECIRCULATION ZONES

FIG. 4 VARIATION OF TRAJECTORY ANGLE

FIG. 1 TEST SETUP

1 Wind tunnel
2 Filter element
3 Swirl chamber nozzle
4 Vane swirler
5 Camera
6 Water supply line
7 Flash unit with stoppers

FIG. 3 VARIATION OF SPRAY CONE ANGLE

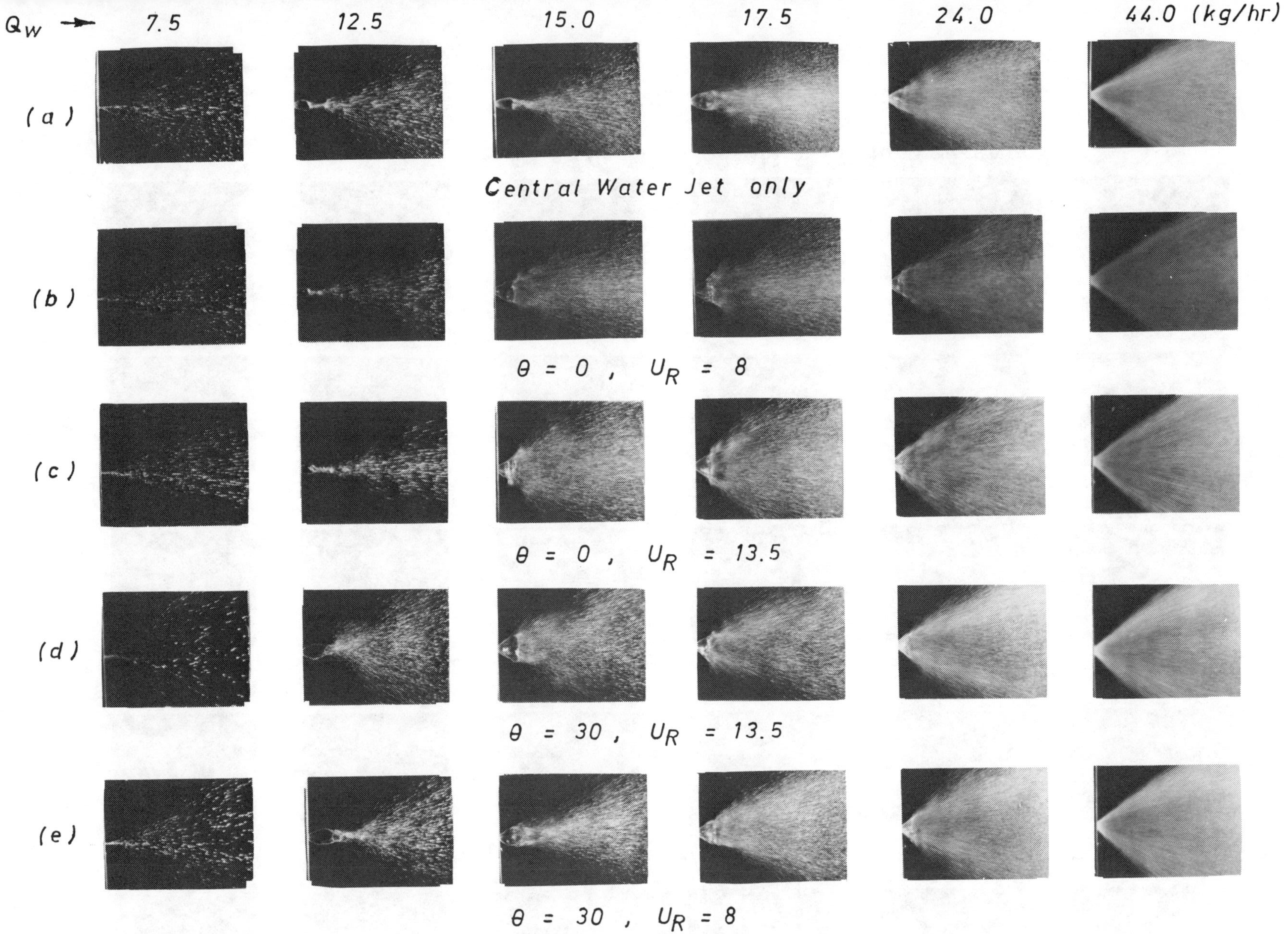

$Q_w \rightarrow$ 7.5   12.5   15.0   17.5   24.0   44.0 (kg/hr)

(a) Central Water Jet only

(b) $\theta = 0$, $U_R = 8$

(c) $\theta = 0$, $U_R = 13.5$

(d) $\theta = 30$, $U_R = 13.5$

(e) $\theta = 30$, $U_R = 8$

FIG. 5

307

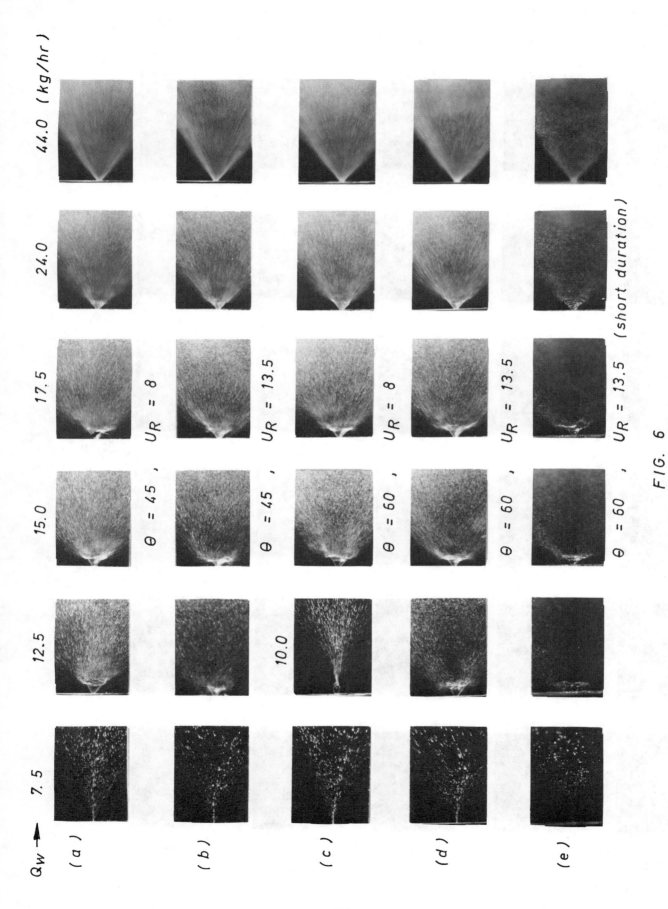

FIG. 6

308

FLOW VISUALIZATION OF AN OSCILLATORY FLUID MOTION IN GLYCERINE TANK

MIKIO HINO* and HARUO FUJISAKI**

It is well known that in oscillatory fluid motions on solid boundaries there result stationary Lagrangian drift currents; e,g. the Stokes drift of water waves and the steady current induced around a vibrating sphere.

Such current velocities are difficult to measure with the usual velocity measurement techniques (pitot tube, hot-wire anemometer and so on), not only because these velocities are very slow but also because these measure only the Eulerian velocities.

The authors have incidentally found that the streak-lines of the stationary Lagrangian drift current in the oscillatory laminar flow over a wavy bottom have been visualized when glycerine is used as a working fluid and lit normally through the side-wall by a flood-light.

INTRODUCTION

Lagrangian Drift Currents

It is well known that in water waves the fluid particles which move on the elliptic orbits are drifted slowly to the direction of wave propagation; i.e. the Stokes drift. Even if, from the Eulerian view point, the fluid velocities are periodic, they are not equally positive and negative from the Lagrangian view point, thus resulting the residual drift current. The existence of this Lagrangian mean velocity (or drift velocity, or mass transport velocity) was pointed out by Stokes (1847) in his classical investigation.

Another example of the Lagrangian drift current is the flow field in a fluid at rest induced by a vibrating sphere or a circular cylinder reciprocating harmonic oscillation. Along the axis of vibration, the outward drift currents are induced and the inward flow toward the equator of the sphere is generated. The problem has been analysed by Schlichting (1932) and others.

Such a Lagrangian drift current is also predicted to form in an oscillatory flow over a wavy boundary. Lyne (1971) presented a theory for cases of the two limiting conditions; i.e. the ratio of amplitude of fluid oscillation and the wave length of wavy wall is very small or very large. The problem was further investigated by the present writers (1974,1975).

These steady streamings are generated by the stresses within the fluids caused by the momentum transport of fluid motion; i.e. a kind of the Reynolds stresses.

---

* Professor, Dept. of Civil Eng., Tokyo Institute of Technology,
O-okayama, Meguro-ku, Tokyo 152, JAPAN

** Graduate Course, Dept. of Civil Eng., Tokyo Institute of Technology, Presently,
Yokohama Office, Bureau of Port and Harbor, Ministry of Transportation.

The Lagrangian and Eulerian Velocities

The position of a fluid element is expressed, in terms of the Lagrangian velocity u_1, as

$$x = a + \int_0^t u_1(a,t')dt'$$

Consequently, the Lagrangian velocity is given, to the second order, by

$$u_1(a,t) = u_E(x,t)$$

$$= u_E(a,t) + \{ \int_0^t u_E(a,t')dt' \}(\frac{\partial}{\partial x_0}, \frac{\partial}{\partial y_0})u_E(a,t) \quad (1)$$

where u_E means the Eulerian velocity at a fixed point (x,t) and a is the initial position of a fluid particle.

To the first order, the Lagrangian velocity is equal to the Eulerian one. Even if the Eulerian velocity field were periodic, there may be a *steady secondary streaming* caused by the second term in the right hand side of eq.(1).

Measurement of Lagrangian Drift Current

Such current velocities are difficult to measure or detect with the usual velocity measurement techniques (Pitot tube, hot-wire anemometer and so on), not only these velocities are very slow but also because these measure only the Eulerian velocities. Generally, the Lagrangian drift currents are detectable by the flow visualization technique.

In this paper, the writters report a simple new method of the visualization technique of a secondary steady streaming by using glycerine as a working fluid.

THEORY OF OSCILLATORY VISCOUS
FLOW OVER WAVY WALL

Two-dimensional viscous oscillatory flow is considered over an infinite wavy wall of small amplitude compared with the wave length

$$y = \alpha \cos\kappa x \qquad (2)$$

The flow oscillation at infinity is described by

$$U = U_\infty \cos\omega t \qquad (3)$$

The equation of motion is given, in terms of a stream function Ψ, as (Lyne 1971)

$$2R^{-1}\frac{\partial}{\partial t}D^2\Psi - \frac{\partial(\Psi, JD^2\Psi)}{\partial(\xi,\eta)} = R^{-1}D^2(JD^2\Psi) \qquad (4)$$

where ξ and η represent the curvilinear coordinate and

$$D^2 \equiv \partial^2/\partial\xi^2 + \partial^2/\partial\eta^2 \qquad (5)$$

$$J = 1 + 2ake^{-k\eta}\cos k\xi + 0(a^2) \qquad (6)$$

$$R = U_\infty\sqrt{2/\nu\omega} \qquad (7)$$

$$a = \alpha\sqrt{\omega/2\nu} \qquad (8)$$

$$k = \kappa\sqrt{2\nu/\omega} \tag{9}$$

$$\tau = \omega t \tag{10}$$

The boundary conditions are

$$\Psi = \partial\Psi/\partial\eta = 0 \qquad (\eta = 0) \tag{11}$$

$$\left.\begin{array}{l} \partial\Psi/\partial\eta \to \cos\tau \\[6pt] \partial\Psi/\partial\xi \to 0 \end{array}\right\} \qquad (\eta \to \infty) \tag{12}$$

The stream function is expanded as

$$\Psi = \Psi_0 + a\Psi_1 + a^2\Psi_2 + \cdots \tag{13}$$

The function $\Psi_1(\xi,\eta,\tau)$ is assumed to be expressible as

$$\Psi_1(\xi,\eta,\tau) = R[\ F(\eta,\tau)e^{ik\xi}\ ] \tag{14}$$

The function $F(\eta,\tau)$ is expanded by the Fourier series, eq.(15),

$$F(\eta,\tau) = \sum_{m=-\infty}^{\infty} f_m(\eta)e^{im\tau} \tag{15}$$

The boundary conditions reduce to

$$\left.\begin{array}{ll} f_m = f_m' = 0 & (\eta = 0) \\[6pt] f_m \to 0,\ f_m' \to 0 & (\eta \to \infty) \end{array}\right\} \tag{16}$$

Substitution of eq.(15) into equation (3) and application of the orthogonality of $e^{im\tau}$ yield a system of simultaneous linear differential equations, eq.(17)

$$\frac{d}{d\eta}\begin{bmatrix} q_{-M,1} \\ q_{-M,2} \\ q_{-M,3} \\ q_{-M,4} \\ q_{-M+1,1} \\ q_{-M+1,2} \\ \vdots \\ q_{M,3} \\ q_{M,4} \end{bmatrix} = \begin{bmatrix} q_{-M,2} \\ q_{-M,3} \\ q_{-M,4} \\ G_{-M} \\ q_{-M+1,2} \\ q_{-M+1,3} \\ \vdots \\ q_{M,4} \\ G_M \end{bmatrix} \tag{17}$$

where $q_{m,1} = f_m$, $q_{m,2} = f_m'$, $q_{m,3} = f_m''$, $q_{m,4} = f_m'''$.

Consequently, the problem reduces to a two-point boundary value problem. However, there is no difficulty encountered, since the problem is linear and solvable by means of the standard technique. A particular solution and the s-set of homogeneous solutions are determined numerically by the Runge-Kutta-Gill method. Finally, as a linear combination of these solutions, the solution which satisfies the boundary conditions are obtained.

EXPERIMENT

Working Fluid

In order to verify the theoretical results for laminar oscillatory flow, and to confirm the phenomenon in an appreciable size, use of extremely high viscous liquid was required. We could not find any alternative other than glycerine.

Experimental Apparatus

An experiment has been performed in a small glycerine tank which is 100 cm in length, 50 cm in width and 35 cm in depth. A 77 cm × 45 cm wavy plate whose wave amplitude and length were α = 1 cm and ℓ = 15 cm, respectively, was installed at the bottom of tank (Fig.2).

Contrary to the theoretical assumption, the bottom wavy wall was driven to oscillate harmonically in the horizontal direction by a motor and crank system. Since the inertial force due to acceleration is negligible, the experiments may be compared with the theory. The period of oscillation was varied between T = 1 to 3 sec. and the amplitude of wavy wall oscillation being 2A = 2.0 cm. The kinematic viscosity of the working fluid (glycerine) was ν = 15.8 cm^2/s and ρ = 1.29 gr/cm^3 at the temperature 8°C. Therefore, the thickness of the Stokes layer was $\delta = \sqrt{2\nu/\omega}$ = 2.2 ~ 3.9 cm.

Flow Visualization Technique

At first, we considered to visualize and measure the very slow velocity field by the dye streak technique. However, a very simple technique, rather a new phenomenon was found. The finding of the technique was quite accidental. When a strong light beam was projected, to take photographs, through a transparent side wall of the tank, a clear pattern of bright and dark streaks appeared on the opposite wall painted white.

(a)   The streak patterns of secondary cellular structure emerged slowly after the initiation of oscillatory motion and gradually they became clearly appreciable (Fig.3).

(b)   By putting a neutrally buoyant particle, it was confirmed that the cellular streak pattern coincide with a Lagrangian particle orbit (Figs.4 and 5).

(c)   When the direction of beam was tilted to the z axis perpendicular to the wall, the cellular pattern became obscured overrapping with each other. This shows that the cellular patters had the axes perpendicular to the side wall.

(d)   The steady cellular stream ascends at the crest of wavy wall and descends at the trough of wavy wall. They are symmetrical with respect to vertical lines through a crest or a trough (Fig.6). The sizes of the cellular pattern are of the order of the Stokes layer $\delta = \sqrt{2\nu/\omega}$.

(e)   After driving the wavy wall in the dark for enough long time, light was suddenly cast through the side wall. The cellular pattern appeared immediately. This fact denies a hypothesis that the flow field was stimulated by some optical effects.

(f)   When a thin metal rod was inserted into the working fluid and it was moved along its surface, a shear plane was visible if light beam was shed with a slight angle with it (Fig.7). The shear trace remained for long enough time.

The detailed mechanism is unknown. However, it may be anticipated that the high shear $\partial V/\partial s$ in the fluid gives rise to the formation of internal refraction surfaces.

(g)   This fact as mentioned in (f), on the contrary, provided means to visualize the flow field. Before the begining of experiment, some vertical or horizontal internal shear planes, in the direction parallel to the beam, were made in a still glycerine by a thin metal rod with sharp edge (Fig.8). These lines when viewed from the side transparent wall served as tracer to measure the velocity field.

Fig.9 shows the gradual deformation of a vertical shear plane which was artificially marked. Fig.10 compares the theoretical and experimental result of the horizontal velocity component of steady secondary flow U_s.

A movie representation of the experiment on an oscillatory glycerine flow over a wavy surface will be given at the Symposium.

REFERENCE

1) Lyne, W̊.H. (1971) : Unsteady viscous flow over a wavy wall, J. Fluid Mech., vol. 50, 33-48.
2) Uda, T. and Hino, M. (1975) : Theory on a laminar oscillatory flow over a wavy wall, Proc. J. Soc. of Civil Engr., No. 237, 27-36.
3) Hino, M. and Fujisaki, H. (1975) : Viscous oscillatory flow over a wavy wall, Proc. 22 Japanese Conference on Coastal Engineering, 35-40.

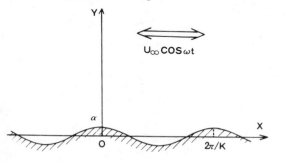

Fig. 1 : Coordinate system

Fig. 2 : Experimental apparatus

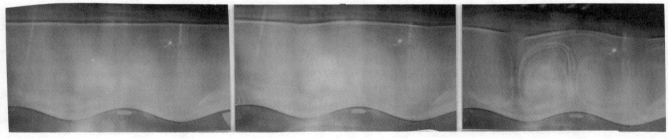

t = 0 min.   t = 8 min.   t = 24 min.

Fig. 3 ; (T = 2.10sec. , A = 1.0cm , θ = 10.5⁰C)

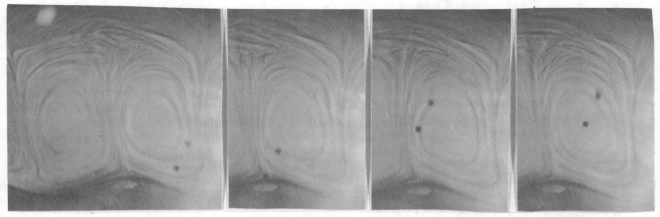

t = 0 min.   t = 2 min.   t = 4 min.   t = 6 min.

Fig.4 : (T = 1.65s, A = 1.0cm, θ = 10°C)

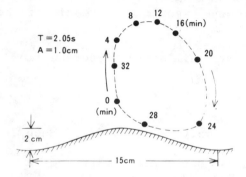

T = 2.05s
A = 1.0 cm

2 cm

15cm

Fig. 5 : Tracing a neutrally buoyant particle.

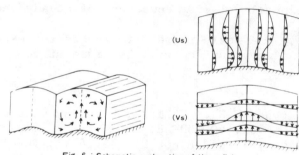

(Us)

(Vs)

Fig. 6 : Schematic explanation of the cellular structure of steady secondary flow field.

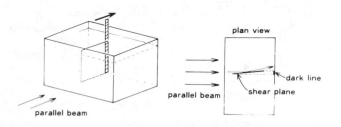

plan view

dark line

shear plane

parallel beam

parallel beam

Fig. 7 : Effect of strong shear in glycerine caused by pulling a thin metal rod along its surface

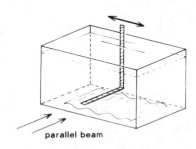

parallel beam

Fig. 8 : Drawing the horizontal streaks as tracer

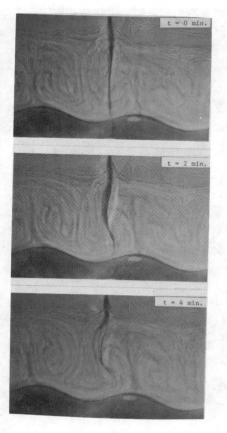

t = 0 min.

t = 2 min.

t = 4 min.

Fig. 9 : (T=3.12sec, A=1.0cm., θ=9°C)

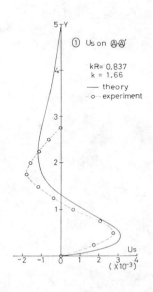

① Us on Ⓐ-Ⓐ'

kR= 0.837
k = 1.66
—— theory
-○- experiment

Fig. 10 ;

314

TECHNIQUES FOR VISUALIZATION OF SEPARATED FLOWS

R. H. PAGE* and C. E. G. PRZIREMBEL**

Several techniques for visualizing regions of separated flows are reviewed. These techniques include color schlieren, shadowgraph, and Wollaston prism schlieren interferometry which are used to visualize the density changes in the gas flow. Also the technique of injecting small talc particles in the gas flow to visualize the streamlines in subsonic regions is discussed. Typical results are presented. A color film shown with the oral presentation of the paper is used to illustrate the results.

1.0 INTRODUCTION

The authors and their colleagues at Rutgers University have carried on extensive theoretical and experimental research with separated flows. As part of their experimental program they have developed several techniques for visualizing regions of separated flows within either supersonic or subsonic flow. These regions of separated flows are of great interest because the flow does not follow the wall or guiding surface and is not susceptible to simple theoretical or experimental analysis. Thus experimental simulation is often required when separated regions are present. The visualization of the flow is essential in order to develop a better understanding of the fluid dynamic phenomena.

Many flow visualization techniques have been used in the study of separated flows. These include surface treatments which optically alter the interaction between a wall (surface) in a fluid as well as those that optically alter a surface between a liquid and a gas.[1-3] The flow visualization of separated flows has also included direct injection methods involving smoke, dies, neutral density beads, etc.[4]

The discovery and development of methods by which an optical system could be used to visualize transparent flows through direct observation of the optical properties of the flowing fluid (without the necessity for instrumentation probes and the introduction of foreign particles) was a major step forward in flow visualization.[5,6] Such methods provide a technique for photographically recording all of the information observed by optical systems.[7] The results reported in this paper highlight the techniques used at Rutgers University.

2.0 VISUALIZATION TECHNIQUES AND EXAMPLES

A number of flow visualization techniques have been utilized effectively for

*Professor of Mechanical and Aerospace Engineering, Rutgers University, USA

**Associate Dean of Engineering, Rutgers University, USA

separated flow studies at Rutgers University. The optical techniques of schlieren, shadowgraph, and interferometry have played a major role as experimental diagnostic tools.

2.1 CONVENTIONAL SCHLIEREN

A conventional schlieren system of the standard Toepler type is shown in Fig. 1. The light from a source is focused by a mirror into a parallel beam which passes through the glass wall of the test section of the wind tunnel. If there are no density gradients normal to the light axis in the wind tunnel test section (for example, when the tunnel is turned off) the light emerges from the test section in an undisturbed parallel beam. It is then focused into an image of the source at the focal plane of the second parabolic mirror. Generally, the knife edge is inserted at this focal plane in such a way that it intercepts about one half of the source image. The image of the object in the test section is brought to focus at the film plane in the camera by an auxilliary lens.

Fig. 2 illustrates the refraction pattern for conventional schlieren systems. The undisturbed light rays converge to form an image at the source of the knife edge. If the knife edge is adjusted for maximum sensitivity, approximately one half of the light from the source is cut off and a uniform grey image is produced at the film plane. Thus the gradients in the test section could cause refraction of light rays as shown by the dotted lines in the lower two sketches of Fig. 2. The refracted rays in the center sketch missed the knife edge entirely thus causing a bright illumination of the upper portion of the film. In the bottom sketch the opposite case of refracted rays being intercepted by the knife edge is shown. The light reinforcement or cancellation on the grey image background produces the familiar black and white schlieren images. It is obvious, from Fig. 2, that density gradients which refract light rays parallel to the knife edge will result in no change in illumination at the film plane. A schlieren system is only sensitive to density gradients having a component normal to the knife edge. Fig. 3 is an example photograph.

2.2 TRI-COLOR SCHLIEREN

The Rutgers color schlieren system[8] is simply a black and white system with a knife edge replaced by a tri-color filter. Fig. 4 shows how the tri-color filter is used to produce color images. The undisturbed light rays of the top sketch all pass through the narrow blue slit and result in a uniform blue illumination of film. When the density gradient is present which causes refraction of some of the light rays a different color appears on the film image. The center sketch of Fig. 4 shows light rays refracted so that they pass through the yellow filter. The bottom sketch shows light rays refracted in the opposite direction so that the refracted rays now pass through the red filter, thus causing a red image on the top half of the film. This color schlieren system, like the black and white system, is only sensitive to density gradients having a component normal to the axis of the blue filter.

2.3 WOLLASTON PRISM SCHLIEREN INTERFEROMETRY

A standard single pass schlieren system may be converted into a single beam schlieren interferometer by replacing the knife edge with a polarizer-Wollaston prism-analyzer combination.[9] Changes in density in the test section are related to fringe shifts. Fig. 5 illustrates the system. Figs. 6 and 7 are example photographs.

2.4 SHADOWGRAPH

The direct shadow method of flow visualization utilizes an optical system identical to that shown in Fig. 1 except that there is no knife edge present. Therefore, the illumination of the image is no longer sensitive to density gradients having a component normal to a particular axis. Indeed, it can be shown, that the change of illumination at the film plane becomes proportional to the rate of change of the density gradient in the test section. Because of its simplicity, the shadowgraph technique is very well known and is ideal for visualizing phenomena involving large changes in the second derivative of the density (for example, shock waves).

2.5 PARTICLE ADDITION

Motion picture films of non-steady separated flow regions utilizing color schlieren and shadowgraph images can be enhanced with tracers. The flow within the separated region may be visualized by the introduction of particles. Particles are selected which follow the recirculation flow paths. The optical superpositioning of the paths of particles or particle groups on the motion picture images greatly improves the visualization of the separated flow regions.

3.0 IMAGE RECORDING

Mercury vapor and Xenon light sources driven by D.C. power supplies have been used for most of the research. A single pulse of a Q-switched ruby laser has been used for short duration photography[10] such as illustrated in Fig. 6, Fig. 7, and Fig. 8.

Photographs and motion picture films have been used to record the flow visualization. Transient flow phenomena involving separated flow regions have been studied at Rutgers utilizing high speed motion picture photography. Cinematographic studies have provided excellent descriptions of various supersonic, separated flow fields.[12] These high speed motion pictures are often viewed on a reduced time scale or on a frame by frame basis.

In all cases the authors have used standard commercial components and standard films for image recording. For single photographs they prefer a 35mm reflex camera body for convenience in film handling. A 16mm Hycam camera was used to obtain high speed motion pictures. This camera, which utilizes a rotating prism for optical compensation, is adjustable from 150 to 8000 frames per second. A timing light which placed small blips of light on the edge of the film, was used to determine the precise frame rate. Sample results have appeared in a film.[13]

4.0 EXPERIMENTAL FACILITIES

The majority of the visualization studies were carried out in two of Rutgers supersonic wind tunnels. Both the Emil Buehler Wind Tunnel (EBWT)[10] and the Rutgers Axisymmetric Near-Wake Tunnel (RANT I)[14] were used. The characteristics of these tunnels are presented in Table 1.

5.0 CONCLUSIONS

Excellent qualitative descriptions of separated flow fields have been obtained by optical methods which have sometimes been combined with trace techniques. In addition to single frame photography the authors have found high speed motion picture photography very helpful in providing a physical understanding of the complex flow fields accompanying separation and reattachment.

6.0 REFERENCES

1. Merzkirch, W. F., "Making Flows Visible". <u>International Science and Technology</u>, N. 58, p. 46-56, Oct. 1966.

2. "Symposium on Flow Visualization", ASME, New York, N. Y., Nov. 1960.

3. Clayton, B. R. and Massey, B. S., "Flow Visualization in Water: A Review of Techniques", <u>J. of Scientific Instrumentation</u>, V. 44, p. 2-11, 1967.

4. Prandtl, L. and Tietjens, O. G., <u>Applied Hydro- and Aeromechanics</u>, Dover Publication, N. Y., N. Y., 1957.

5. Toepler, A., "Beobachtungen Nach Einer Neuen Optischen Methode", Ostwalds Klassiker der Exacten Wissenschaften, N. 157, Leipzig, 1906.

6. Schardin, H., "Toepler's Schlieren Method - Basic Principles for its Use and Quantitative Evaluation", David Taylor Model Basin, U.S.N., Translation 156, July 1947.

7. Merzkirch, W., <u>Flow Visualization</u>, Academic Press, N. Y. and London, 1974.

8. Kessler, T. J. and Hill, W. G., Jr., "Schlieren Analysis Goes to Color", <u>Astronautics and Aeronautics</u>, V. 4, N. 1, front cover and p. 38-40, Jan. 1966.

9. Small, R. D., Sernas, V. A., and Page, R. H., "Single Beam Schlieren Interferometer Using A Wollaston Prism", <u>Applied Optics</u>, Vol. 11, No. 4, p. 858-862, April 1972.

10. Page, R. H. and Sernas, V., "Apparent Reverse Transition in an Expansion Fan", <u>AIAA Journal</u>, Vol. 8, No. 1, pp. 189-190, January 1970.

11. Page, R. H., Kessler, T. J., and Kuebler, A. A., "Color Schlieren Studies of Transient Flows with High Speed Motion Picture Photography", <u>Proceedings 4th Space Congress</u>, Cocoa Beach, Florida, April 1967.

12. Page, R. H. and Przirembel, C. E. G., "Cinematographic Study of Separated Flow Regions", AGARD Conference Proceedings #168, May 1975.

13. Page, R. H. and Przirembel, C. E. G., "Cinematographic Study of Separated Flow Regions", 10 min - 16mm silent color film, Department of Mechanical, Industrial, and Aerospace Engineering, Rutgers University, U.S.A. 1974.

14. Page, R. H., Dixon, R. J., and Przirembel, C. E. G., "Apparatus for Studies of Supersonic Axisymmetric Base Flow", Proceedings of XVIIth <u>International Astro. Congress</u>, Madrid, Spain, October 1966.

TABLE 1

Rutgers Supersonic Wind Tunnels Used with Visualization Studies

	EBWT	RANT
Name	Emil Buehler Wind Tunnel	Rutgers Axisymmetric Near-Wake Tunnel
Test section geometry	Two-Dimensional	Axisymmetric
Test section size	10.16 cm x 11.43 cm	20.32 cm diam.
Mach No. range	1.2 to 4.0	3.0 and 3.88
Stagnation Temperature	Atmospheric	Atmospheric
Stagnation Pressure	$\leq$ 10 atm.	$\leq$ 14 atm.
Compressed Air Moisture Content	1 part per million	1 part per million
Mode of Operation	Intermittent with constant stagnation pressure control	Intermittent with const. stag. press.
Control System	Electro-pneumatic	Electro-pneumatic
Test Time Duration	Programmable by event timer ($\sim$30 seconds)	Programmable by event timer ($\sim$20 seconds)
Model Support	Conventional sting mount with variable angle of attack	Special design with complete support by upstream sting
View Ports	Optical Glass 43.18 cm x 12.06 cm	Optical Glass 22.86 cm x 10.16 cm

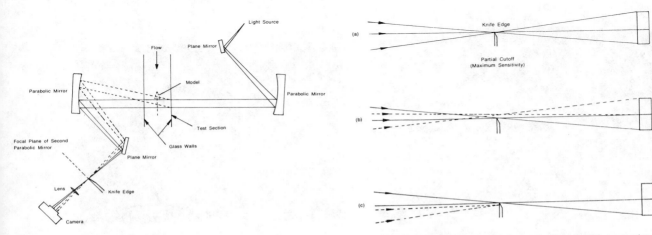

Figure One: Conventional schlieren system schematic.

Figure Two: Refraction pattern for conventional schlieren system.

Fig. 3 Schlieren, M = 2.5

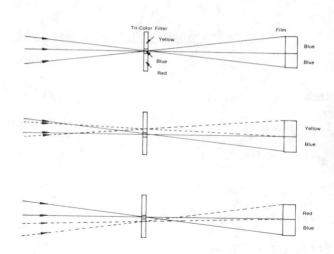

Fig. 4 Refraction with Tri-Color Filter

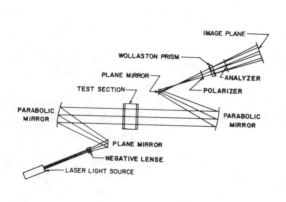

Fig. 5 Wollaston Prism System

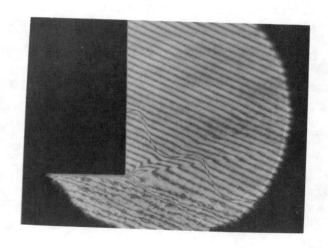

Fig. 6 Corner Flow, M = 3.05

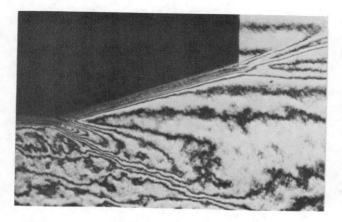

Fig. 7 Beveled Corner Flow

Fig. 8 Turbulent Boundary Layer

SCHLIEREN VISUALIZATION OF THE LAMINAR-TO-TURBULENT TRANSITION IN A SHOCK-TUBE BOUNDARY-LAYER

K. BRACHT* and W. MERZKIRCH**

The laminar-to-turbulent transition in the flow of a shock-tube boundary-layer is measured for shock Mach numbers $1.06 < M_s < 1.5$. Schlieren photographs show a pulsation of the boundary-layer thickness in the transition regime. The observed structure of the turbulent boundary-layer can be related to the burst phenomena known for incompressible flow. The transition is also measured with hot films which are sensitive to the difference in laminar and turbulent heat transfer at the tube wall.

Introduction

In the gas flow behind the shock wave in a shock tube, a boundary layer is formed along the tube wall. Within this layer the gas velocity decreases from the free stream value u_2 to zero at the wall. Since the free stream temperature T_2 usually is higher than the wall temperature T_w (which can be assumed to be the ambient room temperature), a thermal boundary layer is super-imposed on-to the described velocity boundary layer. The assumption of constant static pressure results in a density gradient normal to the wall, with the gas density being higher at the wall than in the free stream. Therefore this boundary layer can be visualized with appropriate optical method, e.g. a schlieren system, provided that the density gradient is large enough for being resolved.

The boundary layer thickness is zero at the position of the shock wave and it increases with increasing distance from the shock (Fig. 1). The actual dimension of the boundary layer thickness is small, in many cases only a fraction of a millimeter. The flow in the layer immediately behind the shock is in any case laminar (Ref. 1,2). The boundary layer flow changes character to become turbulent at a certain distance behind the shock wave (Ref. 3,4). The final flow state would be a fully turbulent channel flow if the tube were long enough. The position at which transition from laminar to turbulent flow occurs and the length of the transition regime depend

---

* Research Assistent
** Professor of Fluid Mechanics; both: Institut für Thermo- und Fluid-dynamik, Ruhr-Universität, 4630 Bochum, Germany

on the parameters of the external flow and on the shape and structure of the wall surface.

The momentum and energy loss in the shock tube boundary layer causes an attenuation of the moving shock wave. The amount of attenuation is different for laminar and turbulent boundary layer flow (Ref. 5). The transition in the shock tube boundary layer also determines the usable test time for shock tube experiments. A further motivation for studying the laminar-to-turbulent transition in a shock tube boundary layer is that no other flow facility than a shock tube reproduce the free stream conditions with such a high degree of precision, and that measurements can be performed with non-intrusive, i.e. optical, techniques due to the density variations occuring in the boundary layer. However, one must be aware of the fact that the boundary conditions in a shock tube flow are different from those in a stationary channel flow.

Visualization with a schlieren system

The test section of the shock tube was provided with viewing windows and served as the test object in a z-shaped schlieren system. A single spark light source was used in this system in order to obtain short exposure ($< 1o^{-6}$ sec) photographs of the flow under study. All schlieren photographs were taken with a horizontal knife edge, i.e. with the knife edge parallel to the plane plate forming the floor of the shock tube. For an adequate spatial resolution it is necessary to have the laminar boundary layer at the beginning of transition as thick as possible. According to Emrich (Ref. 3) this condition can be best met by operating with weak shock waves and low initial pressures. This result, however, is incompatible with a number of experimental requirements. Particularly, the sensivity of the schlieren effect decreases rapidly with decreasing initial pressure or density. As a compromise, shock Mach numbers have been chosen in the range $1.o6 \leq M_2 \leq 1.5$, and initial pressures $o.5 \leq p_1 \leq 1.o$ bar in the low-pressure section of the shock tube.

The visualized pattern of the boundary layer flow is consistent with the earlier results of Emrich (Ref. 3). Fig. 2 shows the smooth pattern of the laminar boundary layer immediately behind the shock wave which appears as the bright streak at the right of the picture. The shock moves from left to right, and the boundary layer thickness is seen to increase with increasing distance from the shock wave. At the far left, at a distance of 7o mm behind the shock, the boundary layer thickness is about $\delta = o.5$ mm. At a greater distance from the shock one may observe that the boundary layer thickness begins to oscillate and that the upper edge of the layer becomes more diffuse (Fig. 3). Emrich interprets this pattern as the beginning of transition. At the end of the transitional regime the boundary layer flow is fully turbulent. The edge of the layer appears diffuse with a large number of thin streaks oblique to the wall (Fig. 4). This structure can be explained with the phenomenon of bursts which have been studied and visualized extensively for incompressible turbulent boundary layers (Ref. 6,7). By means of these bursts, small volumes of cold gas are ejected from the wall region into the upper part of the boundary layer. The traces of these gas ejections are then detected by means of the schlieren system.

A large number of experiments were performed in order to determine the distance x_{tr} between the shock and the beginning of the disturbed region in the flow or the first appearance of the fully turbulent pattern. The values x_{tr} can be correlated through the "critical" Reynolds number (Ref. 3)

$$Re_{crit} = \frac{u_2^2 \cdot x_{tr}}{\nu_2 (u_s - u_2)} ,$$

where u_s is the shock velocity, u_2 the free stream velocity of the gas behind the shock, and ν_2 the kinematic viscosity of the shock heated gas. These Reynolds numbers are shown in Fig. 5 as a function of the shock strength, expressed by the pressure ratio p_2/p_1 across the shock.

Hot Film Measurements

The rate of heat transfer from the gas flow to the (cold) wall of the shock tube can be used as an additional source of information for the laminar-to-turbulent transition in the boundary layer. This heat transfer rate is higher for turbulent flow than for laminar flow. A hot film element imbedded in the wall of the shock tube is a device sensitive to such changes in local heat transfer. A small layer of platinum, o.oo1 mm thick, 2o mm long, and o.5 mm wide, was deposited on a glass surface and served as the hot film element. The platinum layer has an electrical resistance of about 5o Ω; it is connected to wires and built as the measuring element in a constant current circuit. The undisturbed temperature of the "hot" film element without flow is lower than the temperature T_2 of the shock heated gas. A change of the film temperature, due to a variation of the local heat transfer, causes a change of the electrical resistance of the element. The respective change in voltage, which can easily be detected after amplification, is a direct measure of the change in film temperature. Since the mass of the platinum film is extremly small, the temperature changes can be measured without any noticeable delay time.

Fig. 6 is a typical oscilloscope trace of the measured signal for a shock Mach number $M_s = 1.5$. The vertical scale is a measure of the film temperature; the abscissa is the time scale. The first rise of the signal is due to the passage of the incident shock wave. The second steep rise is attributed to the change from laminar to turbulent heat transfer at the tube wall. The respective time multiplied with the shock velocity u_s delivers the transitional distance x_{tr}. The third abrupt increase of the signal is caused by the shock wave reflected from the closed end of the shock tube. The critical Reynolds numbers derived from hot film experiments for two different shock Mach numbers, or for two different pressure ratios p_2/p_1, respectively, are also included in Fig. 5.

Results

From Fig. 5 one may conclude that the measured distances for transition can be correlated satisfactorily with a critical Reynolds number. The results obtained from optical tests and from hot film experiments are in good agreement. These results also agree well with data measured earlier by Emrich (Ref. 3) and Thompson and Emrich (Ref. 8). These data are not included in the diagram of Fig. 5. The increase of the critical Reynolds number with shock strength gives evidence of the stabilizing effect of compressibility or of an increasing flow Mach number. Schlieren photography is an appropriate tool for investigation of this type of boundary layer. The visualization makes clear that the mechanism of transition is very complex and that the structure of the turbulent boundary layer can be related to phenomena which have been observed in the incompressible boundary layer of liquid flows.

References·

(1) H. Mirels, Laminar boundary layer behind shock advancing into stationary fluid, NACA TN 34o1 (1955).
(2) E. Becker, Instationäre Grenzschichten hinter Verdichtungsstößen und Expansionswellen. In: Progress in Aeronautical Sciences, Vol.1 (edited by A. Ferri, D. Küchemann, L.L.G. Sterne), (1961), 1o4, Pergamon
(3) R.J. Emrich, Schlierenoptische Untersuchung des Turbulenzumschlags in der Stoßrohr-Grenzschicht, Z. Flugwiss. 19(1971), 242.
(4) J.A. Bander and G. Sanzone, Shock tube chemistry. 1. The laminar-to-turbulent boundary layer transition, J. Phys. Chem. 81(1977),1.
(5) H. Mirels, Attenuation in a shock tube due to unsteady boundary-layer action, NASA Report 1333 (1957).
(6) S.J. Kline, W.C. Reynolds, F.A. Schraub and P.W. Runstadler, The structure of turbulent boundary layers, J. Fluid Mech. 3o(1967),741.
(7) E.R. Corino and S. Brodkey, A visual investigation of the wall region in turbulent flow, J. Fluid Mech. 37 (1969),1.
(8) W.P. Thompson and R.J. Emrich, Turbulent spots and wall roughness effects in shock tube boundary layer transition, Phys. Fluids 1o(1967), 17.

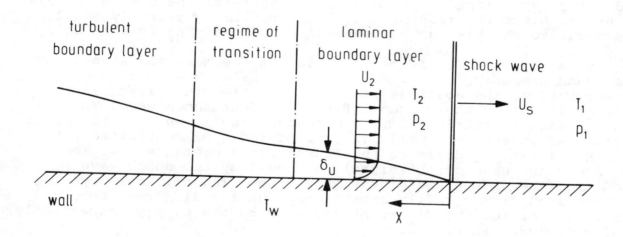

Fig. 1. Geometry of shock-tube boundary-layer

Fig. 2. Schlieren photograph of the laminar shock-tube boundary-layer. The vertical bright streak on the right is the plane shock wave travelling from left to right with M_s = 1.o6. The distance from the shock to the left end of the photograph is about 7o mm.

Fig. 3. Pulsating boundary-layer thickness in the regime of transition; M_s = 1.o6. Distance to shock ~ 1 m.

Fig. 4. Turbulent boundary layer; M_s = 1.41. Distance to shock ~ 8oo mm.

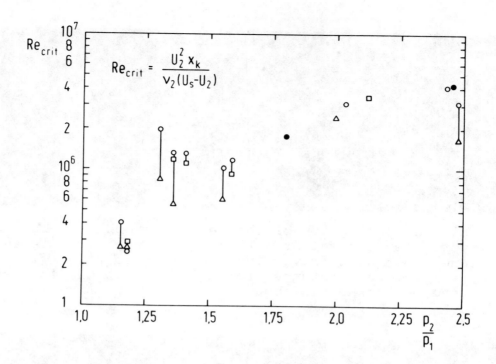

Fig. 5. Critical Reynolds number for different shock strengths p_2/p_1.
Symbols are averaged for a great number of single experiments:
△: first appearance of disturbed pattern,
○: last appearance of laminar flow,
□: first appearance of turbulent flow,
●: hot film experiments.

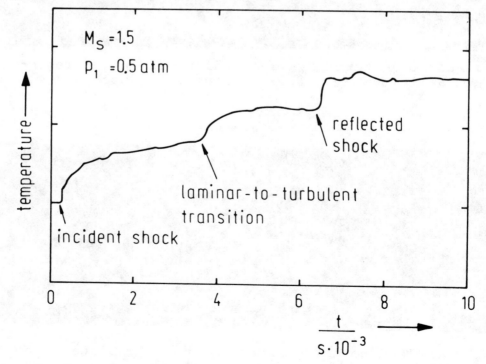

Fig. 6. Signal obtained with hot film element.

FLOW-FIELD VISUALIZATION OF HEAT AND MASS DIFFUSION IN A POROUS TUBE GENERATED FLOW

JOEL M. AVIDOR* and MARY DELICHATSIOS**

In the present study interferometry was used to visualize and investigate the diffusion of heat or mass in the mixing layer that develops at the interface between two dissimilar gas flows generated by two adjacent grids of porous tubes. In the experiment a Mach-Zehnder interferometer with 30 cm diameter optics and a pulsed Xenon laser (λ = 0.5126 μ) serving as the illumination source were used to obtain interferograms of the mixing layer. From typical interferograms presented mass and heat diffusion in the porous tube generated flowfield under various flow conditions have been visualized and compared with hot-wire temperature measurements and species concentration measurements.

Introduction

A major element in the laser fusion program is the availability of scalable nanosecond carbon dioxide laser amplifiers. One of the key technical questions for device scaling and for maximum performance is the development of scalable gain isolation. Flowing of a saturable absorber together with the laser gas in the cavity is a novel scheme to provide windowless scalable gain isolation in large CO_2 laser amplifiers. In order to optimize the operation in such a cavity the diffusion of the saturable absorber into the laser gas is of significant importance.

In this type of device to optimize performance it also becomes desirable to flow the gas parallel to the electric field. To achieve the latter, we use a novel scheme recently developed (Ref. 1) that is based on generating a flowfield in a cavity by injecting gas through a grid of porous tubes. Fig. 1 indicates the flow geometry for the main flow and saturable absorber flow.

The present work deals with results of investigations concerning the distribution of velocity, temperature and species concentration at the interface between two flowfields generated by two adjacent grids of porous tubes. Measurements have been carried out in a cavity where one portion of the flow is thermally tagged or mixed with a tracer gas, with the initial differences in temperature and concentration with respect to the surrounding flow confined to small values.

Two sets of experiments were undertaken. In the first set heat diffusion in the mixing layers was measured using a hot wire anemometer operating as a resistance thermometer. Mass diffusion of the simulated saturable absorber was obtained using a species concentration probe whereby gas was extracted from the mixing layer and the concentration of the saturable absorber was determined from the absorption of an incident IR radiation. In the second set of experiments interferometry was used and heat or mass diffusion in the mixing layers was determined from direct visualization of the deviation of the interference fringes from their unperturbed position.

*Principal Research Scientist   Avco Everett Research Laboratory, Inc.
**Senior Research Scientist   Everett, Massachusetts 02149 USA

Experimental Apparatus

A schematic of the apparatus that has been constructed and used for our experiments is shown in Fig. 2. The porous tubes used here are manufactured by Mott Metallurgical, Farmington, Conn. The tubes are fabricated from metallic powder (S.S. 316) to yield tubes with controlled porosity. The powder is precompressed into a form by the application of a controlled pressure, and subsequently heated to the sintering temperature. Very low-porosity tubes were used here, designated by the manufacturer as having 0.5μ filtration rating. They were eccentric in their cross section with a 0.040" thin wall, 0.250" I.D., 0.375" O.D. and 13" long.

The porous tubes were arranged in a parallel grid 2.5 cm apart and connected to two large diameter pipes (2" I.D.). The tubes were fed from two high pressure nitrogen manifolds, one serving as a flow source plenum for the room temperature nitrogen and the other for the slightly heated nitrogen. The parallel porous tube array (12 inc. by 13 in. in overall dimensions) was enclosed in a 12 inc. high plexiglass enclosure to prevent entrainment of ambient air into the near-field of the porous tube generated flow.

Compressed nitrogen was supplied to the grid from a high pressure bottle bank. The supply pressure was regulated using a dome pressure regulator which ensured constant plenum pressure during each test run. No filtration of the gas was used and no degradation of the low-porosity tube flow capability was observed over a large number of tests.

Temperature Measurements

We have investigated the structure of the temperature field in the interface between the simulated laser gas flow and the saturable absorber flow. For this purpose the "saturable absorber" gas was heated slightly ($\Delta T \approx 15^{\circ}C$) utilizing a pebble bed as a heat exchanger and was then introduced into three porous tubes (Fig. 1). These tubes were equipped with thermocouples glued to the surface at various axial positions, which served to monitor the heated gas temperature during each test to ensure that data was taken under temporally stable heating conditions. The splitter plates shown in the figure were used in order to possibly reduce the growth of the turbulent mixing layers.

For measuring mean temperature distributions both very fine thermocouples (wire size 0.0005") and a hot wire as a resistant thermometer were used. The hot wire was also used to measure temperature fluctuations and in conjunction with a true RMS voltmeter (TSI) yielded RMS temperature fluctuation data.

Hot wire anemometry was also used for mean flow velocity and RMS velocity fluctuation measurements. The constant temperature hot wire system (Thermo-Systems, Inc.) used for this purpose consisted of a constant temperature anemometer in conjunction with a linearizer yielding a linear net calibration curve. The velocity flowfield displays a wake-like behaviour close to the tubes (Ref. 1). The large velocity nonuniformities rapidly decay and at a distance of about 10 tube spacings the flow is quite uniform.

Fig. 3 shows a typical profile of the temperature distribution in the mixing zone and the corresponding velocity flowfield at a distance $X = 10$ cm above the porous tubes. The wake-like behaviour of the flowfield is evident. However the temperature distribution on either side of the mixing layers is quite uniform. This figure also includes the definition of the mixing zone width that we determined from the maximum slope of the temperature profile.

The rate of the spreading of the temperature field is shown in Fig. 4. In this figure we present data of the mixing layer growth taken with and without the presence of a splitter plate. The interesting observation here is that initially the mixing layer is smaller when the splitter plate is used, however, the mixing layer rate of growth is about the same as without the splitter plate further downstream of the porous tube grid.

Species Concentration Measurements

Mass diffusion in the turbulent mixing layers between the laser gas and saturable absorber flow has been measured using an IR absorption technique. The scheme involves seeding the saturable absorber region with SF_6, which is used as a tracer in the mixing region. A small gas extraction probe samples various locations in the mixing layers. The concentration of SF_6 is then determined from the absorption of radiation from a low power CO_2 laser beam.

The block diagram of the arrangement for the absorption measurement is shown in Fig. 5. The laser beam from a low power (4 watts) GTE-Sylvania 941E CO_2 laser enters a 50% beam splitter generating the reference beam and the beam that enters the absorption cell. The intensities of the reference beam, I_o, and the transmitted beam, I_t, are recorded by two pyroelectric detectors (Molectron P003). A balance amplifier (not shown) is used for the transmitted beam to ensure that the detector output is equalized before introduction of SF_6 into the cell.

Figure 6 shows the species concentration probe. The probe is made of plexiglass and consists of two parts. A screen is inserted between the two parts to ensure that the diverging flow of the sampled gas fills the test section of the probe. Two 1" diameter optical quality germanium windows are inserted in the test section of the cell to allow the transmission of the CO_2 laser radiation through the cell. The sampled gas flowing through the probe is sucked with a 1/2 mm diameter, 10 cm long hypodermic tube.

When gas flows into the cell the transmitted CO_2 radiation will be equal to the difference between the incident radiation on the cell and the absorbed radiation as follows:

$$I_a = I_o - I_t = I_o (1 - e^{-aPx})$$

where

$\quad\quad a$ = absorption coefficient
$\quad\quad P$ = the partial pressure of the absorbing gas (SF_6)
$\quad\quad x$ = the absorption cell length

Measurements of the mixing layer characteristics and growth for mass diffusion were almost identical to the measurements of temperature diffusion presented in Figures 3 and 4. This result has been well known and will be shown again in the interferometric measurements right below.

Interferometry

Figure 7 shows a schematic of the Mach-Zehnder interferometer (Ref. 2) that was next used to visualize the diffusion of heat and mass in the porous tube generated flowfield.

The system consisted of 30 cm diameter optical elements which were mounted on a vibration isolation table to ensure mechanical stability. A pulsed xenon laser radiating at 5126 Å with a pulse length of 0.3 μsec was used as the illumination source. The laser beam was expanded using a 40:1 beam expander to fill the 30 cm diameter optical elements. The beam was then split into two beams at plate P_3 by partial reflection and transmission forming the reference beam r_1 and the beam r_2 that passes through the test section. They were both gathered by a concave mirror and focussed on the screen where the fringes were formed.

A point that warrants some discussion here is the problem of focussing both the test section and the fringes on the screen. For clarity only two (coherent) rays r_1 and r_2, are shown in Fig. 7a. If the plates P_1, P_2, P_3, P_4 are adjusted exactly parallel to each other, the two rays overlap again when they leave plate P_4. In this case both of them appear to have come from the same source, S. Now if the splitter plates P_3 and P_4 are rotated, the rays appear to come from different virtual sources as illustrated in Fig. 7b. A rotation of plate P_3 about an axis normal to the plane of the rays does not affect the transmitted ray r_1, but makes r_2 appear to come from S_2. On the other hand, a rotation of plate P_4 does not affect r_2 but inclines r_1 in such a way that its virtual source is S_1. The two rays no longer overlap when they leave P_4, but it is necessary only that they coincide at the screen, where the fringes are to be formed. In other words, they must be brought to focus there. On the other hand, the optical arrangement must be such that the test section is also focussed on the screen. To accomplish both, the two rays r_1 and r_2 must appear to cross there. Such an intersection may be arranged by giving the splitter plates suitable rotations.

The test section is shown in Fig. 8 with the optical beam parallel to the porous tubes. Half of these tubes were fed with room temperature nitrogen (at $T \sim 280^o K$) and the rest of the tubes with lower density gas which was either nitrogen slightly heated ($T \sim 295^o K$) or nitrogen with a small concentration of helium (of about 3% by volume).

Direct visualization of the mixing layers between the two gas streams was achieved with this arrangement and some typical interferograms are shown in the following figures. Fig. 9a is a tare shot (an interferogram without flow in the test section) with the fringes almost horizontal. Fig. 9b shows a typical interferogram of the mixing layer between two gas streams with a small temperature difference.

One notices the mixing layer region and the characteristics of its spreading. The fringe shift in the mixing region is proportional to the density difference which in turn is proportional to the temperature difference from the reference stream (i. e., the room temperature nitrogen stream on the right of the photograph). Close to the tubes, in the lower part of the photograph the mixing layer thickness is small and steep temperature gradients are apparent. The gradients diffuse, that is, the mixing layer grows as the gases move upwards. The two vertical wires in the photograph are 2.5 cm apart and the horizontal wire is 18 cm above the porous tubes. The mixing layer thickness grows to about 3 cm at a distance of 25 cm above the porous tubes, and this result is consistent with hot wire temperature measurements presented earlier in this paper in Figures 3 and 4.

Figure 9c shows a similar interferogram of heat diffusion. In this case a splitter plate was inserted between the porous tubes at the mixing region, and the thickness of the mixing region near the tubes at the lower part of the photograph is smaller than it was in Fig. 9b. The growth characteristics of the mixing layer are as in Fig. 9b.

Figures 10a, 10b, 10c are similar to Figs. 9a, 9b, 9c and show the mass diffusion characteristics between the gas streams. In this case the lower density gas was nitrogen with a small concentration of helium (of about 3% by volume) The interferograms without a splitter plate (Fig. 10b) and with a splitter plate (Fig. 10c) exhibit the same behaviour as the heat diffusion interferograms of Figs. 9b and 9c.

Summary and Concluding Remarks

In the present study hot wire anemometer measurements, species concentration measurements and interferometry were used to investigate and visualize the diffusion of heat and mass in a porous tube generated flowfield.

The experimental results obtained by all methods show that the rates of spreading of heat and matter are mutually equal. An approximately linear growth of the mixing layer equal to 0.10 X was measured.

Some advantages of interferometry as compared with the other methods are worthwhile pointing out.

1. The method is direct and relatively simple.
2. Being an optical method, it does not disturb the flow.
3. It provides us with instantaneous visualization of the whole flowfield rather than point wise results.

References

(1) J.M. Avidor, N.H. Kemp, C.J. Knight, "Experimental and Theoretical Investigation of Flow Generated by an Array of Porous Tubes", AIAA J. Vol. 14, No. 11, Nov. 1976 pp. 1534-1540.

(2) Born, M., and Wolf, E., "Principles of Optics", Pergamen Press (1964) p. 463.

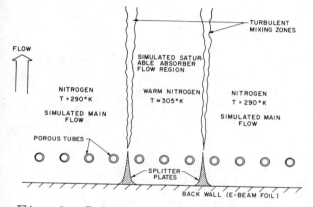

Fig. 1   Schematic of the Simulated Saturable Absorber Flow

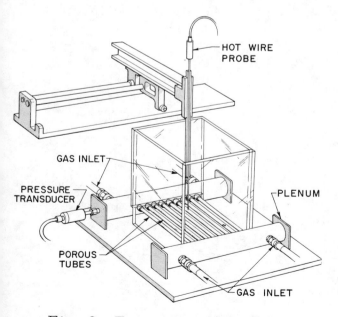

Fig. 2   Experimental Set-up

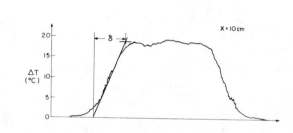

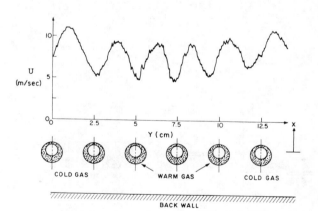

Fig. 3   Temperature Distribution and Velocity Flowfield at 10 cm above the Porous Tube Array

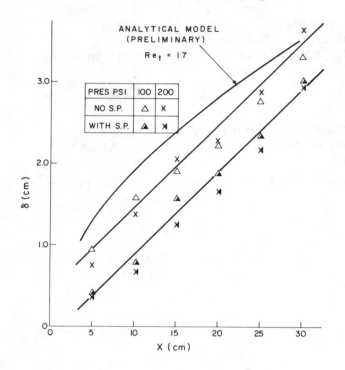

AXIAL SPREAD OF MIXING LAYER

ANALYTICAL MODEL
(PRELIMINARY)
$Re_t = 17$

PRES PSI	100	200
NO S.P.	△	X
WITH S.P.	▲	✖

Fig. 4   Axial Spread of the Mixing Layers without and with Splitter Plates Between the Warm Gas - Cold Gas Tube Arrays

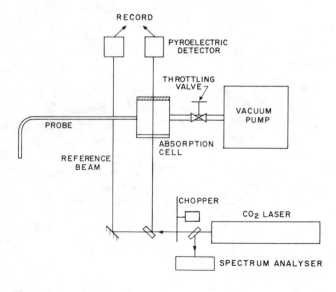

Fig. 5   Block Diagram of the Species Concentration IR Absorption Measurement

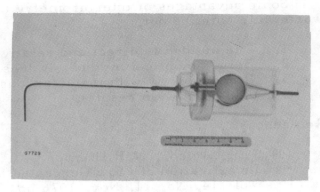

Fig. 6   The Species Concentration Probe

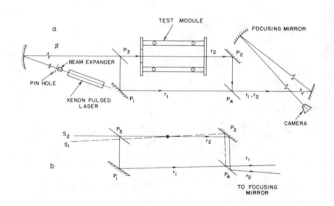

Fig. 7   The Mach-Zehnder Interferometer

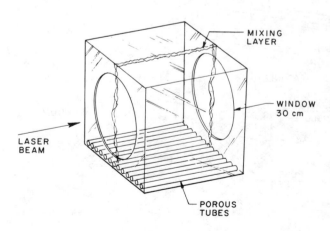

Fig. 8   Optical Set-up in Rectilinear Test Section

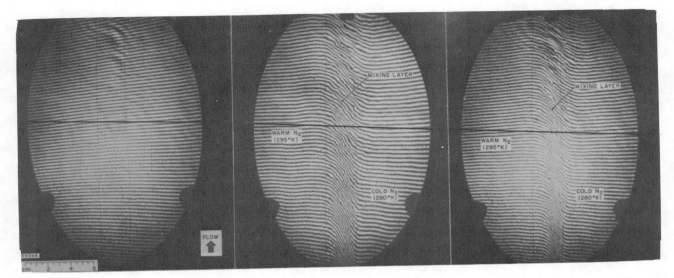

Fig. 9   Interferograms of the Mixing Layer Between the Two Gas Streams of Fig. 8 with a Different Temperature. Vertical wires 2.5cm apart, horizontal wire 18cm above porous tubes.   a. Tare shot   b. $\Delta T = 15^{o}K$ c. Same as b with the addition of a splitter plate between the tubes at the mixing region.

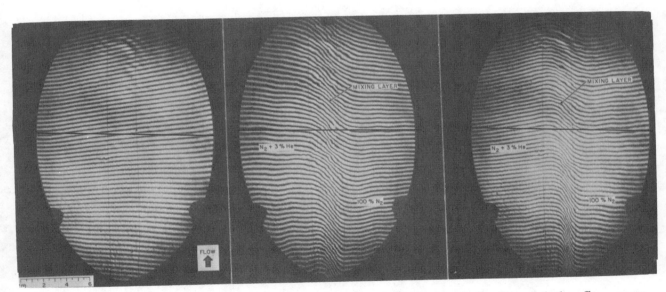

Fig. 10   Interferogram as in Fig. 9 with Helium Addition into One of the Streams. Vertical wires 2.5 cm apart, horizontal wire 18 cm above porous tubes. a. Tare shot   b. Helium concentration 3% by volume.   c. Same as b with the addition of a splitter plate between the tubes at the mixing region.

(Work supported under ERDA Contract EY-76-C 02-4-55*000)

AERODYNAMIC FLOW VISUALIZATION IN THE ONERA FACILITIES

CLAUDE VERET,* MICHEL PHILBERT,** JEAN SURGET,**
and GUY FERTIN**

Shadowgraphy, schlieren and interferometry are used in the ONERA facilities for transonic and supersonic aerodynamic flow visualization. Apparatus equipping several wind tunnels are described and results shown. Studies of aerodynamic flows in turbomachinery compressors require special visualization set-ups : optical systems with cylindrical lenses concentric to the hub carrying the blades were realized, one for a supersonic annular blade cascade, another for an axial compressor. Other systems were made for centrifugal compressors. The paper also describes a holographic interferometry set-up realized for specific applications in wind tunnels the improvements and shows brought about by this new interferometry technique.

1. SHADOWGRAPHY AND SCHLIEREN INSTRUMENTS FOR WIND TUNNELS

An example of shadographic image is given on Fig. 1. It was obtained at the Modane-Avrieux S2 wind tunnel, by use of a shadograph apparatus of 800 X 600 mm view field size (Ref. 1).

Fig. 2 shows a shadowgram obtained at Mach number 6 in the Modane-Avrieux S4 wind tunnel. This hypersonic wind tunnel is fitted on one side with a 800-mm-dia. window but, due to practical requirements, the other side is not available for visualization ; so, a high efficiency retroreflective screen is placed on this other side : the shadow of the flow is projected on this screen and recorded by still or movie cameras.

Several ONERA wind tunnels are equipped with schlieren instruments. In particular, Z type, single pass, schlieren benches, with a field of view of 350 mm in diameter, are fitted on the R2 and R3 supersonic wind tunnels of the Chalais-Meudon Center. In order to eliminate vibration effects appearing when the flow is started, a self-controlled device has been adapted to the knife edge on each of these benches.

Figures 3 show two aspects of the aerodynamic configuration around a space shuttle model, visualized by the schlieren method.

The S3 wind tunnel of Modane is equipped with a coaxial, single pass, schlieren system (Ref. 2). The view field, 800 mm in diameter, is obtained by means of two parabolic mirrors of the same size. The width of the test chamber (distance between the internal faces of the windows) is 760 mm for three-dimensional model studies, and only 560 mm for two-dimensional models in the transonic domain. In these conditions, it is difficult to use the normal optical circuit. Actually, because the large width of the chamber, the light rays are much deflected in the flow regions where the density gradients are high, mainly near the leading edge, the trailing edge and the shock wave ; deflections of more than one degree are reached. Deviated rays are rejected far from the knife edge and they do not reach the image, so that dark zones are associated with the high density gradient regions. Such effects are shown on Fig. 4a where a dark circle around the leading edge can be observed, looking like a thickening of the model and of the shock wave.

---

* Head, Optics Division, Office National d'Etudes et de Recherches Aérospatiales (ONERA) - 92320 Châtillon (France)

** Research scientist, ONERA

In order to avoid this spurious effect, the receiving unit of the schlieren set-up has been modified. The use of an objective associated to a bicolor filter schlieren mask makes it possible to collect all the rays, including the largest deviations. The picture of Fig. 4b, though printed in black and white, shows clearly the improvement of the image quality due to this device.

2. FLOW VISUALIZATION IN TURBOMACHINERY

Optical processes have also been used to study the flows within supersonic compressors developed at the Palaiseau Center. Still photographs and motion pictures reveal the shock wave evolutions and the turbulent regions inside and outside the interblade channels of fixed vanes or rotating blade cascades.

2.1. Supersonic axial compressor with short blades

Fig. 5 shows a schlieren image taken with an exposure time of 100 ns in a supersonic axial compressor with short blades. We can see, on the left side, the external branches of the bow shocks attached to the blade leading edges and, in the middle, between the blades, the internal branches followed by lambda shocks. This compressor includes a closed tubular circuit inside which freon gas is set in motion by an experimental rotating annular cascade (rotor diameter 440 mm, rotation speed 12,000 r.p.m., blades height 10 mm). In order to allow the passage of a test light beam, the external casing is fitted with a cylindrical window. The optical device used here for flow visualization is a coaxial schlieren system working by reflection on the hub, previously polished, acting as a cylindrical mirror (Ref. 3). The exposure time must be less than 1 μs in order to photograph the shock wave when the rotor is at its maximum speed. The flash lamp can also be replaced by a stroboscopic source in relation with a movie camera. Films were taken with a normal speed camera (24 frames per second) (Ref. 4).

2.2. Supersonic radial compressor

Another schlieren system has been developed for flow visualization in the fixed straight blades diffuser of a radial compressor (Fig. 6). A plane steel mirror is fixed on the back casing (seen as a black part, partially covered by blades, on Fig. 6). A window is sealed on the upper cover of the casing.

An autocollimation schlieren device is used for sending a parallel light beam on the mirror, through the window. The field of view at the blade level is approximately rectangular and 180 $\times$ 135 mm in size.

On Fig. 6a can be seen from right to left : the bow shocks attached to the leading edges, oblique shocks and their reflections on the walls in the interblade channels and recompression shocks ; the flow becomes subsonic and turbulent behind these last shocks. On Figures 6b, c, d can be seen the recompression shocks travelling gradually upstream, up to the throat section. When counterpressure increases further, a surge phenomenon appears accompanied by the generation of a wave network moving with the rotor (Fig. 6e). This regime is highly unsteady (Ref. 5).

2.3. Axial supersonic compressor with long blades

In this compressor, the visualization method used is the shadowgraphy (Ref. 6) ; this method is perhaps less comprehensive than schlieren but it can be used even if the hub carrying the blades has an irregular shape. The surface between the blades is covered with a retroreflective coating, and a window is sealed in the casing in front of the blade tips.

3. INTERFEROMETRY

This optical process is well adapted to study aerodynamic phenomena as it allows, in addition to visualization, a quantitative analysis when the flow is two-dimensional or axisymmetrical.

Interferometry has been used of the ONERA wind tunnels for about twenty years. The biggest Mach-Zehnder interferometer built at ONERA has an elliptical field of view 100 $\times$ 150 mm in size (Ref. 7).

Fig. 8 shows an example of an interferogram obtained by means of one of the Mach-Zehnder interferometers equipping the Chalais-Meudon ONERA fluid mechanics laboratory (Ref. 8).

Classical interferometers are very expensive because they require the use of highly accurate optical and mechanical components. When new equipment is needed, it is at the present time more likely to be holographic interferometers. Actually, they are cheaper to implement and easier to operate for the same accuracy and they offer increased possibilities. Several holographic benches, of modular design, were built and operated for a broad range of experiments on various wind tunnels (Ref. 9, 10). The largest field of view is 200 mm in diameter, with overall dimensions as small as 1.5 m in length.

In practice, two main methods are used : stored beam and double exposure. With the former, the interferogram appears in real time as with a classical interferometer, changing with the flow inside the test chamber.

With the second method, it is necessary to wait for the processing time of the holographic plate to observe the interferogram of the phenomenon. However this technique is the best when the interferogram has to be processed for quantitative measurements.

The data processing is performed from a negative photographic reproduction of this interferogram. The use of modern processors makes this operation, which was previously time consuming, simpler and faster (Ref. 11). The working process, now entirely automated, uses a Joyce-Loebl 3 C S microdensitometer coupled with a Hewlett Packard 2100 microprocessor connected to a disk memory, where the programs and intermediate data are stored.

The processed results, such as density values along various scanning lines or velocity profiles within the isentropic flow regions, are numerically printed, and diagrams are traced on a tracing table.

The holographic procedure vastly improves the use of the interferometry in the aerodynamics domain, providing new experimental possibilities, which are not accessible to classical interferometry(Ref. 12).

4. CONCLUSION

The visualization optical techniques have been used since the beginnings of experimental aerodynamic research. They are appreciated because, on the one hand, they don't induce disturbances within the studied flows, and, on the other hand, they provide a general view of a flow field which, even in the case where they remain qualitative, improve highly the comprehension of complex physical phenomena and make much easier the interpretation of the point-to-point measurements performed by means of other physical techniques.

Shadowgraphy and schlieren techniques are, at the present time, mainly associated with industrial wind tunnels. Interferometry, offering easier implementation and operation, new possibilities, improvement in the data processing by direct connection with a microprocessor, use of less expensive optical devices, will lead in the near future to a more extensive use and possibly to its application in industrial facilities.

REFERENCES

1. C. VÉRET, Appareil d'ombroscopie de la soufflerie S2 Modane, Rech. Aéron., no. 90 (Sept.-Oct. 1962), p. 33-39.
2. C. VÉRET, Appareil de strioscopie de la soufflerie S3 Modane, Rech. Aéron., no. 78 (Sept.-Oct. 1960), p. 37-40.
3. M. PHILBERT and G. FERTIN, Schlieren systems for flow visualization in axial and radial flow compressors, ASME paper 74-GT-49 (1974).
4. G. FERTIN, Recording of schlieren pictures for flow visualization in rotating machines, in "High Speed Photography" P-J. Rolls, Ed. Chapman and Hall 1975, p. 375-379.
5. Y. RIBAUD and P. AVRAN, Visualisation par strioscopie dans un diffuseur de compresseur centrifuge supersonique, in "Proceedings of the Second Symposium on Air Breathing Engines", published by the Royal Aeron. Soc., London (1974).
6. G. FERTIN, Visualisation de l'écoulement dans un compresseur axial supersonique à aubes longues, La Rech. Aérosp., no. 1975-3, p. 165-168.
7. R. CHEVALÉRIAS, Problèmes particuliers posés par la mise au point d'un interféromètre strioscopique sur une soufflerie supersonique, Revue d'Optique, t. 35 (1956).
8. J.L. SOLIGNAC, Etude interférométrique de l'écoulement à Mach 5 autour d'une sphère, Rech. Aérosp., no. 125 (1968), p. 31-39.
9. M. PHILBERT and J. SURGET, Application de l'interférométrie holographique en soufflerie, Rech. Aérosp., no. 122 (1968), p. 55-60.
10. J. SURGET, Etude quantitative d'un écoulement aérodynamique par interférométrie holographique, Rech. Aérosp., no. 1973-3, p. 161-171.
11. J. DELERY, J. SURGET and J.-P. LACHARME, Interférométrie holographique quantitative en écoulement transsonique bidimensionnel, Rech. Aérosp., no. 1977-2, p. 89-101.
12. J. SURGET, Two references beam holographic interferometry for aerodynamic flow studies, in International Conference on "Applications of holography and optical data processing", Jerusalem, August 1976 - Pergamon Press.

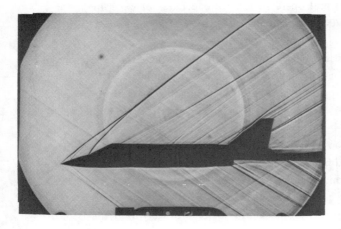

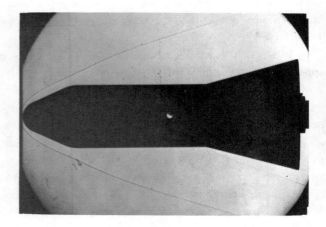

Fig. 1 — Schadowgram of a supersonic flow in the S2 Modane wind tunnel.

Fig. 2 — Shadowgram obtained on the S4 Modane wind tunnel (Mach number 6).

Fig. 3 — Schlieren pictures of a space shuttle model in the R2 Chalais-Meudon wind tunnel.

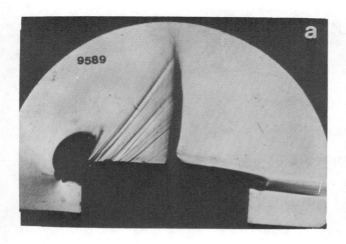

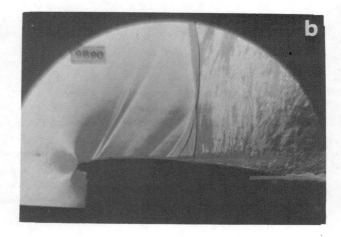

Fig. 4 — Visualization of a two-dimensional flow in the S3 Modane wind tunnel (Mach number 0.76)
a - Classical schlieren with knife edges.   b - Schlieren with a two-color filter.

Fig. 5 — Example of still photograph
in an axial compressor.

Fig. 6 — Rotor and stator of a radial compressor.

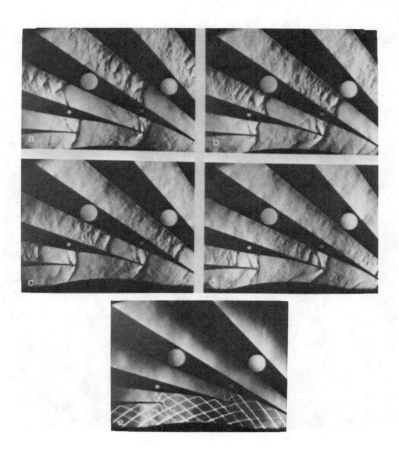

Fig. 7 — Example of schlieren pictures taken in the straight blade diffuser.

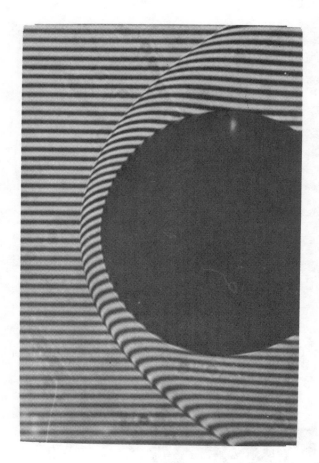

Fig. 8 — Interferogram of a flow around a sphere (Mach number 5).

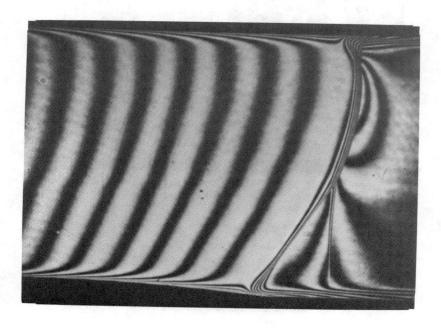

Fig. 9 — Holographic interferogram of a two-dimensional flow near the trailing edge of a bump on a wall.

FLOW VISUALIZATION WITH HOLOGRAPHIC INTERFEROMETRY

F. MAYINGER* and U. STEINBERNER**

The visualization of hydrodynamic processes by the aid of holographic interferometry makes possible the investigation of a great number of flow problems. The applicability of the method requires that the variations of the refraction number, generated in the transilluminated fluid, depend on the flow to be investigated. This is especially true for convection phenomena. The applied interference techniques and the optical arrangement are described. The examples of application mainly illustrate the flow conditions in boundary layers. A laminar and turbulent boundary layer at free convection, the superposition of free and forced convection and the convection in the vicinity of rising gas bubbles are presented. Flows, normally occuring without heat transfer, are made visible likewise.

Introduction

The hydrodynamic and thermodynamic relationships of the convective heat transfer between a fluid and a heat-emitting or heat-absorbing wall are described by a system of partial differential equations. The equations are based on the laws of conversation for mass, momentum and energy. They are presented in detail in the fundamental literature on fluid mechanics (Ref. 1) and heat transfer (Ref. 2,3). The balances show that in general the velocity field and the temperature field are mutually coupled. Therefore it is possible to draw conclusions concerning the flow behaviour by visualizing the temperature field.

In order to observe the temperature field in transparent fluids, optical methods are available, which based on the interference phenomenon of coherent light. For the visualization presented here the holographic interferometry was applied. The observed and recoreded interference fields represent the distribution of lines of constant temperature. The formation and shape of the lines and the direction of movement at temporal and local variation of the field are indicators by means of which qualitative and quantitative statements can be made about the velocity field.

In the most cases of convective heat transfer the temperature distribution and therefore the velocity distribution has boundary character. For this reason only the relationships, in the vicinity of the wall were investigated, i.e. in the boundary layer. The examples of application show the different structures of boundary layers at free convection with internal heat sources in closed spaces resulting from different boundary conditions. To finish up the subject suggestions shall be given for the investigation of flows with which no heat transfer is connected in general.

Applied Interference Methods And Optical Arrangement

Since the development of the laser, the holography invented by Gabor, has been applied in a wide range. With the holography there is an optical recording technique available by means of which the amplitude and phase distribution of a wave front of any given shape can be recorded and reconstructed at a later moment. Applying this principle, a number of new measuring techniques have been developed during the last few years (Ref. 4). Thereby the holographic interferometry has gained particular importance.

For a better understanding of the used interference techniques a clear presentation of the holography shall be given briefly. The general theory is very complex and can be taken from literature (Ref. 5,6). The conditions, occurring at the recording and reconstruction of a parallel light wave, as was applied in the experiments also, are presented in fig. 1. For the recording the object wave is superposed with a second wave. If the object and reference wave are coherent to each other a stationary interference field occurs. The resulting fringe system can now be recorded on a suitable light sensitive material, e.g. a photographic plate. After the development the photographic plate contains the whole information about the object wave because the amplitude and phase distribution are stored in form of a photographic density distribution. If the plate, now called hologram, is illuminated with a coherent light wave the interference pattern functions as a diffraction grating with a variable grating constant. Thereby waves of different order of diffraction occur at the reconstruction. One of the waves of the first order travels in the same direction as the original object wave and also possesses the same amplitude and phase distribution if the reconstruction wave equals to the reference wave. This directly reconstructed object wave is used with the interference techniques described subsequently.

For the visualization of flow processes it is advantageous to apply a method by means of which the relationships in the fluid can continuously be observed and recorded. For this purpose the real-time-method has been developed. The basic idea of this method is to record the object wave at known conditions in the fluid by the aid of the holography, to reconstruct it subsequently and to superpose if with the momentary existing object wave at the conditions to be investigated. The waves will interfere and the differences between the wave fronts will continuously be visible and measurable. Fig. 2 shows the principle of the real-time-method at the example of a heated horizontal plate. In the first step the object wave is holographically recorded at constant temperature distribution in the fluid. After the development and the repositioning of the photographic plate the original object wave is continuously reconstructed with the reference wave and superposes with the momentary object wave. If the conditions in the test section have remained constant and the hologram has exactly the same position as at the recording no interference fringes are visible. This zero field serves as a control. Now the flow conditions which are to be investigated can be produced by heating the plate. The local differences of the density of the fluid in the boundary layer cause a phase shift of the wave fronts of the momentary object wave. The interference field, resulting from the reconstructed and deformed object wave can continuously be observed behind the hologram and photographed or filmed. If flow conditions exist which temporally change rapidly the application of a high speed camera is advisable in order to obtain a good temporal resolution of the processes.

Should such a phototechnical equipment not be available it is possible to obtain informations about the flow with the double exposure method. This method makes use of the property of recording several object waves on a hologram successively. Fig. 3 shows this principle, again at the heated horizontal plate. With a first exposure the object wave is recorded as with the real-time-method at a constant temperature distribution in the fluid. Then the plate is heated. The produced temperature field deforms the object wave which is now recorded with a second exposure on the same photographic plate by superposition with the reference wave. With fast processes the second exposure time is to be chosen correspondingly short. This, however, can be achieved by a short unblocking of the focussed laser beam with relatively simple devices. After the development, the hologram is illuminated with the reference wave. Both wave fronts are now simultaneously reconstructed and will interfere. The interference picture can be observed and photographed. The disadvantage of this method is that a hologram respectively records only one single temporal state of the interference field, i.e. a continuous observation is impossible. In addition it should be noted that the double exposure method using a double pulse laser as a light source allows extremely short exposure times of approximately 20 ns. Thereby it is for example possible to investigate super sonic flows.

The main difference between the described methods and the classical interferometry is that with the holographic interferometry the object wave is compared with itself. This results in considerable simplifications in the experiments. Optical imperfections are eliminated because the interfering waves have passed through the same test section.

A commonly used optical arrangement for both holographic interference methods is presented in fig. 4. A laser is used as a light source. The laser beam is divided into an object and reference wave with a beam splitter. Both beams are expanded to parallel waves with the help of lens systems and are brought to coincidence on the photographic plate with mirrors. The test section is in the expanded part of the object wave. Recording the interference pictures it should be considered that, applying parallel light the maximum size of the interference picture that can be recorded equals the diameter of the front lens of the camera. If the diameter of the object wave is larger than the lens of the camera an additional lens has to be used for refocussing so that the illuminated cross-section can be photographed.

Examples For Application

As mentioned in the beginning, particularly those flow relationships are investigated which occur under different conditions at convective heat transfer in the zone near the wall. The interference pictures which respresent the temperature fields in all examples supply the following informations about the boundary layer of the flow:

- thickness of boundary layer
- transition from laminar to turbulent flow
- position of flow separation

- visualization of flow conditions
  from
    a) shape of the stationary interference field
    b) shape, dimension, position, frequency and direction of motion
       of
       the time variing interference field
       as a result of
       fluctuations, instabilities or superposed flow disturbances.

Free convection in horizontal fluid layers with internal heat sources occurs very frequently at chemical reactions and shall be the theme of the first example. The flow conditions, occurring in the layer and the resulting temperature distribution were theoretically and experimentally investigated by the aid of the holographic interferometry (Ref. 7,8). A temperature field typical of the convection is presented in fig. 5. The test fluid was water which was electrically heated to simultate uniform volumetric heat sources. At the top and at the bottom the layer as limited by cooling channels which provide constant wall temperatures. The interference picture shows that an unstable boundary layer occurs in the upper part, due to differences of density; here the cold fluid is above the warmer fluid. Out of this boundary layer cold fluid flows in narrow zones to the bottom, recognizable at the separations of the isotherms. In contrast to that the warm fluid rises in wide zones. At the lower part the warm fluid is above the colder fluid and leads to a stable boundary layer. The separations at the upper boundary layer temporally vary in size, position and number. The separations produce eddies in the area between the two boundary layers without essentially influencing the lower boundary layer and cause a considerably better heat transfer at the top. The heat transfer is proportional to the temperature gradient at the wall and can be estimated qualitatively by the density of the isotherms likewise.

A partially similar behaviour can be obtained if the free convection with internal heat sources occurs in a semicircular cavity at the same boundary conditions (fig.6). For reasons of symmetry only one half of the cross section is presented. At the upper part and in the lower part of the bottom the boundary layers show the same characteristics as in the horizontal layer. In the upper part of the curved bottom the fluid flows down in a thin boundary layer at the wall of the chamber and forms a big stationary tip vortex. The direction of movement can be observed when a separation is seized by the vortex and is drawn into the laminar boundary layer.

In connection with a lot of chemical processes the relationships during the inflow of a fluid into the just described fluid layer are interesting. In order to investigate this mechanism fluid in form of a jet was supplied at the top (fig. 7). With that a superposition of free convection and forced convection exists fluid mechanically. A theoretical solution of this problem can only be achieved by comprehensive numerical calculation methods (Ref. 9). If the momentum flux of the jet is larger than the inertia forces of the fluid, resulting from the free convection, the beam penetrates the boundary layer of the bottom up to the wall and forms a stagnation point flow. The fluid, transported by the jet, flows towards the free convection at both sides of the stagnation point and then separates from the wall. The points of separations are clearly recognizable at the distribution of the isotherms. The shape of the isotherm at this point shows that a pair of vortices rotating with opposite direction, occur in the area of the stagnation point flow. The jet itself shows a slight widening up to the reversal. At the borders of the jet the mixing with the surrounding fluid increases with the pass of the jet which is visible from the isotherms. The thermal load of the wall compared to the pure free convection, significantly changes only in the area of the stagnation point due to the superposition of the forced convection. For the verification of the observed flow conditions the velocity field, measured with a Laser-Doppler-Anemometer, is presented in fig. 8. The arrows only serve as indicators of direction.

In a third example the vertical boundary layer shall be observed at free convection with internal heat sources. Fig. 9 shows several sections which are typical of the development of the turbulent boundary layer and their position at the lateral, cooled wall of the used rectangular test chamber. The beginning of the turbulent movement is characterized by the first occurence of smooth waves at the border of the until then laminar boundary layer. For the definition of the transition from laminar to turbulent flow a critical value is normally given for the Rayleigh number, describing the convection process. This value is obtained if the local Ra-number is calculated with the observed distance between the transition point and the starting point of the boundary layer. In the case of the boundary layer, presented here, a critical Ra-number of approximately 10^9 was obtained. The result corresponds to the investigations by Eckert (Ref. 10) and Cheeseright (Ref. 11) at a heated vertical plate. Below the transition area the waves are fully developed and flow periodically downwards. The observed wave frequency was 0,25 1/sec. With increasing distance to the transition area the waves become more irregular and begin to turn over and to form eddies. In this area a fully developed turbulent flow occurs. The direction of rotation of the eddies corresponds to the flow direction of the fluid, which flows down in the boundary layer. The thickness of the flow boundary layer can be calculated according to an estimation by Schlichting (Ref. 12). The ratio of the thickness of the velocity boundary layer δ_V to the thickness of the temperature boundary layer δ_T is proportional to the square roof of the Pr-number. Fig. 10 shows a comparison between a temperature boundary layer measured holographically and the velocity profile of the same position in the laminar range, obtained with a Laser-Doppler-Anemometer. The thickness of the velocity boundary layer, obtained from the temperature field shows good agreement with the direct measurement.

The next example is concerned with flow processes in two-phase fluids. The convection movement in the vicinity of a wall, where gas bubbles come out and rise in the surrounding fluid was investigated. Such processes appear at two-phase heat and mass transfer with and without chemical reactions, where a mixing effect is generated in a liquid by means of gas bubbles. Similar relationships occur with bubble boiling in subcooled fluids, where steam bubbles rise at a heated wall (Ref. 13). The starting situation of the investigation was the horizontal fluid layer with internal heat sources, described in the first example whose stable boundary layer at the bottom of the test chamber is shown once more in fig. 11a. If air is blown in through a nozzle at the bottom now a periodical growth and separation of the bubbles takes place at the wall. The bubble mechanism causes a convection movement which transports cold fluid with the bubble from the wall to the top and simultaneously for reasons of continuity brings warm fluid to the cooled wall at both sides of the bubble. The direction of movement of the thereby induced eddies is cearly recognizable (fig. 11b) at the shape of the interference lines in the vicinity of the bubble. Connected with the convection movement a relatively thin boundary layer occurs at the bottom. This results in a considerably higher heat transfer. In the upper part of the fluid layer then rising bubbles cause an intensive intermixing of the fluid.

The investigation shows that the holographic interferometry is very suited for the observation of fast and strongly fluctuating processes. The exposure time of the interference picture produced according to the real-time method was 1/500 sec. A 4 Watt Argon Laser was used as a light source whose power of about 514 nm is also sufficient for high speed photographs up to 3000 frames/sec and more.

Flows, normally occuring without heat transfer can be made visible with the holographic interferometry likewise. In these cases an arbitrary temperature field is generated in the flow in order to obtain the wanted informations. Applying this method the flow behaviour may not essentially be influenced by the thermic buoyancy. The latter can generally be supposed with the forced flow of the fluids air and water if the flow velocities are not too small. As an example for this kind of visualization an air flow, drawn in by the rotor of a ventilator, was chosen. The aim of the investigation was to calculate the width and position of the drawn in air flow in order to arrange best the intake ports in the housing of the ventilator. For the production of the interference picture in fig. 12 the double exposure method was used. The first exposure was performed at stationary rotor and a surrounding air of room temperature. Then the rotor was turned on and the air was electrically heated in some distance above the rotor. During the second exposure the warm air, drawn in with the flow to be investigated, yields the change of the optical pass which is necessary for the interferometrical visualization. Form the interference lines the air flow which is entering the rotor and which is sharply sett off to the surrounding is clearly recognizable. To supplement the last example it should be mentioned that the flow behaviour within the boundary layer of fluids flowing around bodies can be made visible with the described method also. In this case it is advantageous to cause the interference field by heating the body under investigation and to have the applied fluid at the state of the surrounding.

Concluding Remarks

The holographic interferometry proves to be a valuable method for the visualization of flows in many fields of fluid mechanics. The application requires diffraction differences in the transilluminated fluid which depend on the flow to be investigated. Therefore the method is particularly suited for the observation and analysis of flow processes which are simultaneously superpositioned by a heat transfer. The interferometrical investigations can be performed relatively fast if the optical arrangement is available. A high optical quality of the test section is not required. In addition, the experimental technique is considerably simplified in comparison to the classical interferometry. The presented examples show a few possibilities of application at free and forced convection. Even flows, normally occurring without heat transfer, can be made visible under particular special conditions. Therefore the holographic interferometry is suited for the investigation of a large number of flow problems ranging from creeping flow up to very rapidly changing processes.

References

(1) R.B. Bird, W.E. Stewart, E.N. Lightfoot, Transport Phenomena, (1966) John Wiley and Sons
(2) Gröber, Erk, Grigull, Die Grundgesetze der Wärmeübertragung, (1963), Springer
(3) E.R.G. Eckert, Wärme- und Stoffaustausch, (1966), Springer
(4) F. Mayinger, W. Panknin, Holography in Heat and Mass Transfer, 5th Int. Heat Transfer Conference, Tokyo 1974
(5) H. Kiemle, D. Röss, Einführung in die Technik der Holographie, (1969), Akademische Verlagsgesellschaft Frankfurt/Main
(6) R.J. Collier, C.B. Burckhardt, L.H. Lin, Optical Holography, (1971), Academic Press
(7) M. Jahn, Holographische Untersuchung der freien Konvektion in einer Kernschmelze, Dissertation, T.U. Hannover (1975)
(8) H.H. Reineke, Berechnung der freien Konvektion mit inneren Wärmequellen, Dissertation, T.U. Hannover (1974)
(9) H.H. Reineke, R. Schramm, U. Steinberner, Reaktortagung Mannheim, (1977), 221 - 224, Deutsches Atomforum e.V. (DAtF)
(10) E.R.G. Eckert, T.W. Jackson, Analysis of turbulent free convection boundary layer on flat plate, (1950), NACA TN 2207
(11) R. Cheeseright, Turbulent natural convection from a vertical plane surface, (1968), ASME Publication 67-HT-17
(12) H. Schlichting, Grenzschichttheorie, (1965), Verlag G. Braun
(13) F. Mayinger, D. Nordmann, W. Panknin, Chem.Ing.Techn. 46, (1974), 209

+ Professor and Director of the Institut für Verfahrenstechnik, Technical University of Hannover, Hannover, FRG

++ Co-worker of Prof. Dr.-Ing. F. Mayinger, Institut für Verfahrenstechnik, Technical University of Hannover, Hannover, FRG

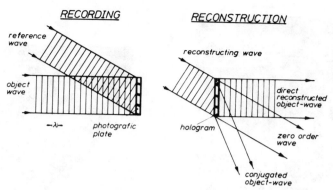

Fig. 1 Principle of holography

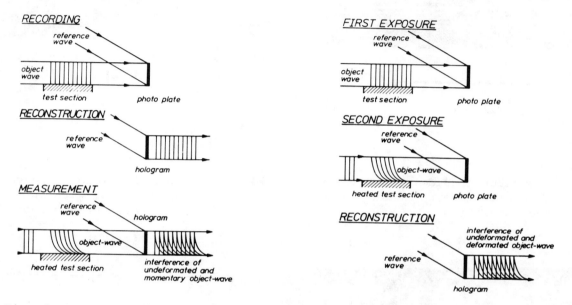

Fig. 2 Real-time-method

Fig. 3: Double exposure technique

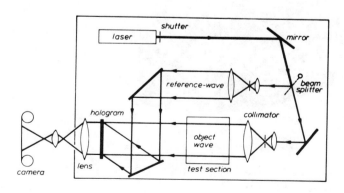

Fig. 4 Holographic interferometer

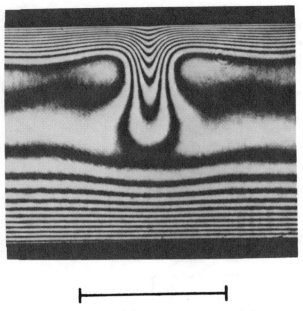

10 mm

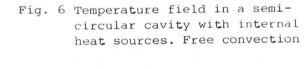

10 mm

Fig. 5 Temperature field in a horizontal
layer with internal heat sources.
Free convection

Fig. 6 Temperature field in a semicircular cavity with internal
heat sources. Free convection

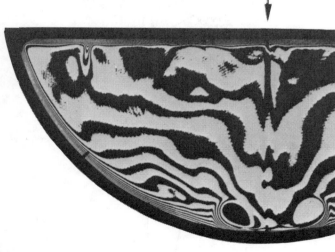

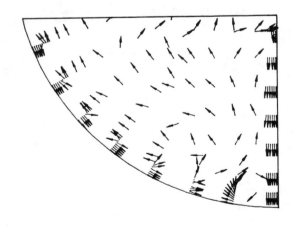

Fig. 7 Temperature field in a
semicircular cavity with inflowing
yet. Superposition of free and
forced convection.

Fig. 8 Velocity field in a semicircular
cavity at the same conditions as
in Fig. 7

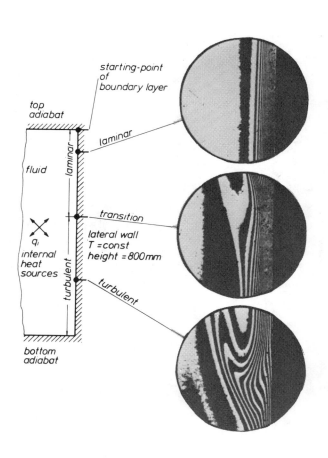

Fig. 9 Temperature field of a
vertical turbulent boundary
layer

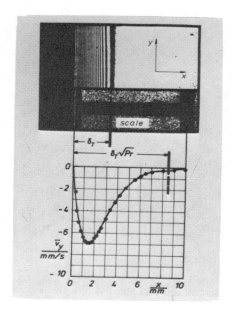

Fig. 10 Velocity-profile and
temperature-boundary
layer

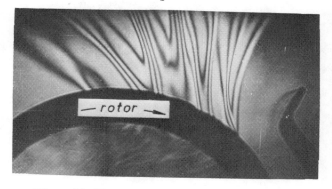

Fig. 12 Temperature field of a flow
drawn in by a rotor

Fig. 11 Lower part of
the
a) temperature field
in a horizontal
layer with inter-
nal heat sources
b) temperature field
of the same layer
with rising gas
bubbles

10 mm

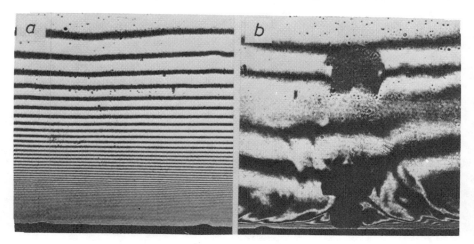

HOLOGRAPHIC FLOW VELOCITY ANALYSIS

M. YANO* and I. FUJITA**

Instantaneous velocity profiles in arbitrary section of water flow have been measured by holographic interferometry.

Each directional component of velocity is analyzed from the reconstruction images of the holograms. In this method, the measuring range of velocity is limited by Doppler effect.

Another limitation is caused by the seeded particles which are contained either naturally or artificially in flow.

The quantity and quality of the seeded particles are related to the quality of the hologram.

The complicated analysis of the velocity evaluation from the hologram is occured by the difference of the indices of refraction between water of flow and surrounding air.

Introduction

Holography becomes a familier technique in wide range of research fields in science and technology. We have known many examples of the application of that in fluid mechanics. However, there is few study to measure directly the velocity in flow.

In this paper, a measuring method of velocity in flow by holographic interferometry is described.

The cloud of colloidal micro suspended particles have been used as the tracers of flow in the measurement. When the measuring section in flow is illuminated by laser beam expanded only uni-direction, the scattering beam from the tracers is recognized as a brilliant membrane in flow. (Ref.3)

Instantaneous movement of the cloud of tracers in the membrane is seemed to be similar to the deformation of a elastic thin plate. The movement of the tracers is concerned with directly the velocity of flow in short time.

When the hologram of the membrane is made by the double exposure method, the velocity of flow is analyzed from the interferometric fringe pattern of the reconstructed image of the hologram and the time interval of double exposure.

Obviously, the other methods of real time and time averaging of holographic interferometry can not be applied to the measurement.

As the previous works, the experiments concerned with seeded tracers in flow have been tried. At the first work, only city supplied water is used in flow, which is naturally contained micron order air babbles and other particles confirmed by the observation of microscopic view. The tracers of flow are seemed to be the air babbles. (Ref.4)

At the second work in the method the tracers are seeded artificially in flow. The tracers have been used the particles of different quality, concentration, and diameter. The satisfied holograms have been obtained from the work in flow of wide range of Reynolds number.

* Professor of Civil Engineering, The Kobe University, Kobe, Japan

** Graduate Student of Civil Engineering, The Kobe University, Kobe, Japan

From the obtaining fringe pattern on the membrane, the velocity components of flow have been calculated with trial and error method because of the difference of the indices of refraction between water and air.

The calculating values of the local velocity and total discharge of the flow are compared with the direct measuring values and confirmed to be coincident each other.

Measuring Range of Velocity

In double exposure holography for moving object, the phase and angular velocity of the scattering beam from the object are both shifted from those of the original illuminated beam. (Ref.5)

Therefore, it is necessary to neglect either one of the information. This method is developed under the condition to neglect Doppler effect. In one exposure, on the recording plate of holography, the intensity of holographic information concerning with m particles of the tracers in the membrane is represented by the theory of partial coherent light.

$$\langle R_e (E_0 \cdot E_i^*) \rangle = \frac{1}{T} \int_0^T R_e \left\{ \sum_m A_0 A_{mi} \exp i \left[(\phi_0 - \phi_{mi}) + \omega_{Dm} t \right] \right\} dt \qquad (1)$$

where T is the exposure time, E_0 and E_i are respectively illumination and scattering beams ($E = A \exp i (\phi + \omega t)$), phase ($\phi_0 - \phi_{mi}$) represents the information of m particle, and angular velocity ω_{Dm} is Doppler shift generally unconsidered in static holography. The equation (1) is rearrenged

$$\langle R_e (E_0 \cdot E_i^*) \rangle = \sum_m \sqrt{A_0^2} \sqrt{A_{mi}^2} \cos(\phi_0 - \phi_{mi}) \sin \left(\frac{2\pi}{\lambda} |V_m| \cdot T \right) \Big/ \left(\frac{2\pi}{\lambda} |V_m| \cdot T \right) \qquad (2)$$

where the phase ($\phi_0 - \phi_{mi}$) is assumed to be independent of time, also the angular velocity of Doppler shift ω_{Dm} is assumed to be the maximum value $\omega_{Dm} = \frac{4\pi}{\lambda} |V_m|$, and the velocity $|V_m|$ is assumed to be constant during the exposure time T.

According to the time variation of the equation (2), the effect of Doppler can be neglected when the velocity $|V_m| \cdot T \ll \frac{\lambda}{2}$.

In the method, a ruby laser has been used for the source of beams. The exposure time T is equal to the puls width $T = 30 \times 10^{-9}$ sec. and the wave length $\lambda = 0.6943 \mu$ m..

As the result, the measurable range of velocity is determined as $0 < |V_m| \ll 10 \, m/s$.

The interval time of double exposure has to select shorter than the period of the fluctuation of flow and the passing time of the particle in the membrane.

The thickness R of the membrane has been determined

$$R = 2.44 \lambda \frac{F}{D} \qquad (3)$$

where F is the forcal length of the cylindrical lens used to expand the beam in uni-direction, and D is the diameter of the incident beam to the lens. The thickness of the membrane is approximatry R=0.1 mm. determined by the characteristics of the instruments and is smaller than the micro scale of the turbulence in flow.

When the interval time of double exposure is increased and moving distance of the tracers beyonds the thickness of the membrane, fringe pattern is not obtained on the image of the membrane.

As the results of the experiments, actual freedom of the time interval of the double exposure is rather narrow as described in the following section.

Seeding Tracers

In general measurement of flow velocity, it is undesiable condition to contain suspended particles in flow. However, in this method, the tracers are necessary for the measurement.

Therefore, at first, city supplied water has been used for the flow. The tracers are air babbles contained naturally in flow.

All experimental results of the first work have shown the images of the brilliant membrane in flow. However, the interferometric fringe patterns showing the movement of the tracers have not been obtained in all runs of the experiments.

In this work, the latitude of the interval time of double exposure is too narrow to obtain the fringe localized on the image of the membrane. And the runs of the experiment have been only successful in flow of low Reynolds number. (Ref.4)

In order to expand the method in high Reynolds number of flow, as the second work, the tracers of different kind of materials have been seeded artificially in flow.

The materials of seeded particles are coupper, soil, nylon, and carbon particles. These particles with different diameter and concentration by the precipitation method have been prepared.

An example of carbon particles is shown in Photo.1. The order of the concentration of tracers is 10^{-6} g/ℓ in flow.

The images of the brilliant membrane in flow have been also obtained from the holograms of all experimental runs of the second work. However, the interferometric fringe patterns have only obtained when soil and carbon particles have been used as the tracers.

In successful runs of the experiment, the interval time of the double exposure is seemed to be related to the diameter and the concentration of the particles as shown in Fig.1. The distribution of the successful runs at the fixed concentration of the tracers is a Gauss type as shown in Fig.2.

Analysis of Velocity

The interferometric fringe pattern on the recording image of the membrane in flow is represented the moving distance $|d\ell|$ of the tracers in the membrane during the interval time of the double exposure.

$$N\lambda = d\ell \cdot (|k_o - |k_i) \tag{4}$$

where the number N is the counting value of the fringe from the base fringe, and $|k_o$ and $|k_i$ are unit vectors of illuminated and scattered beams.

The velocity $|V_m|$ of the tracers is assumed to be constant during the interval time $T\ell$ of double exposure. The moving distance of the tracers is also represented

$$d\ell = V_m \cdot T\ell \tag{5}$$

The number of the fringe pattern is proportional to the local velocity of the flow in the membrane.

When three holograms are constructed at the same time, the fringe pattern from each hologram is

$$N_\ell \lambda = d\ell \cdot (|k_o - |k_{i\ell}) \qquad (\ell = 1, 2, 3) \tag{6}$$

The directional components of the moving distance $|d\ell|$ of the tracers are calculated from the equation (6) whenever the unit vector $|k_{i\ell}$ of the scattering beam is determined. (Ref.2)

The unit vector $|k_o$ of illuminated beam is known from the arrengement of the optical system in the holo-camera.

The unit vector $|k_{i\ell}$ of the scattering beam is determined by the calculation of the refraction as shown in Fig.3.

Refering Fig.3, considering the refraction laws, the vector $|k_{i\ell}$ is determined by

$$|y_o| \tan \psi + |y_i| \tan \theta = \sqrt{x_i^2 + (z_i - z_o)^2} = const. \tag{7}$$

where $n \sin \psi = \sin \theta$

The coordinate of the point A in Fig.3 is given. The point B is the center position of the eye or the lens used to take a photograph of reconstructing image of the membrane in flow. The coordinate of the point B is determined by the direct geometric measurement of the optical arrengement in the method.

The number N of the fringe pattern is evaluated from the photograph to take a reconstruction image of the membrane of double exposure.

The difference of the indices of refraction between water of flow and surrounding air of the channel is the reason of the deformation of the constructed image in the hologram from the original.

Refering Fig.3, the point A in the membrane is corresponded to the point A' recorded in the hologram. The photograph of the reconstruction image of the hologram corresponds to the apparent membrane OA' in the Fig.3 as the results of the refraction.

The coordinate of the point A'(x,y,z) is determined from the following equations.

$$\frac{s - l_{i,t}}{-y} = \tan \theta$$

(8)

$$(s + ds - l_{i,t}) \cos(\theta + d\theta) = -y \sin(\theta + d\theta)$$

(9)

Using the equations (8) and (9), the relation between the real and apparent membranes is obtained.

In the method, the light source of the reconstruction of hologram is used an argon laser. The wave length λ_r of the reconstruction beam is different from that of the construction beam of hologram.

As the result of the difference of the wave length, the reconstruction image is further deformed from the original.

Considering the method depended on Fresnel holography, the phase of holographic information is a function of the distances of the beams and wave length.

After the calculation of the process of reconstruction, the beam phase of the reconstruction image is obtained. Comparing with the both beam phases of the information and reconstructing images, the ratio M of the dimensions between the original and the reconstruction images of the membrane is obtained as the following. (Ref.1)

In longitudinal direction z' to the recording plate of the hologram, the ratio M_ℓ is

$$M_\ell = \left| - \frac{\lambda_r}{\lambda} \left(\frac{z_i'}{z_r'} + \frac{\lambda_r}{\lambda} \frac{z_i'}{z_s'} - \frac{\lambda_r}{\lambda} \right)^{-2} \right|$$

(10)

In transverse direction x' y' to the recording plate of the hologram, the ratio M_t is

$$M_t = \left| 1 - \frac{z_i'}{z_s'} + \frac{\lambda}{\lambda_r} \frac{z_i'}{z_r'} \right|$$

(11)

where the suffix i, s, and r are respectively represented as information(scattered), reference, and reconstruction beams of the holography.

Using the equations (8)-(11), the coordinate on the photograph of the reconstruction image corresponds to the coordinate on the original membrane in flow. The number N of the fringe pattern is evaluated on the original membrane.

Finally, using the equation (5), the directional components of the velocity in the section of flow are obtained.

Instead of the calculation using the equations (8)-(11), a marked grid plate having the same dimension of the membrane is immersed in still water of the channel, and is constructed the hologram. From the photograph of the reconstruction image of the hologram, the corresponding coordinate on the original membrane is easy determined on the photograph of the membrane.

Conclusion

Examples of the photographs taken the reconstruction images of the membranes of double exposure are shown in Photo.2 and 3.

The experiment of the Photo.2 is shown laminar flow using only city supplied water in a horizontal section of flow. The calculating results of the velocity is shown in Fig.4.

The experiment of Photo.3 is used the tracers of carbon particles in vertical section of flow of turbulence. The obtaining directional components of velocity and the averaging velocity are shown in Fig.5. The flow in the channel is seemed to be slightly meandering.

Direct measuring the velocity and the quantity of discharge of the flow, the accuracy of the experimental results of the velocity is confirmed.

When the particles of carbon having the diameter less than one micron meter are used to the tracers of flow, the best results showing the clear fine fringe pattern on the image of the membrane are obtained in wide range of Reynolds number of flow.

The rough calculation of the velocity profiles in the section of flow is not necessary to calculate the equation (8)-(11) for the coordinate of the membrane. However, it is necessary to use the equation (8)-(11) for the calculation of the velocity in detail of flow.

The equation (10) and (11) are surely simplified when the well known condition is used in the optical system of the holography. (Ref.1)

In future work, extending the method, when two cross membranes are used in flow, the velocity gradient may be obtained for the analysis of the homogeneity and the isotropy of flow.

References
(1) T. Numakura, Holography, (1974), 23, CORONA.
(2) K. Shibayama and H. Uchiyama, Measurement of Three-Dimensional Displacements by Hologram Interferometry, Appl. Opt., 10-9 (1971-9), 2150.
(3) M. Yano and H. Nishida, Flow Visualization Around a Bridge Pier, Symp. on Flow Visual. ISAS, 3 (1975-7), 105.
(4) M. yano and Y. Kisa, Flow Measurement with Holography, Symp. on Flow Visual. ISAS, 4 (1976-7), 141.
(5) M. Zambuto and M. Lurie, Holographic Measurement of General Forms of Motion, Appl. Opt., 9-9 (1970-9), 2066.

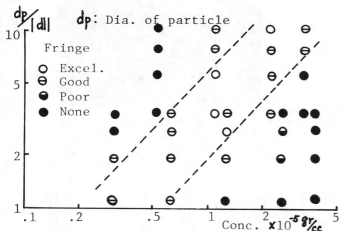

Fig.1 Successful Runs of the Method

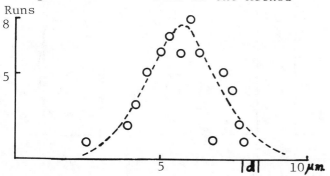

Fig.2 Successful Runs of the Method

Photo.1 Microscopic View of Carbon Particles

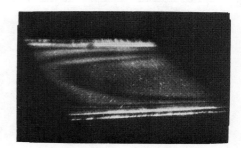

Photo.2 Interferometric Fringe Pattern of Laminar Flow in Horizontal Section

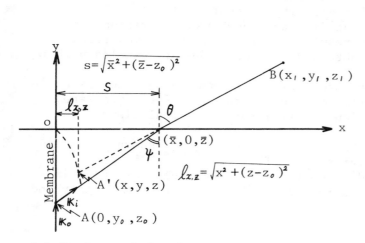

$s = \sqrt{\bar{x}^2 + (\bar{z} - z_o)^2}$

$l_{x,z} = \sqrt{x^2 + (z - z_o)^2}$

(a) Horizontal Section

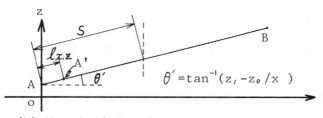

(b) Vertical Section

$\theta' = \tan^{-1}(z_i - z_o / x)$

$\bar{x} = |y_o| \tan\psi \cos\theta'$

$\bar{y} = 0$

$\bar{z} = |y_o| \tan\psi \sin\theta' + z_o$

$x = l_{x,z} \cos\theta'$

$y = -(|y_o|/n) \cos^3\theta \times (1 - \sin^2\theta / n^2)^{-3/2}$

$z = l_{x,z} \sin\theta' + z_o$

Fig.3 Refraction of Scattering Beam from the Membrane

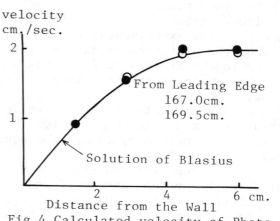

Fig.4 Calculated velocity of Photo.2

Photo.3 Interferometric Fringe Pattern of Turbulent Flow in Vertical Section

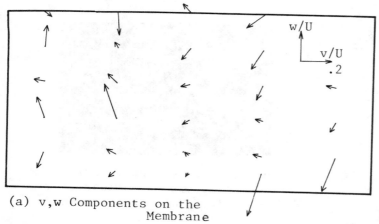

(a) v,w Components on the Membrane

$U = 20.6 \, \text{cm/s}$

(b) Vertical Velocity Profiles

Fig.5 Calculated Velocity of Photo.3

FLOW VISUALIZATION OF BOUNDARY LAYERS IN WATER BY IN-LINE HOLOGRAPHY

J. H. J. VAN DER MEULEN* and J. H. RATERINK**

A new method is presented for the visualization of boundary layers in water. The method is based on the application of in-line holography. Visualization of the boundary layer flow is effected by the injection of a 2 percent sodium chloride solution. Flow observations on two axisymmetric bodies are presented, showing laminar boundary layer separation on one body and transition to turbulence on the other body.

Introduction

Until recently, visualization of boundary layers in water has received little or no attention. At low velocities crude results can be obtained by using dye injection or oil film techniques, but these methods fail at high velocities. In 1962, Bland and Pelick (Ref. 1) introduced the schlieren method for the purpose of flow visualization in a water tunnel. They pointed out that schlieren pictures of the flow in water will have a greater resolution than those in air. Schlieren observations of the flow could be made, even without heating. However, the authors did not pursue the method any further.

In 1973, the schlieren method was successfully adopted by Arakeri (Ref. 2) to visualize boundary layers on axisymmetric bodies in water. The density gradients required for the schlieren effect were created by internal heating or cooling of the test bodies. It was found that two bodies, frequently used in cavitation research, exhibited laminar boundary layer separation. Unawareness of this important flow phenomenon had obscured the results of comparative cavitation studies, made in the past.

In the present paper a new method is presented for the visualization of boundary layers in water. The method is based on the application of in-line holography. Visualization of the boundary layer flow was effected by injecting a fluid with a slightly different index of refraction from the surrounding fluid.

---

* Netherlands Ship Model Basin, Wageningen, The Netherlands

** Institute of Applied Physics TNO-TH, Delft, The Netherlands

Holographic method

The flow visualization experiments were made in the high speed water tunnel of the Netherlands Ship Model Basin. This tunnel is provided with a 50 mm square test section with rounded corners; the maximum speed in the test section is 30 m/s. Boundary layer phenomena on two axisymmetric bodies have been studied. The first body, a hemispherical nose, is known to possess laminar boundary layer separation (Ref. 2). For the second body, a blunt nose, laminar separation was not predicted theoretically (Ref. 3). Both bodies were made of stainless steel, the diameter being 1 cm.

In-line holography has been used for making records of the flow patterns. Holography has become one of the most important areas of modern optics since the invention of the laser as a new light source. Holography is usually described as a method for storing wavefronts on a record from which the wavefronts may later be reconstructed. The record, formed in photo-sensitive material, is called a hologram. In forming holograms two sets of light waves are involved: the reference waves and the subject waves. In the present case of in-line holography, only one set of waves is used basically. The undeflected light waves from this set of waves act as reference waves, the light waves deflected by the subject act as subject waves.

A schematic diagram of the applied optical system is shown in Fig. 1. The light source is the Korad K-1QH pulsed ruby laser of the Institute of Applied Physics TNO-TH. To improve the resolution of the system, the red light from the ruby laser is converted to ultraviolet light, with a wavelength of 0.347 µm, in a KDP crystal. The pulse duration is 25 nanoseconds and the maximum energy 4 mJ in the TEM_{00} mode. A telescopic system (L_2 and L_3) is used to obtain a laser beam with a diameter of 30 mm. A mirror reflects the beam into the test section of the tunnel. In the walls of the plexiglass test section, two optical glass windows are inserted. The location of the body in the test section is such that the nose is illuminated by the laser beam over a length of about 20 mm, and the body contour is imaged on the hologram. A shutter is placed on the first window. The camera containing the holographic plate is located close to the second window. Agfa-Gevaert Scientia Plates 8 E 56 and 8 E 75 with a resolution up to 3000 lines/mm were used as recording material. Reconstruction of the holograms was made with a continuous-wave He-Ne gas laser ($\lambda = 0.633$ µm). The reconstructed image was studied with a microscope with a magnification between 40x and 200x. Further details of the system are given in (Ref. 4).

The main advantages of the present method of in-line holography are: (1) the required holographic system is simple, (2) three-dimensional information on the subject is stored in the hologram, which permits detailed analysis afterwards, using the reconstructed image, and (3) high speed flows can be investigated by applying a pulse laser. Also time information can become available if the laser pulse contains a number of pulses. This application of multiple exposure holography has been successfully used to study the type of cavitation occurring on the blunt nose (Ref. 4).

Flow visualization technique

Flow visualization can be effected by artificially changing the index of refraction of a small isolated volume in the flow. The velocity of light c is related to the index of refraction n by the relation $c = c_o/n$, where c_o is the velocity of light in vacuum. If there is a gradient of index of refraction normal to the light rays, the rays will be deflected since, according to the above relation, light travels more slowly when the index of refraction is larger. The deflection of the light rays is recorded in the hologram, and upon reconstruction the isolated volume with a different index of refraction can be detected.

Several methods have been attempted to visualize the boundary layer flow about the axisymmetric bodies (Ref. 4). The ultimate method consisted of injecting a sodium chloride solution into the boundary layer from a hole located at the stagnation point of the body. The diameter of the hole is 0.08D, where D is the diameter of the body. The effect of the concentration of sodium chloride on the index of refraction of the solution is given in Fig. 2. In this figure the temperature effect on the index of refraction of water is included. Thus, the effectiveness of adding NaCl relative to creating a temperature difference is well demonstrated. Two parameters have been varied systematically to find the optimum conditions for flow visualization: (1) the ratio of the injection velocity V_i to the velocity in the test section V_o, and (2) the sodium chloride concentration. This parameter study was made with the hemispherical nose. A stable flow was found for $V_i/V_o < 0.25$; above this value the flow was unstable. These stability observations are in agreement with those reported earlier by Gates (Ref. 5). The optimum value for V_i/V_o appeared to be 0.1-0.2. The optimum concentration of NaCl was found to be about 2 percent (for $V_i/V_o = 0.2$). Below this concentration the details of the boundary layer flow become less pronounced, as shown in the photographs of Fig. 3(a) and (b).

A possible improvement of the present visualization technique may be obtained by injecting a fluid with a lower index of refraction than the surrounding fluid. In that case the light rays will be deflected away from the body, instead of deflected into the body, and interference with the undeflected light is promoted. These considerations are confirmed by the results of experiments performed by Arakeri (Ref. 6) during his stay at the Netherlands Ship Model Basin. Arakeri studied the boundary layer flow around a 1.5 caliber ogive axisymmetric body. Gradients of index of refraction were created by heating or cooling. By heating the test body, and making use of the present holographic set-up, Arakeri was able to visualize laminar separation on the 1.5 cal ogive. This is shown in the photograph presented in Fig. 4. However, when cooling was applied, laminar separation could not be observed. Earlier attempts to visualize laminar separation on the 1.5 cal ogive by the application of the schlieren technique were not successful (Ref. 7).

Boundary layer flow observations

Fig. 3(b) shows laminar boundary layer separation on the hemispherical nose. Reattachment of the shear layer and transition to turbulence are both observed. In the transition region the flow is still
visualized by the NaCl, but further downstream, where the turbulence
is more developed, mixing of the NaCl prevents further observations.
The angular position of separation γ_S was found to be 85.5°. This
value is equal to the one predicted theoretically. Nor the length of
the separated bubble, nor the position of separation were effected by
the NaCl concentration of the injected fluid or by V_i/V_o. The influence of polymer additives on laminar separation is shown in Fig. 3(c).
It is found that the polymer additive suppresses the occurrence of
laminar separation by inducing "early turbulence". Mixing of the NaCl
in the turbulent boundary layer of polymer solutions is less pronounced
than in pure water (Ref. 4). Fig. 5 shows transition from laminar to
turbulent boundary layer flow on the blunt nose. Laminar separation
does not occur, which is in agreement with theoretical predictions
(Ref. 3). The influence of polymer additives on the free shear layer
of water jets is well demonstrated by the photographs presented in
Fig. 6. These photographs were made by using the same experimental
set-up as given in Fig. 1. The velocity of the 0.8 mm diameter jet
is 8 m/s, while the fluid in the test section is stagnant. The jet
fluid is identical to the fluid in the tunnel. However, for flow
visualization, 2 percent NaCl is added to the jet fluid.

References

(1) R.E. Bland and T.J. Pelick, The Schlieren Method Applied to Flow
    Visualization in a Water Tunnel, Trans. ASME, J. Basic Engng.,
    Vol. 84, Dec. 1962, pp. 587-592.

(2) V.H. Arakeri, Viscous Effects in Inception and Development of
    Cavitation on Axisymmetric Bodies, Cal. Inst. Tech. Rep. E-183-1,
    Jan. 1973.

(3) E. Silberman, F.R. Schiebe and E. Mrosla, The Use of Standard
    Bodies to Measure the Cavitation Strength of Water, St. Anth.
    Falls Hydr. Lab. Rep. 141, Sept. 1973.

(4) J.H.J. van der Meulen, A Holographic Study of Cavitation on
    Axisymmetric Bodies and the Influence of Polymer Additives,
    Ph. D. Dissertation, Enschede, 1976.

(5) E.M. Gates, Visualization of Drag-Reducing Fluid Flows in a
    Water Tunnel, ASME Cavitation and Polyphase Flow Forum, 1976,
    pp. 9-11.

(6) V.H. Arakeri, Experimental Investigation of Laminar Boundary
    Layer Stability with Surface Heating and Cooling, N.S.M.B.
    Rep. 01054-2-CT, June 1977.

(7) V.H. Arakeri, A Note on the Transition Observations on an
    Axisymmetric Body and Some Related Fluctuating Wall Pressure
    Measurements, Trans. ASME, J. Fluids Engng., Vol. 97, March
    1975, pp. 82-86.

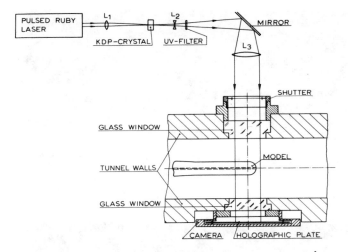

Fig. 1. Schematic diagram of optical system for making holograms of flow phenomena in test section of tunnel.

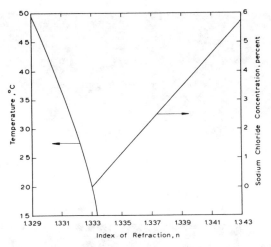

Fig. 2. Index of refraction of water as a function of the temperature and the concentration of sodium chloride.

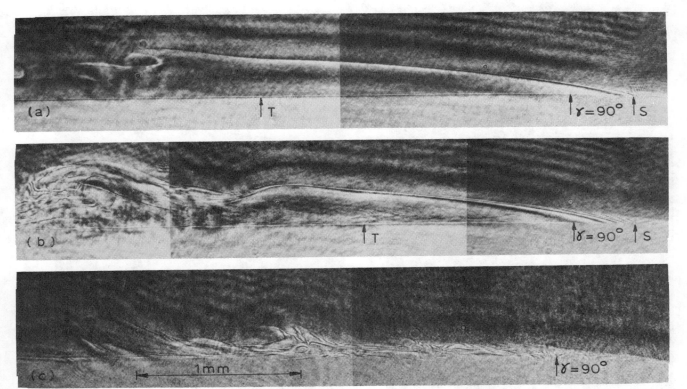

Fig. 3. Photographs showing laminar separation (S) and subsequent transition to turbulence (T) on hemispherical nose (a, b), and influence of polymer additives (c). The flow is from right to left.
(a) Injection of 0.5 percent NaCl; V_i/V_o= 0.2; V_o= 4.3 m/s.
(b) Injection of 2 percent NaCl; V_i/V_o= 0.2; V_o= 4.0 m/s.
(c) Injection of 2 percent NaCl + 500 ppm Polyox WSR-301; V_i/V_o= 0.2; V_o = 4.0 m/s.

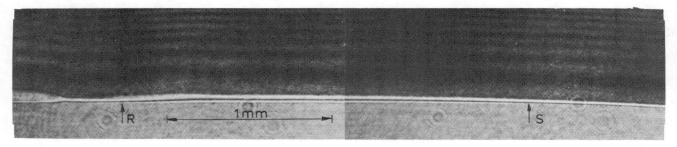

Fig. 4. Photograph showing laminar separation (S) and reattachment (R) on 1.5 cal ogive ($s_S/D = 1.24$). The flow is from right to left. $T_{tunnel} = 30.8°C$; $T_{body} = 36.7°C$; $V_o = 15$ m/s. (From Arakeri (Ref. 7)).

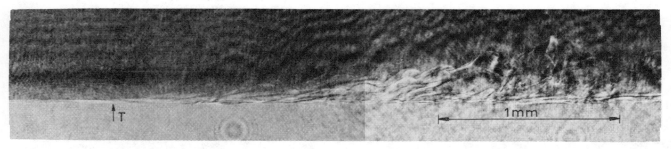

Fig. 5. Photograph showing transition (T) from laminar to turbulent boundary layer on blunt nose ($s_T/D = 1.68$). The flow is from left to right. Injection of 2 percent NaCl; $V_i/V_o = 0.12$; $V_o = 8.0$ m/s.

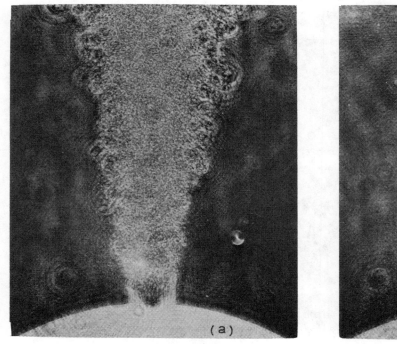
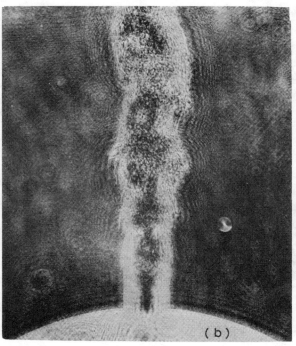

Fig. 6. Photographs showing influence of polymer additives on jet flow. The velocity of the 0.8 mm dia. jet is 8 m/s.
(a) Water + 2 percent NaCl.
(b) Water + 2 percent NaCl + 50 ppm Polyox WSR-301.

OPTIMIZATION OF FLOW BIREFRINGENCE MEASUREMENTS BY RATIONAL CHOICE OF SPECTRAL FREQUENCY OF BIREFRINGENCE-DETECTING RADIATION

JERZY T. PINDERA*

Flow birefringence is often used for visualization and determination of liquid flow. However, all three major models of mechanism of shear strain rate-induced birefringence, do not contain information on influence of wavelength of birefringence-detecting radiation on major parameters of birefringence.

Paper presents evidence that to optimize the determination of flow patterns using flow birefringence, it is necessary to know the relationships between the wavelength of radiation, the amount of birefringence and the optic axis orientation. As an example, the response of aqueous solution of Milling Yellow is chosen. Some other parameters of optical response are discussed.

Introduction

The phenomenon of the flow-induced birefringence has been known for more than a hundred years; the concepts of stress birefringence, form birefringence, and orientation birefringence in liquids were developed and accepted already at the turn of the century; the concept of dynamic orientation of anisometric particles has been introduced long ago. However, there is still a strong tendency to develop and to accept very simple phenomenological models of all involved phenomena.

In all theories of flow birefringence the dynamic character of interaction between radiant energy and matter is neglected and it is believed that the amount of birefringence does not depend on the wavelength of birefringence-detecting radiation.

Consequently, it is believed that the relationships between the shear stress or shear strain rate and the major parameters of birefringence as the amount of birefringence (separation of wavefronts), and the extinction angle (direction of optic axis with regard to streamline direction), can be described by means of constants which do not depend on the spectral band of radiation, and that the limits of linear mechanical and linear optical responses of liquids are the same.

As a result, some particular phenomenological models and pertinent relations are very often developed and recommended as a foundation of experimental techniques for flow determination, without any proof that the assumptions made are justified, sometimes without any explicit information on the assumptions made, and usually without a discussion of the range of applicability of the derived relations. For instance, it is common to consider the coefficients in the flow birefringence relations to be constants, independent of the frequency of the birefringence-detecting radiation. Consequently, it is rather customary to present experimental results on birefringence response of liquids without any information on the dominant wavelength or bandwidth of the birefringence-detecting radiation.

---

* Professor, Department of Civil Engineering, University of Waterloo, Waterloo, Ontario, Canada   N2L 3G1

To introduce the flow-induced birefringence as a reliable engineering tool for flow visualization and determination of flow fields, it is necessary to investigate the degree of correlation between such phenomenological mathematical models and derived relations, and the actual mechanical and optical responses of real liquids.

Obviously, the optimization of flow determination by means of flow birefringence measurements is not possible when the basic patterns of interaction between radiation used and flowing real medium are not known or are neglected.

Flow Determination By Means of Flow Birefringence

Determination of the velocity field is one of major problems of studies of flow patterns of real liquids in conduits and around solid bodies. The velocity vector can be determined indirectly, e.g., on the basis of information on the shear strain rate tensors for Newtonian liquids in a steady state, when boundary conditions are known.

It has been known for more than a hundred years, that some relationships exist between the direction of principal strain rate axis and the related shear strain rate $\dot{\gamma}$ on the one hand, and the optic axis direction and the amount of flow-induced birefringence Δn, on the other. Because of this, it is easily possible to obtain the light intensity modulation pattern related to the velocity field in a suitable liquid by using the well known polarization techniques.

A sample of such light intensity modulation patterns produced by velocity field in liquid flowing through a rectangular conduit is given in figure 1. Two kinds of light intensity modulation are presented: classical modulation of transmitted radiation, and modulation of radiation scattered from a primary laser beam. A second kind of modulation of scatter radiation related to the modulation of primary beam is easily obtainable when the scattered radiation is not filtered by a polarizer, [1,2].

Using a narrow band of radiation of the width of 1-10 nm, it is possible to measure the amount of flow-induced birefringence in the transmitted radiation with a high precision; measurement errors are usually less than 1%.

The accuracy of the birefringence measurement using scattered radiation depends on whether or not the mathematical model of Rayleigh scattering simulates satisfactorily the actual pattern of light scatterings in chosen liquid. However, even if the actual light scattering deviates significantly from Rayleigh scattering it is possible to measure accurately birefringence by measuring the scattered light intensity distribution, using suitable recording techniques, [1, 2].

The second parameter of birefringence, namely the direction of optic axis, can be measured with an error less than 1°, using a narrow band radiation.

In case of a steady state laminar flow within Newtonian range, the dependence of the amount of birefringence Δn on the shear strain rate $\dot{\gamma}$ is often presented by a linear relation
$$\Delta n = C_1 \dot{\gamma} \tag{1}$$

Depending on the assumed mechanism cf birefringence the dependence of the optic axis direction on the shear-strain rate $\dot{\gamma}$ or the strain rate $\dot{\varepsilon}$ and the principal strain rate axis direction is presented in various forms. Two typical relations, the first given by Natanson, Zaremba i Zakrzewski at the turn of the century and the second given by Wayland are:
$$\cot 2\chi = \pm \tau \dot{\gamma} \qquad \text{and} \qquad \Lambda_o - \chi = C_2 \dot{\varepsilon} \sin 2\Lambda_o \tag{2a, 2b}$$
where τ is relaxation time and C_1 and C_2 are constants.

With regard to linearity it is usually assumed that the mechanical and optical ranges of linear response of real liquids are the same.

The question is whether all these assumptions presented above represent satisfactorily the actual responses of real liquids.

Flow Birefringence Response of Real Liquids

Aqueous solution of Milling Yellow (NGS 1828) has been chosen as a representative sample of a birefringent liquid, since it seems to exhibit all three mechanisms of flow-induced birefringence: stress, form, and orientation birefringence.

In this paper and in the listed references, the birefringence responses of real liquids are analyzed both qualitatively and quantitatively with respect to pertinent typical mathematical models of responses, i.e., the validity and range of applicability of assumptions is analyzed, and the values of parameters of mathematical models are determined and presented as functions of pertinent physical parameters.

The following assumptions are investigated:
- amount of birefringence does not depend on wavelength of radiation;
- optic axis direction does not depend on wavelength of radiation;
- ranges of linear mechanical and optical responses are the same.

These assumptions seem a priori to be incorrect, since the known photoelastic responses of solids exhibit dependence on wavelength of radiation in the elastic, visco-elastic, and plastic ranges of deformation [3, 4]. Principles of optimization of experiments, including the transfer function of measurement system, are outlined in [5].

Obviously, the first optical response to be determined is the spectral transmittance, Fig. 3, [3]. The spectral transmittance carries indirect information on the linear ranges of optical response, [4].

Relations between the shear stress and shear strain rates for various concentrations supply information on the range within which the mechanical response of liquid is practically linear, Figure 4, [1].

The spectral frequency dependence of induced birefringence parameters has been presented and discussed in [3, 4]; particular case of the dependence of flow birefringence on the band of radiation has been discussed in [1, 2, 5, 6].

The fact that the amount of birefringence depends on the wavelength of radiation is illustrated by Fig. 5. This figure illustrates qualitatively the fact that the linear range of birefringence response depends on the spectral band of radiation, since the measurements were performed for rectangular conduit flow using transmitted light technique. Results presented on Fig. 5 are obviously influenced by the fact that the ratio of the sides of the rectangular conduit, b/a, was equal to 2.9.

In Fig. 6, the dependence of the amount of birefringence on the shear strain rate is presented. This relation has been determined in a steady state simple shear flow, using as a measure the modulation of primary laser beam indicated by intensity distribution of unfiltered scattered radiation.

Figure 7 presents the normalized spectral birefringence of Milling Yellow in the visible and near infrared bands-(see papers [3,4])-for shear strain rate values up to 4 sec^{-1}. Obviously, in this range of shear strain rates, the birefringence response is linear within the red and infrared spectral bands, and is practically linear within the visible band of radiation.

Consequently, the relation between the amount of birefringence and the shear strain rate, ought to be presented in a general form as

$$\Delta n = C_1(\lambda)\dot{\gamma} \qquad \text{where} \qquad \frac{C_1(\lambda_1)}{C_1(\lambda_2)} \neq \frac{\lambda_2}{\lambda_1} \neq \frac{\lambda_1}{\lambda_2} \qquad (3, 4)$$

since $\qquad C_1 = C_1(\lambda) \quad$ is a nonmonotonic function. $\qquad (5)$

The orientation of optical axis with respect to the streamline direction or to the principal strain rate axis, has been an object of discussions during the last hundred years. It has been known already at the end of the last century, that the optic axis direction is not collinear with the principal strain rate axis, and therefore it has been called "anomalous", since it also depends on shear strain rate:

$$\chi = \chi(\dot{\gamma}) \qquad (6)$$

It is interesting to note that already at the beginning of this century, this dependence was an object of an animated discussion reflected in papers published at that time by the members of Polish Academy of Sciences in Cracow.

However, in the subsequent years a tendency has developed to simplify the mathematical models of flow birefringence as far as possible and - as one of the results of this tendency - it was believed that the optic axis direction (or the extinction angle), does not depend on wavelength of radiation as in the case of stress-induced birefringence in elastic solids.

A sample of the actual dependence of optic axis direction represented by extinction angle χ on wavelength of radiation for several values of shear strain rate produced in aqueous solutions of Milling Yellow is given in Fig. 8. More extensive information on this subject is given in [2].

According to Fig. 8, the general form of dependence of the extinction angle on the shear strain rate and the wavelength of radiation, ought to be in form
$$\chi = \chi(\gamma, \lambda) \text{ where the function } \chi = \chi(\gamma = const, \lambda) \text{ is nonmonotonic.} \qquad (7,8)$$

Conclusions - Optimization of Flow Birefringence Measurements

There exists a relationship between spectral transmittance, dispersion of birefringence, and dispersion of optic axis. Thus, information on spectral transmittance is needed to optimize the experiment. As a general rule the spectral bands containing absorption bands should be avoided, [1-6].

Relations presented in Figures 5-8 show that the wavelength of birefringence-detecting radiation is a major parameter of flow-birefringence experiments. For instance, when aqueous solution of Milling Yellow is used as birefringent liquid, the optimal wavelength is within the infrared band, between 800 and 900 nm.

Since light scattering by liquids is not necessarily compatible with the Rayleigh model, precautions must be taken to assure that the recorded intensities of scattered radiation are related unequivocally to corresponding birefringence.

To determine reliably and with a desired accuracy the velocity field in a liquid by using birefringence transmission technique or birefringence integrated technique [1,2,5],-as a direct technique or as a model technique,-it is necessary to optimize the experiment rationally, choosing the most suitable dominant wavelength of birefringence-detecting radiation.

References

[1] Pindera, J.T., Krishnamurthy, A.R., Foundations of Flow Birefringence in Some
     Liquids. Proceedings of the International Symposium on Experimental Mechanics:
     Experimental Mechanics in Research and Development, University of Waterloo,
     June 12-16, 1972. Edited by J.T. Pindera, H.H.E. Leipholz, F.D.J. Rimrott, D.E.
     Grierson. SM Study No. 9, Solid Mechanics Division, University of Waterloo,
     1973, pp. 565-599.
[2] Pindera, J.T., Krishnamurthy, A.R., Characteristic Relations of Flow Bire-
     fringence. Part 1: Relations in Transmitted Radiation. Part 2: Relations in
     Scattered Radiation. Experimental Mechanics - to be published.
[3] Pindera, J.T., Cloud, G.L., On Dispersion of Birefringence of Photoelastic
     Materials, Experimental Mechanics, Vol. VI, No. 9, (1966), pp. 470-480.
[4] Pindera, J.T., Straka, P., On physical measures of rheological responses of some
     materials in wide ranges of temperature and spectral frequency. Rheologica Acta,
     Vol. 13, No. 3, (1974), pp. 338-351 (846-859).
[5] Pindera, J.T., Alpay, S.A., Krishnamurthy, A.R., New Developments in Model
     Studies of Liquid Flow by Means of Flow Birefringence. Transactions of the
     CSME, Vol. 3, No. 2 (1975) pp. 95-102.
[6] Krishnamurthy, A.R., Pindera, J.T., On the Dependence of Flow Birefringence on
     Spectral Frequency and Shear Strain Rates. Proceedings of the Seventh Inter-
     national Congress on Rheology. Gothenburg, Sweden, August 23-27, 1976, pp.
     614-615.

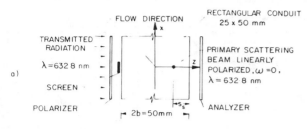

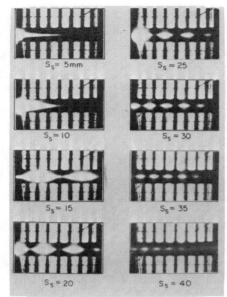

Fig. 1 – Sample of integrated photoelastic measurement of rectangular conduit flow using transmission isochromatics and scattered secondary isochromatics.

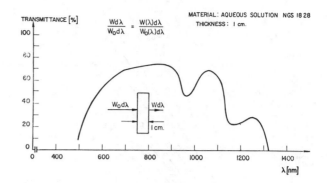

Fig. 3 – Spectral transmittance of aqueous solution of Milling Yellow

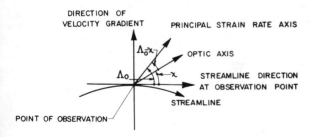

Fig. 2 – Typical mathematical model used for analysis of flow bire-fringence: mechanical and optical parameters.

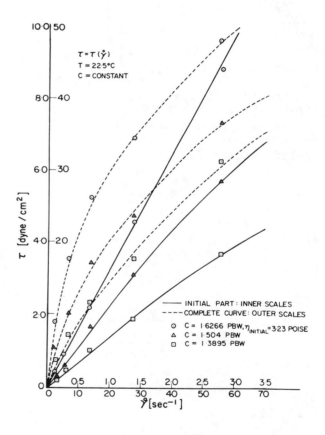

Fig. 4 – Mechanical response of aqueous solution of Milling Yellow: dependence of shear stress τ on shear strain rate $\dot\gamma$ for various concentrations C.

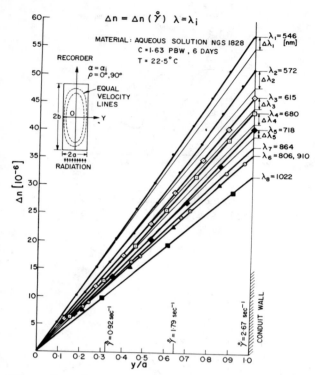

$\Delta n = \Delta n (\overset{\circ}{\gamma}) \quad \lambda = \lambda_i$

MATERIAL : AQUEOUS SOLUTION NGS 1828
C = 1·63 PBW , 6 DAYS
T = 22·5°C

RECORDER
$\alpha = \alpha_i$
$\rho = 0°, 90°$

EQUAL VELOCITY LINES

$\lambda_1 = 546$
$\Delta\lambda_1$ [nm]
$\lambda_2 = 572$
$\Delta\lambda_2$
$\lambda_3 = 615$
$\Delta\lambda_3$
$\lambda_4 = 680$
$\Delta\lambda_4$
$\lambda_5 = 718$
$\Delta\lambda_5$
$\lambda_7 = 864$
$\lambda_6 = 806, 910$
$\lambda_8 = 1022$

CONDUIT WALL

$\overset{\circ}{\gamma} = 0.92 \ sec^{-1}$
$\overset{\circ}{\gamma} = 1.79 \ sec^{-1}$
$\overset{\circ}{\gamma} = 2.67 \ sec^{-1}$

Fig. 5 – Spectral optical response of Milling Yellow solution to shear strain rate, evaluated from transmission isochromatics produced in rectangular conduit flow.

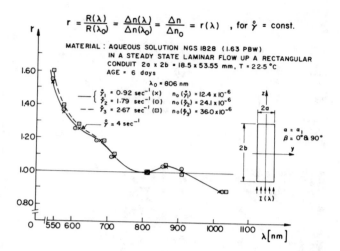

$r = \dfrac{R(\lambda)}{R(\lambda_0)} = \dfrac{\Delta n(\lambda)}{\Delta n(\lambda_0)} = \dfrac{\Delta n}{\Delta n_0} = r(\lambda) \quad , \text{ for } \overset{\circ}{\gamma} = const.$

MATERIAL : AQUEOUS SOLUTION NGS 1828 (1.63 PBW)
IN A STEADY STATE LAMINAR FLOW UP A RECTANGULAR
CONDUIT 2a x 2b = 18.5 x 53.55 mm , T = 22.5°C
AGE = 6 days

$\lambda_0 = 806 \ nm$

$\begin{cases} \overset{\circ}{\gamma_1} = 0.92 \ sec^{-1} \ (\times) & n_0(\overset{\circ}{\gamma_1}) = 12.4 \times 10^{-6} \\ \overset{\circ}{\gamma_2} = 1.79 \ sec^{-1} \ (\odot) & n_0(\overset{\circ}{\gamma_2}) = 24.1 \times 10^{-6} \\ \overset{\circ}{\gamma_3} = 2.67 \ sec^{-1} \ (\square) & n_0(\overset{\circ}{\gamma_3}) = 36.0 \times 10^{-6} \end{cases}$

$\overset{\circ}{\gamma} = 4 \ sec^{-1}$

$\alpha = \alpha_1$
$\beta = 0° \& 90°$

Fig. 7 – Normalized spectral flow birefringence in visible and near-infrared bands for low shear strain rates.

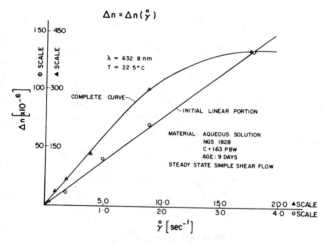

$\Delta n = \Delta n(\overset{\circ}{\gamma})$

$\lambda = 632.8 \ nm$
T = 22·5°C

COMPLETE CURVE

INITIAL LINEAR PORTION

MATERIAL : AQUEOUS SOLUTION
NGS 1828
C = 1·63 PBW
AGE : 9 DAYS
STEADY STATE SIMPLE SHEAR FLOW

$\overset{\circ}{\gamma} \ [sec^{-1}]$

Fig. 6 – Shear-strain-rate dependence of flow birefringence for $\lambda = 632.8$ nm, determined in simple shear flow using scattering technique.

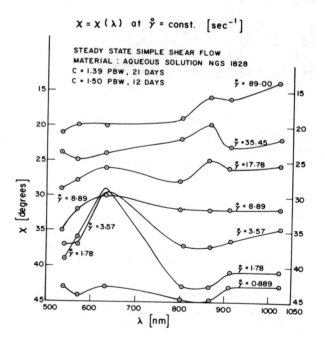

$\chi = \chi(\lambda) \ at \ \overset{\circ}{\gamma} = const. \ [sec^{-1}]$

STEADY STATE SIMPLE SHEAR FLOW
MATERIAL : AQUEOUS SOLUTION NGS 1828
C = 1.39 PBW, 21 DAYS
C = 1.50 PBW, 12 DAYS

$\overset{\circ}{\gamma} = 89.00$
$\overset{\circ}{\gamma} = 35.45$
$\overset{\circ}{\gamma} = 17.78$
$\overset{\circ}{\gamma} = 8.89$
$\overset{\circ}{\gamma} = 3.57$
$\overset{\circ}{\gamma} = 1.78$
$\overset{\circ}{\gamma} = 1.78$
$\overset{\circ}{\gamma} = 0.889$

$\chi \ [degrees]$

$\lambda \ [nm]$

Fig. 8 – Dependence of the extinction angle (optic axis direction with respect to stream line) on wavelength of radiation in wide range of shear strain rates.

THE APPLICATION OF STREAMING BIREFRINGENCE TO THE QUANTITATIVE STUDY OF LOW-REYNOLDS NUMBER PIPE FLOW

M. HORSMANN,* E. SCHMITZ,* and W. MERZKIRCH**

Flows of colloidal solutions exhibiting the effect of streaming birefringence can be investigated with two different visualization methods. Two-dimensional flows can be studied with a transmission polariscope; the scattered-light technique is appropriate to study three-dimensional flow fields. The quantitative evaluation of the flow pictures requires to solve nonlinear, partial differential equations which relate the observed pattern to the flow velocity.

Introduction

The method of streaming birefringence, known for more than one century, has again found increasing interest since it has been detected that an aqueous solution of Milling Yellow dye can serve as a highly sensitive test fluid. In a first approximation this fluid has the properties of a colloidal solution. Therefore the theory of Wayland (Ref. 1) applies when using Milling Yellow (MY) for the visualization of a two-dimensional flow pattern in a transmission polariscope. This theory relates the observed optical pattern to the velocity field of the flow. For low values of the fluid deformation rate, i.e. for low flow velocities, one may use a linearized form of Wayland's theory. The measured "amount of birefringence" (order N of an isochromate) is then proportional to the maximum rate of deformation, $\dot{\varepsilon}_{max}$, which is a function of the velocity components u and v (see below). By introducing a stream function Ψ, Peebles and Liu (Ref. 2) have reduced the problem to the evaluation of a partial differential equation for Ψ. The numerical evaluation of this equation is the key to the application of streaming birefringence for quantitative measurements of two-dimensional flow fields; this problem will be investigated below.

Two-dimensional flows cannot be verified in practice. Even in a channel with a depth-to-height ratio of 1o:1 the deviation from plane or two-dimensional flow is about 5%. The disadvantage of the mentioned method using the transmission polariscope is, that all the optical information is integrated along the path of the light through the test flow and cannot be desintegrated thereafter. Another optical

* Research Assistant
** Professor of Fluid Mechanics; all: Institut für Thermo- und Fluid-
dynamik, Ruhr-Universität, 463o Bochum, Germany

technique suitable for yielding local instead of integrated information has been studied by McAfee und Pih (Ref. 3) and adopted to streaming birefringence. One makes use of the information contained in the light which is scattered from a birefringent fluid flow. The method is examined here with respect to the relation between the optical pattern and the velocity field of the flow.

Flow Facility and Test Fluid

A closed-circuit flow channel filled with the aqueous solution of MY is used for both the light-transmission and the scattered-light experiments. Test sections of different shape, made of Perspex, can be used in the circuit. The flow facility is temperature-controlled because it is known that the optical sensitivity as well as the viscosity of MY are highly sensitive towards changes in temperature. The test fluid is a 1% solution of MY in destilled water (Ref. 4). For the low flow velocities used in these studies this solution can be considered as a Newtonian fluid (Ref. 5). It should be noticed, however, that a knowledge of the fluid's viscosity is not required for the direct evaluation of the optical experiments. The range of Reynolds numbers is $10^{-1} < Re < 10^{-2}$ for all experiments performed within this program.

Two-Dimensional Flow

By using the linearized form of Wayland's theory (Ref. 1) and by introducing a stream function Ψ according to Peebles and Liu (Ref. 2) one is led to the following partial differential equation which relates the flow field (expressed by Ψ) to the observed optical pattern (isochromates):

$$(2\,\frac{\partial^2 \Psi}{\partial x \partial y})^2 + (\frac{\partial^2 \Psi}{\partial y^2} - \frac{\partial^2 \Psi}{\partial x^2})^2 = c^2 \cdot (\frac{\Delta\phi}{2\pi})^2 \qquad (1)$$

$(\Delta\phi/2\pi)$ is the local phase difference between two interfering rays. An isochromate is designated by $(\Delta\phi/2\pi) = N = 0, \pm1, \pm2,\ldots,$ N being the order of the isochromate. The constant c must be calibrated in a known flow field. Here, the calibration is performed in the plane channel flow with constant cross-section. The field to be tested is the flow over a rearward facing step (Fig. 1) Downstream of the step

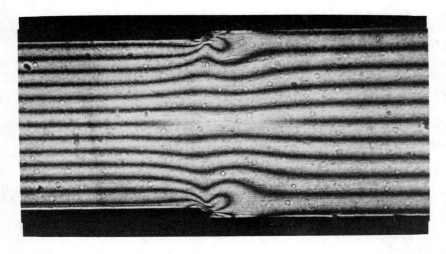

Fig. 1. Isochromates in the flow over a rearward facing step. Height of flow channel 1o mm in front of the step, 11 mm behind the step. Mean velocity $\bar{u} = 0,5$ mm/s in front of the step.

configuration the flow approaches again plane channel flow. This undisturbed situation is specified by horizontal isochromates in Fig. 1.

If one can calculate the stream function Ψ from eq. (1) for the
entire flow field, one may also determine the velocity components u,v
according to the definition u = $(\partial\Psi/\partial y)$, v = $-(\partial\Psi/\partial x)$. In order to
solve eq. (1) it is necessary to replace the right hand side of (1) by
a closed data function D(x,y). This is achieved by means of an
appropriate interpolation procedure with which one determines an
arbitrary number of interpolated points from the discrete data points
obtained on the isochromates. In particular, one may determine the
values of D(x,y) in the mesh points x_i, y_i of a network which is used
to solve eq. (1) by means of a finite difference method. Such methods
when applied to eq. (1) respond very sensitivily to the quality of the
interpolation procedure.

Eq. (1) is written in form of a finite difference scheme; the problem is treated by means of the Newton method. The calculation is
started in the regime of known channel flow and propagates stepwise
along lines x_i = const into the disturbed region. The stream function
is prescribed along the centerline (Ψ=o) and along the wall (Ψ=1). It
was possible to solve the great numerical problems incorporated in
the nonlinear form of eq. (1) in the entire flow field.

Fig. 2 shows the distribution of the stream function for a mean

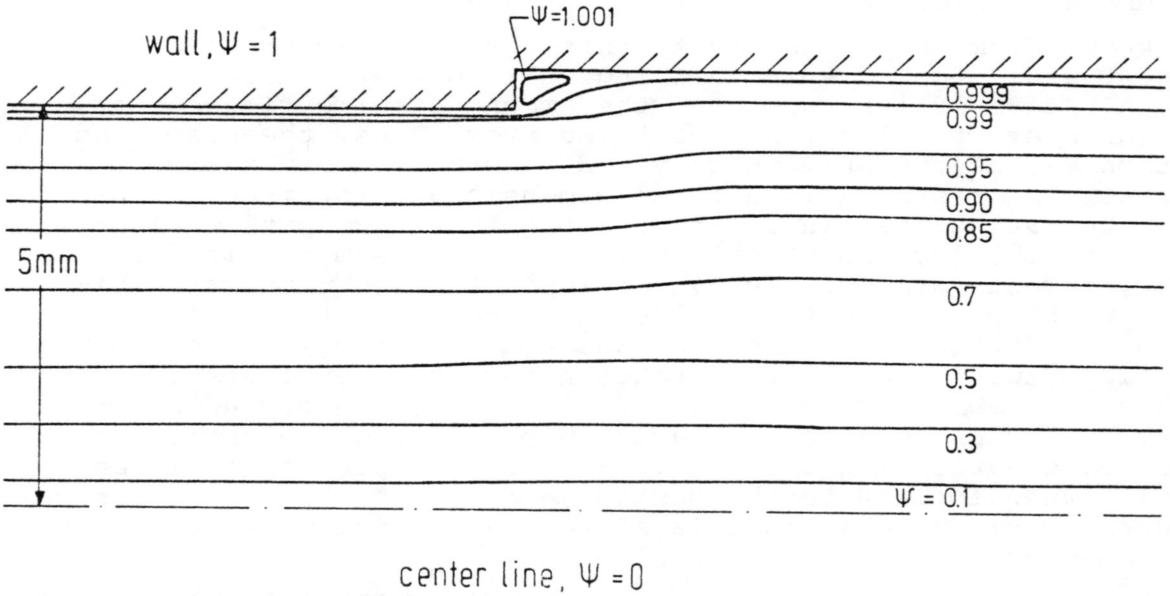

Fig. 2. Pattern of stream function Ψ.

velocity $\bar{u}$ = o,5 mm/s. The values for Ψ increase from Ψ = o along the
centerline to Ψ = 1 along the wall. Closely adjacent to the step is
a limited region with values $\Psi > 1$ which indicates a field of recirculating flow. This is confirmed by the velocity profiles of the u-
component in Fig. 3. The first and the last profile in Fig. 3 are
associated to the undisturbed channel flow upstream and downstream of
the step, respectively. This experiment is a confirmation for the
existence of flow separation and recirculation even at such low Reynolds number (Re = o,6·1o^{-2}). A theoretical prediction of this

phenomenon has been given in Ref. 6.

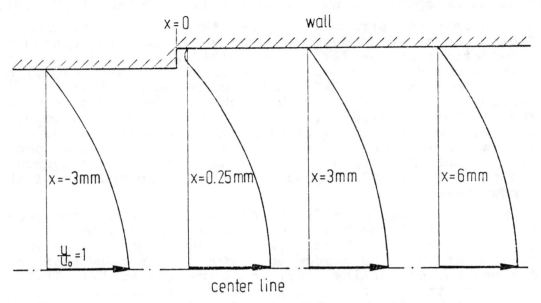

Fig. 3. Non-dimensional velocity profiles u/u_{axis}.

Three-Dimensional Flow: Scattered-Light Method

It is known from photo-elasticity that the light scattered from a birefringent medium may display a fringe-like pattern which depends on the local state of stress or on the amount of birefringence at the position where the light is scattered from. Thus, the scattered light method may be appropriate to yield information on the local values of strain or stress in a three-dimensional distribution instead of integrating this information along the light path. McAfee and Pih (Ref. 3,7) have adopted this method to streaming birefringence. They investigated the fully developed flow in pipes of arbitrary cross-section; these flows have only one velocity component, $u(y,z)$, in the direction of the pipe axis (x-direction). A theoretical relationship between the observed pattern and the local state of flow is not available for this method. Therefore, McAfee and Pih have searched for an empirical relationship to correlate their data. The resulting equation (eqs. 11,12 in Ref. 3) has yielded satisfactory results when applied to the mentioned pipe flow. The question arises whether this relationship is of general validity since it has only been checked for a particular flow case.

Based on a number of experimental observations the following system of equations is proposed for the evaluation of scattered-light experiments:

$$(\partial N/\partial x)^2 = c_1^2 \cdot \dot{\varepsilon}^2_{max_{yz}} + c_2^2(\dot{\varepsilon}^2_{max_{zx}} + \dot{\varepsilon}^2_{max_{xy}})$$

$$(\partial N/\partial y)^2 = c_1^2 \cdot \dot{\varepsilon}^2_{max_{zx}} + c_2^2(\dot{\varepsilon}^2_{max_{xy}} + \dot{\varepsilon}^2_{max_{yz}}) \qquad (2)$$

$$(\partial N/\partial z)^2 = c_1^2 \cdot \dot{\varepsilon}^2_{max_{xy}} + c_2^2(\dot{\varepsilon}^2_{max_{yz}} + \dot{\varepsilon}^2_{max_{zx}})$$

$\dot{\varepsilon}_{max_{ij}}$ is defined in the i-j-plane through

$$\dot{\varepsilon}^2_{max_{ij}} = (\dot{\varepsilon}_{ii} - \dot{\varepsilon}_{jj})^2 + 4\dot{\varepsilon}^2_{ij},$$

$\dot{\varepsilon}_{ij}$ being an element in the respective matrix of deformation velocities. Introducing velocity components u_i and Carthesian coordinates x_i one has

$$\dot{\varepsilon}^2_{max_{ij}} = \left[(\partial u_i/\partial x_i) - (\partial u_j/\partial x_j)\right]^2 +$$

$$+ \left[(\partial u_i/\partial x_j) + (\partial u_j/\partial x_i)\right]^2.$$

A pencil or thin sheet of primary light propagates into the x-direction, and the scattered light will be observed from a direction in the y-z plane. The scattered light displays a pattern of dark fringes of order N and with variable distance Δx between fringes. From the measured values of N, ΔN, Δx one derives the data function $(\partial N/\partial x)$ by means of an appropriate interpolation scheme. The respective procedure applies when the primary light propagates into the y- oder z-direction. The constants C_1, C_2 can be calibrated in a known flow field. The values for C_1 usually are one order of magnitude larger than those for C_2. The constant C_1 is associated to the deformation in the plane normal to the light direction; C_2 is associated to the deformations in the two orthogonal planes which include the light direction. The values of C_1 and C_2 are sensitive towards changes in temperature and concentration of the solution, and they depend on the light wavelenght (see also Ref. 8).

In comparison with the equations proposed in Ref. 3 the system (2) has the following characteristic properties:

(a) The scattered-light effect in streaming birefringence depends on both the diagonal components and the symmetric components of the tensor of deformation velocities. Thus, the system (2) formally agrees with the birefringence theory of Wayland (Ref. 1).

(b) (2) predicts the observed phenomenon that a fringe pattern also exists in the light propagating in flow direction for fully developed channel flow, i.e. in the case studied in Ref. 3.

(c) For the special case studied by McAfee and Pih the system reduces to the equations developed in Ref. 3.

(d) The system (2) should be considered as a first order approximation which applies under the same conditions as the linearized form of Wayland's theory (Ref. 1).

9 unknowns are included in (2), namely the 9 derivatives of the flow velocity, $\partial u_i/\partial x_j$. A general three-dimensional flow problem therefore can only be solved if additional information is available; e.g. the continuity equation can always be taken as one additional equation.

The two-dimensional flow in a divergent channel was used to check the validity of (2). This flow is determined by 4 derivatives: $\partial u/\partial x$, $\partial u/\partial y$, $\partial v/\partial x$, $\partial v/\partial y$. A theoretical solution is available for this case, and scattered-light flow pictures have been calculated by means of the system (2) and with experimentally determined constants C_1, C_2.

The results (Fig. 4) agree well with the visualized flow pattern. The scattered-light pattern usually is of low contrast, and the resulting photographs are not reproduced here.

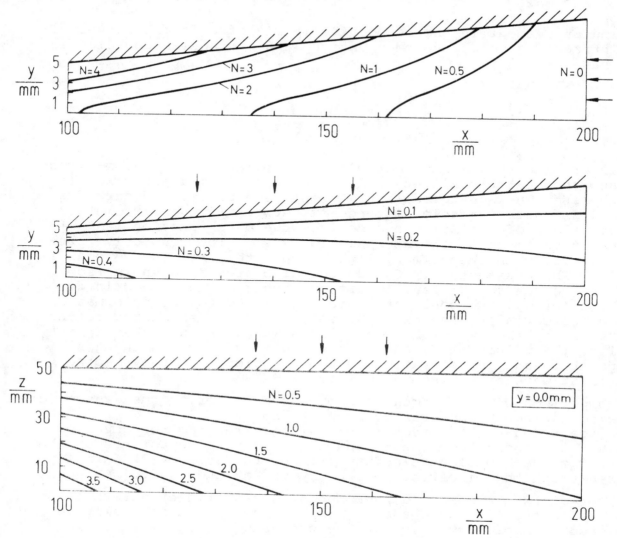

Fig. 4. Scattered-light flow pictures calculated with (2) for the flow in a two-dimensional, divergent channel. Flow in x-direction. Direction of incident light as indicated.

Conclusion

The two methods making use of the effect of streaming birefringence in colloidal solutions and described herin should be applied at low flow velocities or at low Reynolds numbers in order to obtain quantitative flow data. Under these conditions the solution of Milling Yellow dye behaves like a Newtonian fluid. At the same time, first order approximations are valid which relate the observed optical pattern to the flow field. The problem of applying streaming birefringence appears to lie in the numerical evaluation of the respective differential equations.

Acknowledgement

These investigations were supported by a research grant of the Deutsche Forschungsgemeinschaft (DFG).

References

(1) H. Wayland, Streaming birefringence as a rheological research tool, J. Polymer Sci: Part C, No.5(1964),11.

(2) F.N. Peebles and K.C. Liu, Photoviscous analysis of two-dimensional laminar flow in an expanding jet, Exp. Mech., 5(1965), 299.

(3) W.J. McAfee and H.Pih, Scattered-light flow-optic relations adaptable to three-dimensional flow birefringence, Exp. Mech., 14(1974), 385.

(4) M.M. Swanson and R.L. Green, Colloidal suspension properties of Milling Yellow dye, J. Colloid and Interface Sci., 29(1969),161.

(5) W.Merzkirch and M.Horsmann, Non-intrusive testing of low-Reynolds number pipe flows: streaming birefringence, Fluid Dyn. Trans., 8(1976) 61.

(6) E.O. Macagno, T.K. Hung, Computational and experimental study of a captive annular eddy, J. Fluid Mech., 28(1967) 43.

(7) W.J. McAfee and H. Pih, Flow through a semicircular pipe by three-dimensional flow birefringence method, Developments in Mechanics, 6(1971) 277.

(8) J.T.Pindera and A.R.Krishnamurthy, Foundations of flow birefringence in some liquids. Proceed. Int. Symp. Exp. Mech., Univ. of Waterloo (Canada), June 1972, Paper No. 37.

ANALOGICAL STRESS ANALYSIS OF LIQUID FLOW BY THE BIREFRINGENCE DETERMINATION OF JELLY UNDER LOAD

TEIKICHI ARAI* and HIROSHI HATTA**

Stational birefringence patterns of jelly exhibited under fixed loads for samples solidified in two-dimensional ducts with a cylinder obstacle were compared with those reported by Durelli and Norgard in 1972 for the flow of aqueous solution of Milling Yellow NGS dye through a duct of dimensional similarity at low Reynolds number. As a remarkable fact, the results showed that, at some experimental conditions, distribution of the relative intensity and directions of the deviatoric stresses of jelly as well as its relative sites of the set of optically isotropic points gave a complete coincidence with those of the liquid flow, respectively.

Introduction

Most of the past investigations on flow birefringence from the standpoint of physical chemistry have been carried out for the shear deformation in an annular gap of concentric cylinders. They have mainly concerned their applications to the study on size, shape and orientation of solute molecules or the dispersed particles in solution (Ref.1), and partly to the explanation of the first normal stress difference (Ref.2). On the other hand, since the early communication of Humphry (Ref.3),efforts of few researchers have been directed to the application of photoelastic stress analysis to the liquid flowing through ducts fit for the researches of transportation engineering. Among these papers, Durelli and Norgard's one (Ref.4) may be the first where the directions of the principal stresses as well as the relative intensity of the principal stress difference were obtained throughout the duct from precise superimposed map of isoclinics of various extinction angle. Therein aqueous solution of Milling Yellow NGS dye (Ref.5,6) was used as the liquid of Newtonian flow behavior with sufficient sensitivity of birefringence. In the mean time, Arai and his coworkers(Ref.7,8) carried out flow birefringence measurements of polymer and related stress analysis. Successfully and putting the basic research standpoint upon the past experimental results on the solid-like response of the flowing polymer melts oberved in connection with the tube length correction term coefficient, Barus effect (Ref.9,10,11), and flow irregularities(Ref.12,13), they continued experiments on the comparison of the birefringence data of polymer melt flow with those of the rubbery elastic solids of silicone rubber (Ref.14) and jelly(Ref.15) both solidified and polymerized in the same duct for liquid flow, respectively. These rubbery solids gave intrinsically the same birefringences when suitable loads were applied to the same direction of flow as the polymer melts showed. These facts possibly

* Professor of Mechanical Engineering, Faculty of Engineering,Keio University.
  832, Hiyoshi-cho, Kohoku-ku, Yokohama, 223 Japan.
** Assistant of Mechanical Engineering, Faculty of Engineering, Keio University.
  832, Hiyoshi-cho, Kohoku-ku, Yokohama, 223 Japan.

denied the concept of stress relaxation of flowing polymer melt in the course of flow within a duct.

To extend the application and to confirm its validity of the technique of " Analogical stress analysis of liquid flow by the solid birefringence" to Newtonian flow was regarded as impossible when the authors considered the difficulties of molecular orientation of the liquid itself. In the present paper, however, by taking advantage of the use of the before-mentioned Durelli and Norgard's data, the comparisons between the liquid of Newtonian flow and the rubber-like solid of jelly were performed for a duct of dimensional similarity. The results by this analogical method were expected to give some fundamental informations which might be powerful to uncover the basic relations on mechanics between directions of stress and flow of liquid.

Experimental

A duct composed of several steel parts and two window glasses which had been used for flow birefringence measurements of polymer melts (Ref.7,14) was converted to the use of the present work. Its explanatory diagram is given by Fig.1 with indications of the coordinates. The cylinder obstacle was attached within the approach to one of the opposite window glass surfaces with commercial instantaneous adhesive in the direction where the axis was parallel to the light path. Another duct of straight rectangular one with clearance,a,of 10 mm was provided to obtain a result in which the effect of the narrower clearance region was negligible. Used gelatin as the base of jelly was powdery one of Type M-157 produced by Miyagi Chemical Co. Shear modulus of jelly varied with gelatin concentration, temperature and time duration after solidification as well as the temperature it was solved. The modulus of jelly for the gelatin weight percentage concentration of the aqueous solution c of 5,10 and 15 were at 20°C around 1,5 and 10×10^{-2} kg/cm^2,respectively. Gelatin solutions were poured into the ducts provided with stopper at the bottom and left to solidification. Loads,W,were applied by Instron type tester Autograph IM-100 of Schimadzu Seisakusho Co. after removal of the bottom stopper. Birefringences were measured near the cylinder obstacle at the temperature, θ, of the room 20°C. The measuring optical system was the same one used in previous papers. The light source was 100 W high pressure mercury lamp of Toshiba Co.or 500 W super high pressure mercury lamp of Ushio Denki Co. Stress analyses were carried out with photographs taken by Asahi-Pentax ME, F 1.4, camera and Fuji Neopan XXX, ASA 200,films, and extinction angle, χ, was defined by the anticlockwise rotation angle of the incident polarized light plane from x,z-plane when viewed from the light source .

Basic relations between stress and birefringence

For the analysis of stress and related discussion on flow, following basic equations from (1) to (5) were used in the same manner of photoelasticity analysis(Ref.16):

$$n = \frac{N_{fr}}{B} = C_\sigma (\sigma_1 - \sigma_2) \qquad (1)$$

$$\tau_p = \frac{1}{2} (\sigma_1 - \sigma_2) \qquad (2)$$

$$\tau_p = \frac{\tau_{yx}}{\sin 2\chi} = \frac{\tau_{xy}}{\sin 2\chi} \qquad (3)$$

$$\alpha_{\tau p} = \frac{\tau_p}{\tau_{p-ref}} \qquad (4)$$

$$\alpha_{\tau p} = \frac{n}{n_{max}} \qquad (5)$$

where n is fringe orde per unit optical path, N_{fr} observed fringe order, C_σ stress optical coefficient, B the duct width and identical with the optical path, ($\sigma_1 - \sigma_2$) principal stress difference, τ_p principal shear stress usually called in literature maximum shear stress, τ_{xy} shear stress exerted in x direction on the plane of constant y, $\alpha_{\tau p}$ non-dimensional principal shear stress, τ_{p-ref} reference principal shear stress and n_{max} maximum value of n within the duct. It must be notified that under the foregoing definition of χ , eq(3) prove valid only for the case where the direction of the incident polarized light plane is coincident with that of either one of the isostatics.

Results and discussion

At first the following points must be noticed: (1) Both birefringence and stress comparisons between solid and liquid were carried out under the assumption that the deformation and the flow were commonly two-dimensional. (2) Deformation of the duct and the volume change of jelly by load applicaton were neglected. (3) Optical responses to loads were instantaneous and there observed no practical time-dependent change. (4) The direction of extensional isostatics was regarded as identical with the average direction of the longer axis of the orientated dye molecules. (5) Any kind of accidental slippage or asymmetrical set of cylinder obstacle in duct was easily discriminated by the birefringence response from the regular one.

Fig.2 shows a typical example of isochromatic fringe patterns in the dark field with indication figures of N_{fr} for jelly of c=10 wt.%. Therein, all the points of C_1, C_2 and C_3 and their symmetrically corresponding ones with x,z-plane,C_1', C_2' and C_3', N_{fr} was zero. At the points of C_s and C_s' on the cylinder surface at y=0, N_{fr} was also zero. The largest fringe order denoted by N_{fr-max} was observed at the points S_m and S_m' on the cylinder surface and at x=0, and was calculated graphically by extrapolation in the superposed isochromatic map of dark and light fields patterns. When raising load, onset of fracture of jelly took place at or close near to either one of these points. By taking the value as for the reference fringe order per unit optical path, n_{fr}, non-dimensional principal shear stress, $\alpha_{\tau p}$ of eq.(5) through eqs.(1),(2) and (4), were obtained on several planes shown by Fig.3 from the same superposed map of isochromatics. Results for jelly of c=10 wt.% on the planes of constant y were shown by Fig.4, and of constant x by Fig.5. In these figures, effect of load difference was not so clear, and in Fig.5 plots for different loads were connected by a single curve. Fig.6 shows a typical isoclinic fringe patterns. From a set of the trajectories for different extinction angles,χ,like this, superimposed maps of isoclinics were obtained as shown by Fig.7. By applying the well-known graphical method (Ref.16) to the map, trajectories of principal shear stress and isostatics were obtained as shown by Fig.8-(a) and -(b), respectively. For clarification, in Fig.8-(b) extensional isostatics were given by solid lines and compressive one by broken lines from empirical reasoning. As is clear by Fig.7, at each point of C_1, C_2 and C_3 and their axially corresponding ones,C_1', C_2' and C_3', all the isoclinics of different extinction angles crossed together, and therefore the directions of stress there were indefinite. When we consider the zero fringe order at the coresponding points in Fig.2, we can understand that these points are well called isotropic singular points, where no deviatoric stress operates. Similarly, the same definition may be applied to the points C_s and C_s', the former one of which,C_s,possibly correspond to the well-known stagnation point in fluid dynamics (Ref.17), and the latter one,C_s',locating at the symmetrical position to C_s' with respect to y,z-plane, to the center of inversely operating stresses as shown by the conjugate isostatics in Fig.8-(b). Effects of load and gelatin concentration in jelly will be seen by Fig.9, where the change of the singular points,C_1 and C_3,sites with load was plotted for different gelatin concentrations of jelly. Therein small but clear asymmetries were observed in their relative positions with respect to y,z-plane except for the straight rectangular duct. The cross-section of the straight duct normal to x,y-plane was similar to but one tenth in length of that used by Durelli and Norgard (Ref.4). When combined with the following description on the stress comparison between solid and liquid, observed effect of the slit portion on the asymmetry of birefringence in x direction suggest the validity of the general concept that flow of liquid can not be free from the downstream restriction. Really, the velocity of shear stress propagation may be the same magnitude of the pressure propagation in liquid, which is by far larger than that of flow. Direct birefringence comparison between solid and liquid was found possible by changing load applied upon jelly of c=15 wt.%. Fig.10 shows the isochromatic patterns, indicating a complete agreement , including, the relative sites of the set of singular points, with those shown in the paper of Durelli and Norgard(Ref.4) and transposed in Fig.11. They made the birefringence determination for the flow of the aqueous solution of Milling Yellow NGS dye (c= about 1.5 wt.%, 1.8poise) through beforementioned duct at Reynolds number of about 0.7 based on the dimension of the cylinder diameter. The chemical name of the dye molecule may be disodium salt of 6,6'-disulfo-o-tolidin-diazo-bis-acetacetanilide (Ref.18). In Fig.12, conjugate principal shear stress lines obtained from the super-posed map of isoclinics for the same test conditions as those of Fig.10 were given together with the streamlines given in the literature as obtained by double exposure photography of electrlytic buble method. As is clear with the curves, directions of streamlines showed sufficient agreement with neither one of the principal shear stress lines. The inconsistency is particularly apparent on and near the x,z-plane and has a tendency to become large as the location approaches the cylinder sur-face. As presented previously (Ref.12,13), when $\theta_{\tau\rho-st}$ is defined as the smaller inter-section angle between streamline and and principal shear stress lines, inelastic shear stress operates at any arbitrary position within the duct becomes dynamically $\tau_\rho \cos 2 \times \theta_{\tau\rho-st}$(Ref.16). Thus the local efficiency for the liquid transportation is given by $\cos 2\theta_{\tau p-st}$, the largest value of which is given when the direction of streamline is parallel to that of either one of the conjugate principal shear stresses. Thus, as variation principle suggests, every liquid tends to flow along the principal shear stress lines .Observed local discrepancies in Fig.12 may be explained by the application of the same principle to the whole flowing system. In the elastic deformation of jelly, the same principle might hold true for the work on stored energy of strain and stress.

Conclusions

From the foregoing results on the birefringence comparison between jelly and aque-ous solution of Milling Yellow NGS dye, the following conclusive interpretations were derived on the mechanical equilibrium of internal slow viscous flow past a circular cylinder: (1) Even for the viscous fluid flow where any practical recoverable shear deformation is scarcely observable, deviatoric stresses operate in the manner as if the overall flowing liquid in the duct were a single continueous elastic solid of a fixed shear modulus. (2) Stresses which contribute to viscous deformation operate only on and along the streamlines. They contribute to the dynamic equilibrium as other elastical-ly operating stresses do. The shear rate on the streamline is defined by Newtonian or non-Newtonian viscosity coefficient, but the time-dependent development of inelastic shear itself does not change the overall stress distribution of the flowing liquid. (3) Intrinsically, every liquid tends to flow along the lines of principal shear stresses. But duct wall restriction, for example, of the shapes of divergence, convergence and curvature as well as of any obstacle set in the flow field can change this status. Thereby, the smaller one of the intersection angles between principal shear stress lines and streamline may define the local dynamic stability as well as the local viscous work efficiency of stress for the liquid transportation.

Acknowledgment

The authors express their thanks to the former students Messrs Y. Hatano, K. Fukui, and T. Yokoyama for their collaborations. Thanks are also due to Hatakeyama Foundation for the research grant of 1976 by which this work was partly supported.

References

(1) R. Cerf and H. Scheraga, Flow Birefringence in Solutions of Macromolecules, Chem. Rev., 51-2(1952-10),185.
(2) H. Janeshitz-Kriegl, Flow Birefringence of Elastico-Viscous Polymer Systems, Fort-sch. Hochpolym. Forsch.,6-2(1969),170.
(3) R. H. Humphry, Demonstration of the Double Refraction due to Motion of a Vanadium Pentoxide Sol, and Some Applications, Proc.Phys. Soc.(London), 35-1(1923), 217.
(4) A. J. Durelli and J. S. Norgard, Experimental Analysis of Slow Viscous Flow Using Photoviscosity and Bubbles, Exp. Mech.,12-4(1972-4), 169.
(5) J. W. Prados and F. N. Peebles, Two-dimensional Laminar-Flow Analysis, Utilizing a Doubly Refracting Liquid, Am. Inst. Chem. Eng. J., 5-2(1959-2),225.
(6) W. Merzkirch, Flow Visualization, (1974), 214, Academic Press.
(7) T. Arai and H. Asano, Analysis of the Flow of Molten Polyethylene through a paral-lel Slit by the Flow Birefringence Method, Kobunshi Kagaku(in Japanese), 29-7 (1972-7), 510.

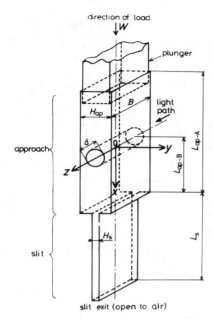

Fig.1

Explanatory diagram for the rectangular duct of birefringence measurement of solid under load.

H_{ap}: 10.0mm, B: 25.0mm, d: 2.5, 4.0, 6,0mm
H_s: 19mm, L_s: 40.0mm, L_{ap-A}: 77.0mm
L_{ap-B}: 19.0mm

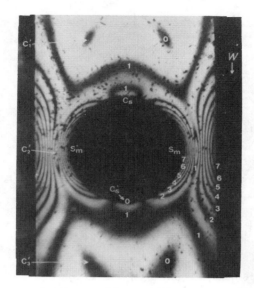

B: 25.0mm, a: 9.8 mm, d: 6.0mm
θ: 20°C, W: 880g

Fig.2

Isochromatic fringe pattern near a circular cylinder for the deformation of jelly (c =10wt.%) with carbon particles dispersed therein.

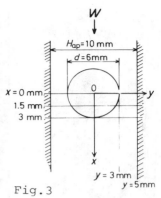

Fig. 3

Cross section for the explanation of deviatoric stress distribution.

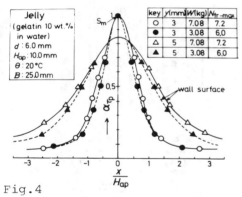

Jelly (gelatin 10 wt.% in water) d: 6.0 mm H_{ap}: 10.0 mm θ: 20°C B: 25.0mm	key	y(mm)	W(kg)	N_{fr-max}
	O	3	7.08	7.2
	●	3	3.08	6,0
	△	5	7.08	7.2
	▲	5	3.08	6,0

Fig.4

Distribution of non-dimensional principal stress difference $\alpha_{\tau p}$ on the planes of two different values of y.

Jelly (gelatin 10 wt.% in water) θ: 20°C d: 6.0mm H_{ap}: 10.0mm B: 25.0mm	key	x(mm)	W(kg)	N_{fr-max}
	─○─	0	7.08	7.2
	─●─	0	3.08	6.0
	─□─	1.5	7.08	7.2
	─■─	1.5	3.08	6.0
	─△─	3.0	7.08	7.2
	─▲─	3.0	3.08	6.0

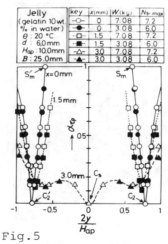

Fig.5

Distribution of non-dimensional principal stress difference $\alpha_{\tau p}$ in y direction on the three planes of different values of x.

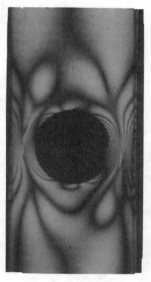

H_{ap}:10.0mm, d:6.0mm, W:5.08 kg, θ: 20°C.

Fig.6   Isoclinics of jelly (c= 10 wt.%) at χ =20°.

(8)   T. Arai, H. Asano, H. Ishikawa, J. Mizutani and S. Murai, Flow Patterns and Principal Stress Lines for Molten Polyethylene in the Entrance Region of a Parallel Slit, Kobunshi Kagaku(in Japanese),29-10(1972-10),743.

(9)   W. Philippoff and F. H. Gaskins, The Cappilary Experiment in Rheology, Trans. Soc. Rheology, 2(1958), 263.

(10) T. Arai and H. Aoyama, Die Wall Restriction on the Elastic Shear Deformation in Viscoelastic Flow of Polymer Melt, Trans. Soc. Rheolgy,7(1963), 333.

(11) T. Arai, The Entrance Effect and the Barus Effect, Proceedings of the 5th International Congress on Rheology, Vol.4(1970), 497, University of Tokyo Press.

(12) T. Arai, Possible Shear Fractures on the Maximum Principal Shear Stress Lines in Polymer Extrusion, paper presented at the Joint Meeting of the U.S. and Japan Societies of Rheology, Honolulu, Hawaii, June 6, 1975.

(13) T. Arai, Flow Instability, Shear Break Strength and Maximum Principal Shear Stress

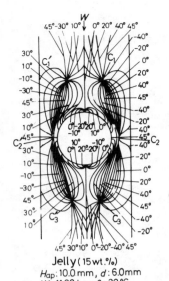

Jelly (15 wt.%)
H_{ap}:10.0 mm, d:6.0mm
W:11.08 kg, θ:20°C

Fig.7

Superimposed map of isoclinics.

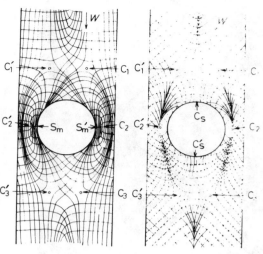

Jelly (15wt.%)
d:6.0 mm, H_{ap}:10.0 mm, W:11.08 kg, θ:20°C
(a)   (b)

Fig.8

Maps of principal shear stresses (a) and of principal stresses (b) with the indications of the set of singular point sites by open circles with arrow

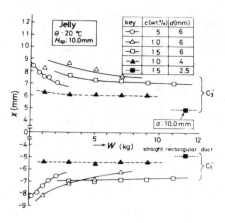

Fig.9

Load and gelatin concentration dependences of the sites of zero points C_1' and C_3' for the deformation of jelly under fixed load.

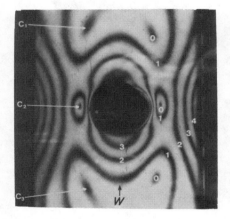

θ:20°C, a:9.8mm, B:25mm
d:2.52 mm, W:ca.12kg, in dark field

Isochromatic fringe pattern near a circular cylinder for the deformation of jelly(c=15wt.%)

Fig.10

Dark-field isochromatics in photoviscous flow about a cylinder

[after A.J.Durelli and J.S.Norgard. Exp. Mech., 12, 169 (1972).]

Fig.11

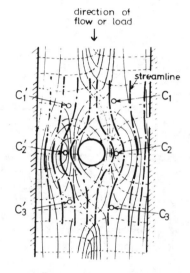

direction of
flow or load

Map of conjugate principal shear stress lines of jelly (c=15wt.%) near a circular cylinder in comparison with the streamlines of Milling Yellow N.G.S. water solution [A.J.Durelli and J.S.Norgard. Exp. Mech. 12 (4), p.16 (1972).] drawn in reduced scale-fit for the superposition of the facts of dimensional similarity.

Fig.12

tor Polymer Melt Flow, Preprint of Japan Soc. Mech. Eng.(in Japanese), No.750-8 (1975-8),101.

(14) T. Arai and H. Ishikawa, Birefringence of Rubber-like Solid in Slit duct under Deformation, Kobunshi Ronbunshu(in Japanese), 31-8(1974-8), 501.

(15) T. Arai, H. Hatta and T. Ikeda, Birefringences of Rubber-like Solids in Slit Ducts under Loads As Compared with Those of Polymer Melts Flowing through the Same Ducts, Proceedings of the 7th International Congress on Rheology, (1976-8),442.

(16) A. Kuske and G. Robertson, Photoelastic Stress Analysis, (1974),5, 162,Academic Press.

(17) H. Schrichtung, Boundary Layer Theory, 4th Ed., J. Kestin, transl.,(1960),33, 78, 112, 221, McGraw Hill.

(18) Society of Dyers and Colourists, Colour Index, 3rd Ed., Vol.4(1975),4198, Bradford, West Yorkshire.

CAVITATION

VISUALIZATION OF JET CAVITATION IN WATER AND POLYMER SOLUTIONS

J. W. HOYT* and J. J. TAYLOR**

The effect of polymer additives on underwater jet cavitation has been studied using a special camera. The drag-reducing additive, polyacrylamide, at a concentration of 25 ppm, decreased the cavitation inception index and greatly changed the appearance of the cavitation bubbles. In pure water, the cavitation appearance resembles ragged groups of small bubbles with the overall impression of sharpness and roughness, but in polymer solution the bubbles are larger, rounded, and of completely different appearance.

Introduction

The study of cavitation in water has been stimulated by the recent discovery that small amounts (parts per million) of high molecular-weight, long-chain polymers of the type which reduce turbulent fluid friction are also effective in lowering the cavitation inception index (i.e., it becomes more difficult to cause cavitation in a given flow situation). Recent work in this area has been summarized by Acosta and Parkin (Ref. 1).

Cavitation in the shear layer of an underwater jet is profoundly affected by the presence of polymer additives as discussed in Ref. 2 and 3. While the effect of polymers in lowering the cavitation inception index can be rationalized to some extent by considering the viscoelastic properties of polymer solutions (Ref. 4 and 5), actual observations indicate little difference in growth rate of spark-generated bubbles (Ref. 6) or gas-bubbles (Ref. 7) in polymer solutions as compared with water.

However, in spite of theoretical analysis (Ref. 8 and 9) which indicate the bubble collapse rate should be less in polymer solutions than in water, experimental data (Ref. 10) show the cavitation erosion rate to be enhanced in polymer solutions. Underwater jet cutting of metals, which may also be a cavitation effect, is greatly enhanced by polymer addition to the jet fluid (Ref. 11). Ting (Ref. 12) discusses these experiments in view of the polymer relaxation times and polymer solution parameters.

Further, the radiated noise from cavitation bubble collapse has been found to be greater in jet cavitation (Ref. 13) and in cavitation around a rotating disk (Ref. 14) for polymer solutions compared to water. All these conflicting experimental and theoretical studies make it desirable to visualize polymer solution flow during cavitation events so as to have a better basis upon which to carry out further studies and evaluations of the effect of polymer solutions on cavitation.

* Professor, U.S. Naval Academy, Annapolis, Maryland, 21402, USA
** Consultant, Naval Ocean Systems Center (Code 8101), San Diego, California, 92152, USA

Experimental

The apparatus consisted of a water nozzle 0.115 inches (.45 mm) in diameter which discharged into a covered box having optical glass walls approximately 2-1/2 inches (7 cm) apart. The box was arranged to provide carefully controlled recirculation of the jet fluid, so the ambient surroundings of the jet were nearly stationary at all times. A slight negative pressure was measured at the top of the box over the jet; this negative pressure was almost exactly balanced by the static head of fluid at the jet centerline and hence the static pressure at the jet centerline was taken as atmospheric.

The inlet pressure to the nozzle was measured by a calibrated pressure gage, and the water leaving the apparatus was sampled. Air content of the samples was measured by a van-Slyke apparatus, while the polymer content of the samples was estimated with a turbulent-flow rheometer.

The apparatus received fluid from a large pressure tank. Water or polymer solution contained in the tank could be held under vacuum overnight, then re-pressurized, as described in Ref. 2, to provide a deaerating effect. The polymer solution used was a 25 ppm solution of polyacrylamide, Calgon Corp., (Pittsburg, Pa.) type TRO-375.

The most critical part of the apparatus was the camera and lighting equipment, all of which was specially designed for this investigation. The camera used a rotating mirror to provide image-motion compensation of the predominantly axially-moving jet stream, and was fitted with dual lenses which provided two images, separated vertically, on the same film frame. The basic ideas involved in the dual-camera design are cited in Ref. 15.

The illumination method used is sketched in Figure 1 and is of the type termed Köhler illumination in microscopy or "trans-illumination" in high-contrast enlargers. As shown in Fig. 1, light from the point source (electronic flash) is imaged by the condensing lens at the diaphragm of the camers lens and thus on the film. The system is very efficient, i.e., there is no lost light and every point on the film sees all the point source and only that. Cavitation bubbles interfere with the light transmission by refracting and reflecting the light so that they appear black in the photos. Flashes from the two flash-heads shown in Fig. 1 can be separated in time, electronically, so that two images with a known time separation are produced on the same film frame. The illumination system is so free from diffused light that the camera body does not need a separator, and there is very little "ghosting" of one image on the other. The electronic flash units are operated at low power because of the high illumination efficiency; the image duration is thus on the order of 1 to 4 μ sec. Verichrome and Plus-X roll film of ASA rating 125 was used.

In order to locate one edge of the flow from the nozzle in the photographs, an 0.016 inch hole was drilled in the side of the nozzle at the top, approximately 0.09 inches from the nozzle end. A dye solution could be released through this hole into the top surface of the jet. The dye streak was controlled so that dye was emitted only for a few seconds before and after each photograph, in order to avoid a dye background level in the recirculating portion of the water.

Results

Figure 2 shows a typical scene of cavitation events in water having a high air content. The photographs are taken 10 μ sec apart; the upper scene was taken first. Air bubbles in the surrounding water appear in the background; measurements indicate that the air content ratio α/α_s is 1.12, where α is the measured air content of a sample of the water and α_s is the saturated air content for the water temperature according to the International Critical Tables. The cavitation index σ is defined as:

$$\sigma = \frac{P - P_v}{\frac{1}{2} \, p \, V^2}$$

where P = the local pressure around the jet (taken as atmospheric)

P_v= vapor pressure of water at test temperature

σ = density of water

V = computed jet velocity

For the conditions of Figure 2, σ is 0.22, i.e., well into the cavitation regime, according to the results of Ref. 2. Due to the high air content of the water (above saturation) the cavitation shown here could be termed "gaseous" cavitation, or cavitation based on finite gas bubbles as nuclei. From the flow visualization standpoint, the camera has largely discriminated against the background bubbles, these appearing as dimly lighted streaky objects, while the illumination provides clear and sharp profiles of cavitation bubbles moving with the jet velocity. Note that the cavitation appears to be masses of smaller bubbles with rather sharp and irregular overall contours, and that significant changes in the bubbles have occurred in the 10 µ sec interval between photographs.

Figure 3 shows cavitation in water having a relatively low air content, α/α_s = 0.65, at a cavitation number of 0.18, with 10 µ sec between the photos. One would expect the cavitation to be of the "vaporous" type, i.e., based on infinitesimal discontinuities in the water rather than finite gas bubbles as in Fig. 2. However, the appearance of the cavitation bubble masses is quite similar to those shown earlier. Of special interest in Fig. 3 is the dye streak showing vortex formation at the jet boundary, with cavitation events centered at the regions where vortices are pairing together.

Figure 4 shows cavitation in 25 ppm polyacrylamide polymer solution, at air content α/α_s of 0.82 and cavitation number of 0.14. Only sporadic cavitation can be seen at this cavitation number; a considerable inhibition of cavitation is thus evident in polymer solutions as noted in Ref. 2. Of particular interest is the marked change in bubble shape in the polyacrylamide solution; the bubbles appear much larger and smoother than cavitation bubbles in pure water. The same comments apply for Figure 5, taken at a still lower cavitation number, 0.13. In both Fig. 4 and Fig. 5 the scenes were taken 80 µ sec apart, and considerable changes in bubble shape are evident in this interval.

Discussion

The image-compensation features of the camera and the illumination technique provide a useful flow visualization tool for detailed study of cavitation events. The technique has been especially helpful in pointing out the significant changes in cavitation bubble appearance in polymer solution as compared with their appearance in pure water.

The generally larger bubble size evident in polymer solution cavitation could offer a partial explanation of the higher radiated noise from cavitating polymer solutions (Ref. 13 and 14) as well as increased erosion (Ref. 10 and 11) compared with water, in that a larger volume is involved as the bubble collapses. Since theoretical studies offer no hint as to larger cavitation bubble size in polymer solutions, these photographs may be useful as a basis for further analyses. And of course, the photographs confirm once again that cavitation inception is inhibited in polymer solutions.

Finally, the dual-photo technique shown here permits analysis of rapidly-occurring events such as cavitation in an inexpensive way.

Acknowledgement

This work was carried out at the Naval Ocean Systems Center under Office of Naval Research, U. S. Navy sponsorship, and continued at the U. S. Naval Academy under a Naval Sea Systems Command Research Professorship.

References

1. Acosta, A. J. and B. R. Parkin, "Cavitation Inception – A Selective Review." J. Ship Research, 19 (1975) 193-205.
2. Hoyt, J. W., "Effect of Polymer Additives on Jet Cavitation." ASME Trans. J. Fluids Engineering, 98 (1976) 106-112.
3. Baker, C. B., J. W. Holl, and R. E. A. Arndt, "Influence of Gas Content and Polyethylene Oxide Additive upon Confined Jet Cavitation in Water," 1976 Cavitation and Polyphase Flow Forum, (1976), 6, ASME, New York.
4. Ting, R. Y., "Effects of Viscoelasticity on Cavitation in Drag Reducing Fluids," in The Role of Cavitation in Mechanical Failures, (1974) 100-105, U.S. Bureau of Standards Special Publication 394.
5. Ting, R. Y., "Viscoelastic Effect of Polymers on Single Bubble Dynamics," AIChE J. 21 (1975) 810-814.
6. Ellis, A. T. and R. Y. Ting, "Non-Newtonian Effects on Flow-Generated Cavitation and on Cavitation in a Pressure Field," in Fluid Mechanics, Acoustics, and the Design of Turbomachinery. NASA SP-304 (1974) 403-421.
7. Ting, R. Y. and A. T. Ellis, "Bubble Growth in Dilute Polymer Solutions," Physics of Fluids, 17 (1974) 1461-1462.
8. Folger, H. S. and J. D. Goddard, "Collapse of Spherical Cavities in Viscoelastic Fluids," Physics of Fluids, 13 (1970) 1135-1141.
9. Shima, A. and Y. Tsujino, "The Behavior of Bubbles in Polymer Solutions," Chemical Engineering Science, 31 (1976) 862-869.
10. Ashworth, V. and R. P. M. Proctor, "Cavitation Damage in Dilute Polymer Solutions," Nature, 258 (1975) 64-66.
11. Kudin, A. M., G. I. Barenblatt, V. N. Kalashnikov, S.A. Vlasov, and V. S. Belokon, "Destruction of Metallic Obstacles by a jet of Dilute Polymer Solution," Nature Physical Science, 245 (1973) 95-96.
12. Ting, R. Y., "Polymer Effects on Microjet Impact and Cavitation Erosion," Nature, 262 (1976) 572-573.
13. Hoyt, J. W., "Effect of Polymer Additives on Jet Cavitation," Trans 16th American Towing Tank Conference, Sao Paulo, Brazil, 1 (1971) 7.0.
14. Walters, R. R., "Effect of High-Molecular Weight Polymer Additives on the Characteristics of Cavitation," Advanced Technology Center, Inc., Dallas, Texas, Report No. B-94300/2 TR-32 (1972).
15. Taylor, J. J., "Camera Apparatus for Making Photographic Images on Moving Cut Film Pieces," U. S. Patent 3,925,796 (1975).

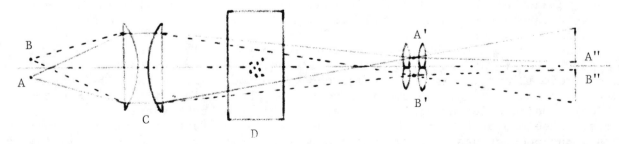

Fig. 1. Sketch of illumination set-up for dual-lens camera. A,B: electronic flash units; C: condenser; D: end view of tank with cavitating water jet; A', B': camera lens; A",B": film plane.

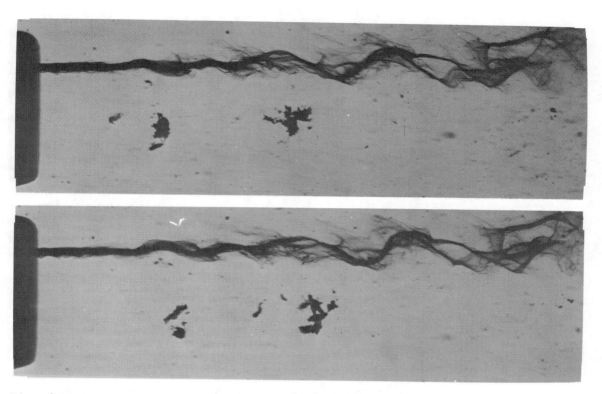

Fig. 2. Jet cavitation in water (dye streak is in top boundary of the jet).
The lower photo was taken 10 μsec after the upper. $\alpha/\alpha_s = 1.12$; $\sigma = 0.22$.

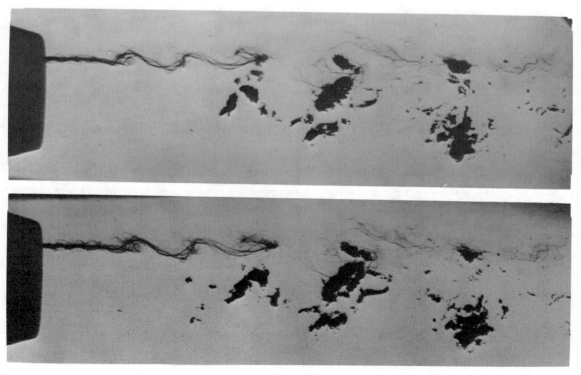

Fig. 3. Jet cavitation in deaerated water. The lower photo was taken
10 μsec after the upper. $\alpha/\alpha_s = 0.65$; $\sigma = 0.14$.

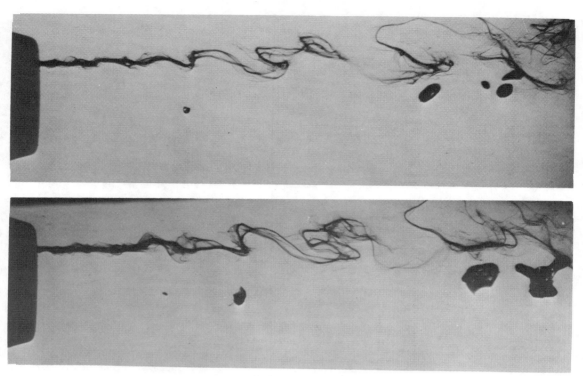

Fig. 4. Jet cavitation in 25 ppm polyacrylamide solution, with 80 μsec delay in lower photo. $\alpha/\alpha_s = 0.82$; $\sigma = 0.14$.

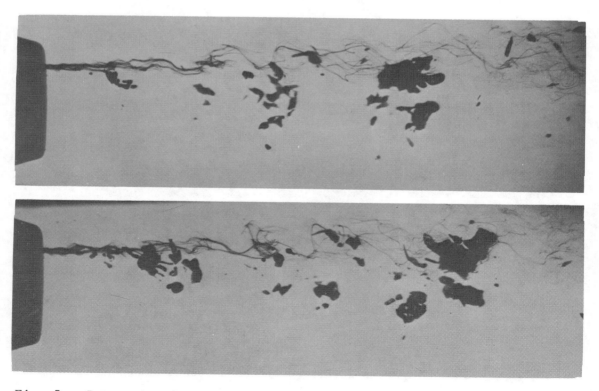

Fig. 5. Jet cavitation in 25 ppm polyacrylamide solution; delay of 80 μsec between photos. $\alpha/\alpha_s = 0.82$; $\sigma = 0.13$.

FLOW VISUALIZATION BY CAVITATION ON AXISYMMETRIC BODIES

H. MURAI* and A. IHARA**

H. Murai* and A. Ihara**

By using the flow visualization technique by cavitation, unsteady flow past spheres and a hemispherical nosed body is visualized. Vortex shedding from spheres is studied in the Reynolds number range $0.9 \times 10^5 < Re < 7.8 \times 10^5$, at free stream turbulence levels Tu=0.1%, 1.35% and 1.93%. At the low turbulence level, beyond the upper critical Reynolds number periodic vortex sheddings are observed, but at the high turbulence level periodic vortex sheddings are not observed. At the Reynolds number less than the upper critical Reynolds number, the Strouhal number and the wake cofiguration coincide with those observed by Achenbach. The separation bubbles on both models are visualized.

Introduction

The use of the flow-Visualization technique by cavitation is favorable for the high speed flow of liquid because of following reasons; (i) it is not necessary to injet the other tracer than operating liquid, (ii) it is possible to deduce the static pressure and the stream velocity at the place where the cavitation takes place, (iii) it is suited to observe the formation of the vortex because cavitation takes place easily at the vortex core. This technique had successfully be used to visualize a internal flow in an axial flow pump by Murai and Takeuchi(Ref.1).

In this experiments by using this technique, unsteady flow past spheres and hemispherical nosed body, which has been the object of numerous examinations, is intended to be clarified. The flow conditions past spheres, in particular, were of great interest with regard to problems of chemical processes. Thus, there is already some information available on properties of the wake at Reynolds numbers less the $Re=10^3$. Experiments carried out at Reynolds number greater than $Re=10^3$, in particular at the vicinty of the upper crtical Reynolds number, are reported by only a few authors; for example, Möller (Ref.2) Cometta (Ref.3), Mujumdar and Douglas (Ref.4), Calvert (Ref.5), Achenbach (Ref.6). The experimental results of several authors seem to be contradictory. While for example, Möller (Ref.2) measured at $Re=10^4$ a Strouhal number of S=2.0, Mujumdar and Douglas (Ref.4) and Achenbach (Ref.6) reported a value lower by one order of magnitude, S=0.2. As to the wake the periodic vortex sheddings were no longler indicated at Reynolds numbers greater than $Re=3.0 \times 10^5$ by Ackenbach[5], and at Reynolds number range $5.6 \times 10^4 < Re < 11.6 \times 10^4$ in highly turbulent free-streams by Mujumdar and Douglas, wake canfigurations in these conditions were not reported because their measurements were

*Professor of Institute of High Speed Mechanics, Tōhoku University, Sendai, Japan

**Lecturer of Institute of High Speed Mechanics, Tōhoku University, Sendai, Japan

carried out by using hot wires. Experimental data at the vicinity of the upper critical Reynolds number thus look like lacking considerably.

The main aim of this experiments was to obtain informations about the periodic boundary-layer separation at high Reynolds numbers and at different turbulence level in free streams, and to find out how the vortexes are generated in a wake by using the flow visualization technique by cavitation. The Reynolds number was varied not only by varying the stream velocity but also by using spheres of different diameter. Experimental results are available in the Reynolds number range $0.9 \times 10^5 < Re < 7.8 \times 10^5$. The turbulence level was varied in the range 0.1-1.93% by introduction of grids upstream of the test section. The pressure distributions for the sphere and hemispherical nosed body were measured at the different Reynolds number and turbulence level.

Test arrangements and measurement techniques

The experiments were carried out in a water tunnel shown in Fig.1 with a $200^{mm} \times 610^{mm}$ test section. The stream velocity at the test section could be varied in the range 4.0-16.0m/s. The turbulence level at the test section in clear condition was about Tu=0.1%. The turbulence level could be varied to Tu=1.35% and 1.93% by introduction of grids upstream in the test section.

Spheres with diameters D=10^{mm}, 20^{mm}, 40^{mm} and 60^{mm} supported from the rear by stings, which had lengths 4.0D and diameters 0.21D and a hemispherical nosed body with D=40^{mm} were tested. The pressure distributions were measured to the sphere with D=60^{mm} and the hemispherical nosed body. The vortex formation by the unsteady boundary separation and wake configurations were visualized by cavitation. The degree of the cavitation could be controlled by varying the static pressure of the water tunnel. Cavitation cavities could be photographed by using a high speed movie camera (Fastax camera No.16-399) at the flaming rate 2000-4500 frame/sec. For certainty, time markes of 2.56×10^{-3} second could be recorded on the film (Eastman 4-x Negative). A high speed stroboscope (Egerton-Germesh hausen and Grier Inc. Type 501) with a illumination time of about 2×10^{-6} second could be used as a source of light. As the cavitation which takes place at the vortex core shed from spheres could easily be ascertained on movie photographs (see Fig.4), the Strouhal number could be calculated by the number of occurrence of this cavitation which was recorded on movie photographs. The vortex shedding frequency was determined by measuring the time necessary for a sequence of 10 vortices to be released. As Shal'nev (Ref.7) had reported that the Strouhal number calculated by the number of occurrence of the cavitation in circular cylinder wakes is considerably affected by the cavitation number, but is nearly same value calculated by the number of shedding of the vortex from circular cylinders at the incentive condition, the effects of the cavitation number to the Strouhal number were examined in this experiments. Fig.2 shows the effects of the cavitation number to the Strouhal number at the Reynolds number of Re=4.1×10^5 and Tu=0.1%. The incipient cavitation number was about Kd=1.1. As the Strouhal number was nearly constant as the cavitation number greater than Kd=0.9 though it abruptly increased at the cavitation number less than Kd=0.8 with decreasing of Kd, photographs of cavitation by the high speed movie camera were taken at the cavitation number of Kd=0.9 in all experiments.

Nomenclature

D : Diameter of the model
f : Frequency of the vortex shedding
Kd: Cavitation number=$2(P-Pv)/\rho U^2$
P : Static pressure in free stream
Pv: Vapor pressure
Re: Reynolds number=$D \cdot U / \nu$

Tu: Turbulence level in free stream=$\bar{u}/U$
S : Strouhal number=$f \cdot D/U$
$\bar{u}$: Root mean square value of longitudinal fluctuating velocity
U : Free stream velocity
ρ : Density of water
ν: Kinematic viscosity of water

Results

Vortex shedding from spheres   Each experimental result of the Strouhal number, which represents the mean value of five runs is shown in Fig.3. The scatter around the mean value was generally less than $\pm 7\%$. In this experiments, the upper critical Reynolds number could be conjectured from the pressure distributions measured to be about Re=4.0×10^5 at Tu=0.1%, Re=2.8×10^5 at Tu=1.35% and Re=2.1×10^5 at Tu=1.93%. The Strouhal number rises gradually up to Re=4.8×10^5, but becomes constant beyond this Reynold number. A remarkable variation of the Strouhal number could not be found by the variation of the free-stream turbulence at Re less than critical Reynolds numbers. For comparsion the experimental results of Achenbach (Ref.6) are also plotted. The present results fit the mean curve representing the results obtained by Achenbach very well at $0.9 \times 10^5 < $ Re $ < 3.0 \times 10^5$. Although Achenbach reported that periodic vortex shedding could not be detected by the hot wire technique beyond the upper critical Reynolds number (in his case Re=3.7×10^5), in this experiments periodic vortex sheddings could be observed beyond the upper critical Reynolds number at Re=7.8×10^5 and Tu=0.1%. In the case of the high turbulence level however, periodic vortex sheddings could not the observed beyond Re=2.1×10^5 at Tu=1.93% and Re=2.5×10^5 at Tu=1.35%. These matters show that the strength of the vortex generated by the rolling up of the shear layer separating from the sphere is strongly affected by the turbulence in free stream.

Wake configurations of spheres   Figs.4, 5, show the behavior of the cavities in time sequence in a wake at the Reynolds numbers of Re less than the critical one. The shear layer separating from the sphere rolls-up downstream apart from the sphere surface. The loops lose individual form immediately after the rolling-up. They then, transform their form to oblique loops more downstream. Oblique loops also can be observed in the process of the rolling-up immendiately behind the sphere surface. In addition to these vortex loops, the different kind of loops which occur after the collape of the rolling-up and are parallel to the free stream can be observed around te sting. The position of the rolling-up of the vortex sheet becomes closer to the sphere surface with increasing Reynolds number and turbulence level (see Fig.7). Beyond the upper critical Reynolds number, the rolling-up took place on the sphere surface. These flow patterns are similar to those described by Achenbach. Fig.6 shows the behavior of the cavities in a wake in time sequences at the Reynolds number greater than the critical one. The position of the rolling-up takes place on the sphere surface. The first oblique loops can be observed to originate from the sphere surface.

Separarion bubbles   The clear film type's cavities regularly to the circumferential direction took place at the separated regions on the hemispherical nosed body at Re $> 3.5 \times 10^5$ and on spheres at Re $> 4.5 \times 10^5$ at Tu=0.1%. In case of the hemispherical nosed body the flow can be considered to separate laminarly (Ref.8) and these film type cavities look like closely to the laminar separation bubble which was photographed by Kline and Runstadler (Ref.9).

Conclusions

The main conclusions to be drawn from the experiments of the flow visualization by cavitation on axisymmetric models can be summarized as follows:
1. The vortex shedding from sphere could be measured beyond the upper critical Reynolds number at the low turbulence level.
2. The effect of the turbulence in a free stream to the Strouhal number could scarcely be observed at Re less than the critical Reynolds number.
3. The flow patterns of sphere wakes could be visualized. The position of the rolling-up of the vortex sheet becomes closer to the sphere surface with

increasing Reynolds number on turbulence level.
4. The separation bubble on sphere surfaces and a hemispherical nosed body surface could be visualized.

Acknowlegements

   The authors wish to thank for the help in experiments of S. Onuma, technician of the Institute, Y. Tsurmi, student of the post graduate course of Tōhoku University, Y. Kimura, technical assistant, and Y. Yamabe, technician of the Institute, who manufactured spheres and the hemispherical nosed body precisely.

References

(1) H. Murai and H. Takeuchi, Observations of Cavitations and Flow Patterns in an Axial-Flow Pump at Partial Discharges, Mem. Instutute of High Speed Mechanics (in Japnese), 24-246 (1968/1969), 315
(2) W. Möller, Experimentelle Untersuchung zur Hydromechanik der Kugel, Phs.Z., 39 (1938), 57
(3) C. Cometta, An investigation of the unsteady flow pattern in the wake of cylinders and spheres using a hot wire probe, Div. Engng., Brown University, Tech. Rep. WT-21, 1957
(4) A. S. Mujumdar and W. J. M. Douglas, Eddy-Shedding From a Sphere in Turbulent Free Streams, Int. J. Heat Mass Transfer, 13 (1970), 1627
(5) J. R. Calvert, Some Experiments on the Flow Past a Sphere, Aero. J. Roy. Aero. Soc. 76 (1972), 248
(6) E. Achenbach, Vortex shedding from spheres, J. Fluid Mech., 62-2 (1974), 209
(7) K. K. Shalnev, I. I. Varga and D. Sebestyen, XXIII. Investigation of the scale effects of cavitation erosion, Phil. Trans. A, 260, 256 (1966)
(8) V. H. Arakeri, Viscous Effects in Inception and Development of cavitation on Axi-Symmetric Bodies, Rep. Fl83.1, Division of Engineering and Applied Science, California Institute of Technology, 1973
(9) S. J. Kline and P. W. Runstadler, Some Preliminary Results of Visual Studies of the Flow Model of the Wall Layer of the the Turbulent Boundary Layer. J. App. Mech., Trans. ASME. 26 (1959), 166

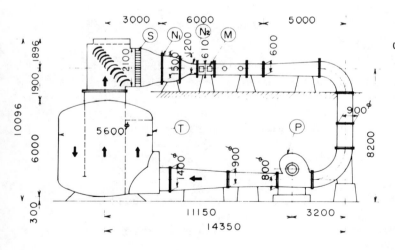

Fig.1 Water tunnel

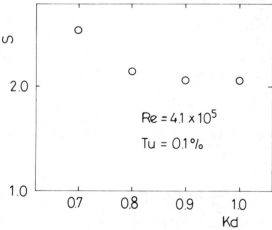

Fig.2 Strouhal number of the sphere as a function of cavitation number. Re=4.1x10^5, Tu=0.1%.

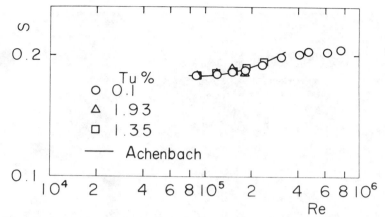

Fig.3 Strouhal number of the sphere as a function
of Reynolds number.

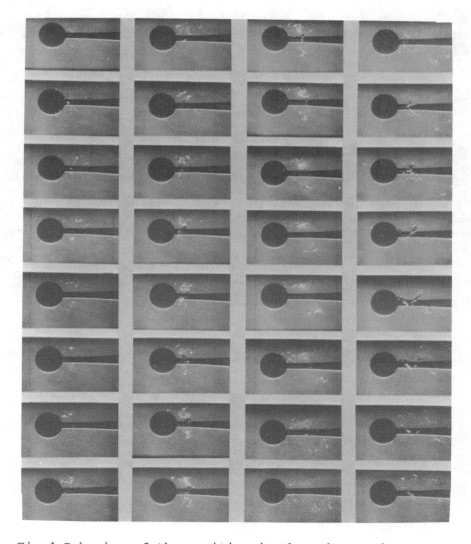

Fig.4 Behavior of the cavities in the sphere wake at
Reynolds number less than the critical one.
Re=1.2x10^5, Tu=1.93%, Kd=0.9, 4300frames/sec, flow
left to right.

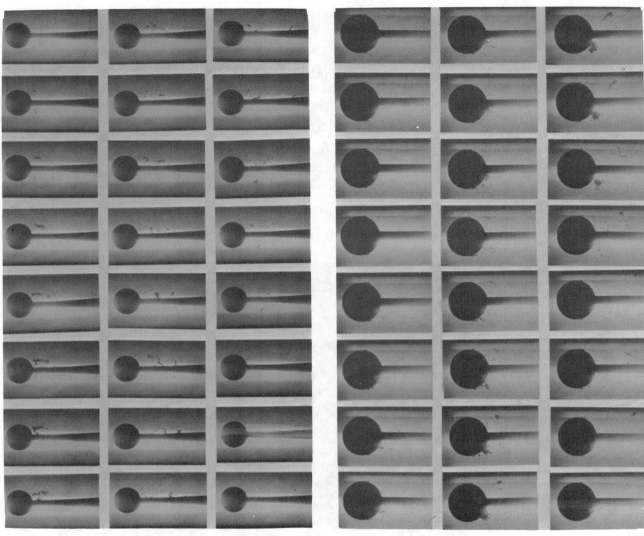

Fig.5 Behavior of the cavities in the
sphere wake at Reynolds number
less than the critical one.
Re=1.2x10^5, Tu=1.35%, Kd=0.9,
4300frames/sec, flow left to right.

Fig.6 Behavior of the cavities in the
sphere wake at Reynolds number
greater than the critical one.
Re=2.9x10^5, Tu=1.35%, Kd=0.9,
4300frames/sec, flow left to right.

Fig.7 The position of
rolling-up of the
vortex sheet as a
function of Reynolds
number.

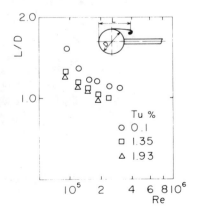

396

VISUALIZATION OF CAVITATION DUE TO WATER HAMMER IN A PIPE LINE

M. SUITA,* M. KAWAMATA,** and R. YAMAMOTO***

It is the purpose of this investigation to visualize the phenomena of cavitation due to water hammer be followed by the water column separation. The water hammer problems treated in many papers are based on the assumptions that vapor pressures do not occur in the systems. If vapor pressures occur in a pipe line, there are many difficult problems to be solved. Based on a few assumptions, the problems of water column separation have been studied by several researchers, but clear explanations have not been made for that. On the other hand, in the field of cavitation, the water column separation does not seem to be treated. Visualization of the phenomena is considered necessary to explain the mechanism of the phenomena.

Introduction

Recently, predictions of various flowing phenomena have been strongly required. The predictions of the flows with the water hammer phenomenon belong to complicated problems. Many papers have been reported on water hammers. Nowadays, we can calculate the unsteady phenomena of pipe flows and so on with fluid friction by a digital computer.(Ref. 1)

But if the water hammer in a pipe line is followed by the cavitation or water column separation, the treatments of the problems are very complicated. Several papers on numerical calculations of unsteady pipe flows with water column separation have been reported up to the present.(Ref. 2, 3, 4) W. H. Li and J. P. Walsh (Ref. 2) studied maximum pressure behind a rapidly closing valve or a pump following power failure with cavitation neglecting the fluid resistance. And in their paper, successive photographs of cavitation behind a closing valve are shown. R. A. Baltzer (Ref. 3) dealt with the column separation assuming that the flow which is occuring beneath the cavity is simply free surface, gravity flow. In his paper, a photograph which indicates the typical vapor cavity accompaning column separation is shown.

As the basic research for the study of pipe flow with cavitation, it is considered necessary to grasp the phenomena. In this paper we describe the visualization of cavitation due to water hammer in a pipe line. The phenomena are photographed by a motor drive camera and a 16 mm movie camera.

---

 * Professor of Mechanical Engineering, Ibaraki University, Hitachi, Ibaraki, 316 Japan
 ** Graduate Assistant of Mechanical Engineering, Ibaraki University
*** Graduate Engineer of Power Reactor and Nuclear Fuel Development Corporation, Tokai, Ibaraki, 319-11, Japan

Experimental apparatus

The phenomena are generated by rapid closure of a butterfly valve which is set up at the upstream end of the pipe line. In order to elongate the time scale of the pipe line, vinyle chloride pipe is used. Fig. 1 illustrates the schematic diagram of the experimental apparatus. Water in the pipe line flows from the high head reservoir to the low head reservoir. The butterfly valve is driven by a oil hydraulic system. Upper part of the pipe line is made of transparent acrylic resin, and has a 44.3 mm square cross section and 0.9 m in length, which is used for photographing. The pipe line has 147 m and 77.4 m long, 0.05 m inner diameter and is set up horizontally. The initial velocities of the flow are 1.0 ~ 0.57 m/s. The time of valve closure is about 0.1 second.

Fig. 2 shows the photographing part of the apparatus which is made of transparent acrylic resin. Water flows from right side to left side. Two iodine lamps (500W × 2) are used for the light sources and they are attached to upper side and lower side of the test section. A motor drive camera is set horizontally. The phenomena are recorded photographically by it at the rate of four sheets per second. The operation of the camera shutter synchronizes with the oil hydraulic system which drives the butterfly valve. And the shutter speed of 1/500 second is used. An extension ring is used. A 16 mm movie camera, driven 48 frames per second, is used for continuous recording.

Fig. 3 shows graphically the pressure variation after the closing of the butterfly valve. At the same time, the shutter sign and the curve of the valve closing are marked in it. The pressures are measured by strain gauge pressure transducers. These phenomena are recorded on an electro-magnetic oscilograph simultaneously.

Results

An example of typical succesive photographs is shown in Fig. 4 . The initial velocity is 0.74 m/s, and the time of valve closure is about 0.1 second. The time intervals between two photographs are about 0.23 second. The first photograph, a) of Fig. 4 indicates the state 0.04 second before the valve closure. The generation of cavitation due to throttling the valve is observed. The second photograph, b) of Fig. 4 shows the state 0.19 second after the valve closure. In this photograph, we can observe a horizontal large cavity situated at the upper part of the channel. It is observed in the third photograph that the horizontal large cavity extends to the down stream side. In the lower part of the section, vigorous boiling at the water temperature is seen. When the pressure waves return from the down stream end, the large cavity suddenly diminishes, in the last photograph of Fig. 4 . The phenomenon repeats a few times, and weakens. The maximum length of the large cavity is about 1 m . For a example, in the position of 7.1 m down stream section from the butterfly valve, we only could find a few small cavities.

As the initial velocity decrease, or the time of valve closure becomes larger, the phenomena weaken.

Fig. 5 and Fig. 6 indicate the influence of the pipe length. The former corresponds to the 147 m long, the latter, 77.4 m long, but the other conditions are the same. The photographs a), b) of Fig. 5 show the state 0.3 second and 0.05 second before the valve closure. The time intervals between two photographs are about o.25 second. The first photograph, a) of Fig. 6 indicates the state 0.02 second after the valve closure. And the time intervals between two photographs are about 0.25 second. In the photographs of Fig. 5 and Fig. 6, the continuous time of the phenomenon is shorter in Fig. 6 . But the boiling phenomena look alike.

To catch the successive change of the phenomena, a 16 mm movie camera is used. Fig. 7 and Fig. 8 show the photographs. Each photograph is selected at the rate of one frame per three frames. Fig. 7 shows the successive growth of the cavitation. Fig. 8 shows the process of the attenuation. The continuous film is projected at the symposium.

Conclusions

The phenomena of cavitation due to water hammer have been photographed. But, in the details of cavitation, we could not catch the mechanism of the bubble growth. It is desirable to take distinct photographs that show the bubble growth.

References

(1) V.L. Streeter and E.B. Wylie, Hydraulic Transients, (1967), McGraw-Hill.
(2) W.H. Li and J.P. Walsh, Pressure Generated by Cavitation in a Pipe, ASCE, EM 6 (1964-12), 113.
(3) R.A. Baltzer, Column Separation Accompanying Liquid Transients in Pipes, ASME Ser. D (1967-12), 837.
(4) J.P.Th. Kalkwijk and C. Kranenburg, Cavitation in Horizontal Pipelines due to Water Hammer, ASCE, 97-HY 10 (1971-10).

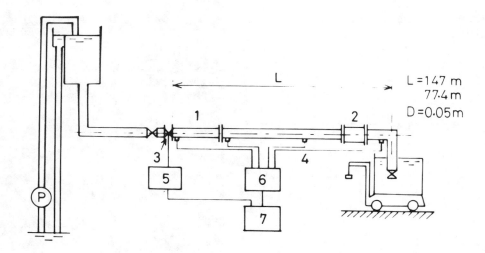

Fig.1 Experimental apparatus.

1: test section,	2: electro-magnetic flow meter,
3: butterfly valve,	4: pressure transducer,
5: oil hydraulic system,	6: amplifier,
7: recorder,	

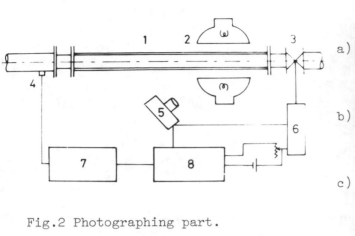

Fig.2 Photographing part.

1: transparent acrylic resion,
2: iodine lamp,
3: butterfly valve,
4: pressure transducer,
5: camera,
6: oil hydraulic system,
7: amplifier,
8: recorder,

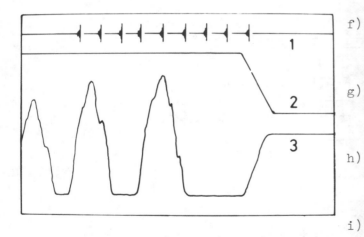

Fig.3 Record of the pressure variation.

1: shutter sign,
2: valve closing curve,
3: pressure variation curve,

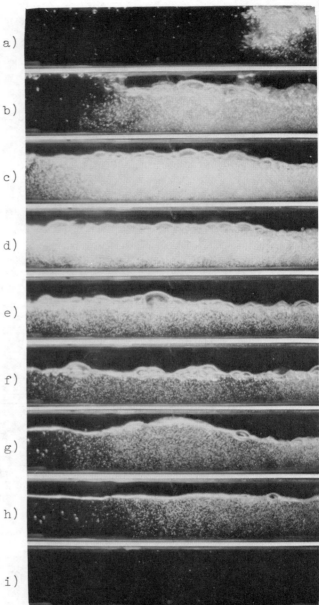

Fig.4 The cavitation due to rapid valve closure.

a)

b)

c)

d)

e)

f)

g)

h)

i)

Fig.5 The cavitation due to rapid valve
closure at the pipe length 147 m.

a)

b)

c)

d)

e)

f)

g)

h)

i)

Fig.6 The cavitation due to rapid vavle
closure at the pipe length 77.4 m.

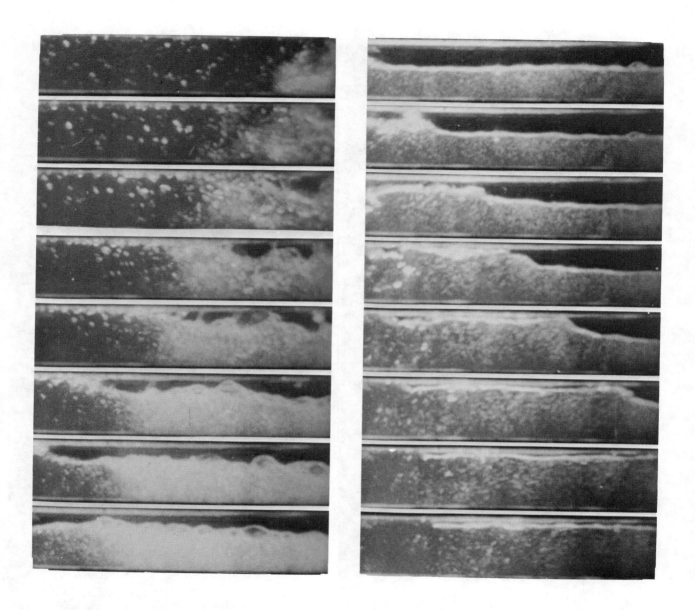

Fig.7 The growth of the cavitation
     by a 16 mm movie camera.

Fig.8 The attenuation of the cavitation
     by a 16 mm movie camera.

INDEX

Nose, flat furnace, and ash erosion, 191–194
Nozzle:
  atomizer, 66, 67
  water, 386
Nozzle flow, 9

Obata, M., 189–194
Ocean thermocline, 43
Ohmi, Y., 161–166
Ohta, Hiroaki, 239–244
Oil dot method, 5
Oil film method, 5
  and boundary layers, 357
  and surface air flow, 31
  and vehicle aerodynamics, 172, 173
  and vortices near a wall, 201–203, illus. 204–206
Oil smoke, 75–77, illus. 78–80
Okajima, A., 143–148
Okamoto, Hisashi, 201–205
Okamoto, Y., 149–154
Okitsu, Shiro, 239–244
Optically activated tracers, 155–158, illus. 159–160
Optic axis, 363–366, illus. 368
Optical methods, 4, 9, 12, 13, illus. 18, 20, 21, 47–79
  and shock tube boundary layers, 321, 323
  and two-phase flows, 291–296, illus. 297–302
  and separated flows, 315
  (See also Birefringence; Holography; Interferometry; Schlieren; Shadowgraph)
Oscillation, vortex excited, 143–145, illus. 146–148
Oscillatory flows, 43, 65
Oshima, Koichi, 195–200
Oshima, Yuko, 81–86

Page, R. H., 285–290, 315–320
Paraffine, 6, 9, 168, 251
Particle suspension method, 6, 119
  (See also Tracer methods, direct injection type)
Particles:
  buoyant, 312
  growth, 129
  luminescent, 7, 161–166
  phosphorescent, 156–157
  radiant, 239–241, illus. 243
  seeded, 351–353
  single, 158, 167, 168, illus. 169
  size, 129
  small, 42–44
  solid, 156
  suspended, 43
Path line method, 6, 136, 144, 157, 273
  and small hydrogen bubbles, 217–218
  and unsteady flows, 209

Path line method (Cont.):
  (See also Tracer methods, direct injection type)
Periodic flow separation, 68
Phase interaction, 291–296, illus. 297–302
Philbert, Michel, 335–340
Phosphorescent particles, 156, 157
Photogrammetry, 30, 281–282
Photography:
  of boundary layer transitions, 88
  of cavitation, 386–388, 398, illus. 400
  convected vs. non-convected views, 120
  and DOP aerosols, 67–68
  in electrochemical systems, 210
  and refractive images, 279–280
  of separated flows, 317
  and smoke wire method, 259
  for three-dimensional views, 117–121
  of water jets, 46
  (See also Cinematography)
Photonics, 29
Pigments as tracers, 6
Pindera, Jerzy T., 363–368
Pipe, 3
  test section in, 102, illus. 106
Pipe flow, 369–375
  cavitation in, 397–399, 400–402
  and hydrogen bubble method, 30
  of polymer solution, 101–105, illus. 106–110
  transient, 161, 166
  turbulent sub-layer of, 46
Pitot tube, 101, 227
Plasma dynamics, 32–33, 41
Plastics, molding of, 44
Polariscope transmission, 369
Polyacrylamide, 386
Polymer melt flow, 377–378
Polymer solution, 101, 360, illus. 362
  and cavitation, 46–47, 385–387, illus. 389–390
Precipitation method, 10
Pressure distribution:
  in boiler model, 194
  in flow around vehicles, 171
  in shock and blast waves, 279, 284
  in shock tubes, 233
Pressure, vapor, and water hammer problems, 397
Prisms, 143
Probes, 30
  avoidance of, 155, 158
  hand-held for smoke generator, 75–77, illus. 78, 257
  and species concentration, 329, illus. 332
  wave, 124
Projectiles, 233
Przirembel, C. E. G., 285–290, 315–320
Pulse light, 7